EXPERIMENTAL DESIGN FOR COMBINATORIAL AND HIGH THROUGHPUT MATERIALS DEVELOPMENT

EXPERIMENTAL DESIGN FOR COMBINATORIAL AND HIGH THROUGHPUT MATERIALS DEVELOPMENT

Edited by

JAMES N. CAWSE
GE Global Research

A JOHN WILEY & SONS, INC., PUBLICATION

Library of Congress Cataloging-in-Publication Data is available.

ISBN 0-471-20343-2

Printed in the United States of America.
αc
10 9 8 7 6 5 4 3 2 1

Results? Why, man, I have gotten lots of results! If I find 10,000 ways something won't work, I haven't failed. I am not discouraged, because every wrong attempt discarded is another step forward. Just because something doesn't do what you planned it to do doesn't mean it's useless. Reverses should prove an incentive to great accomplishment. The only time I become discouraged is when I think of all the things I like to do and the little time I have in which to do them. There are no rules here, we're just trying to accomplish something.

—THOMAS EDISON
(M. A. Rosanoff, *Harper's Weekly,* September 1932, 402)

CONTENTS

CONTRIBUTORS

Erik Agrell
Department of Signals and Systems
Chalmers University of Technology
Goteborg, Sweden
agrell@s2.chalmers.se

Eric J. Amis
Polymers Division
National Institute of Standards and Technology
Gaithersburg, Maryland
Eric.amis@nist.gov

Manfred Baerns
Institut für Angewandte Chemie Berlin-Adlershof e.V.
Berlin, Germany
baerns@aca-berlin.de

David S. Bem
Terial Technologies
Des Plaines, Illinois

Kristin P. Bennett
Department of Mathematics
Rensselaer Polytechnic Institute
Troy, New York
bennek@rpi.edu

Jinbo Bi
Department of Mathematics
Rensselaer Polytechnic Institute
Troy, New York

Curt M. Breneman
Department of Chemistry
Rensselaer Polytechnic Institute
Troy, New York
brenec@rpi.edu

James N. Cawse
GE Global Research
Niskayuna, New York
cawse@crd.ge.com

Steven Cramer
Department of Chemical Engineering
Rensselaer Polytechnic Institute
Troy, New York
crames@rpi.edu

Michael W. Deem
Bioengineering Department
Rice University
Houston, Texas
mwdeem@rice.edu

R. Bruce van Dover
Lucent Technologies
Murray Hill New Jersey
rbvd@mailaps.org

Erik J. Erlandson
NovoDynamics, Inc.
Ann Arbor, Michigan

Mark Embrechts
Decision Sciences and Engineering Systems
Rensselaer Polytechnic Institute
Troy, New York
embrem@rpi.edu

Martha M. Gardner
GE Global Research
Niskayuna, New York
gardner@crd.ge.com

Ralph D. Gillespie
UOP LLC
Des Plaines, Illinois

Fred A. Hamprecht
IWR
University of Heidelberg
Heidelberg, Germany
fred.hamprecht@iwr.uni-heidelberg.de

Laurel A. Harmon
Striatus Inc.
Dexter, Michigan

Martin Holeňa
Institut für Angewandte Chemie Berlin-Adlershof e.V.
Berlin, Germany

Alamgir Karim
Polymers Division
National Institute of Standards and Technology
Gaithersburg, Maryland
alamgir.karim@nist.gov

J. Carson Meredith
School of Chemical Engineering
Georgia Institute of Technology
Atlanta, Georgia
carson.meredith@che.gatech.edu

Steven G. Schlosser
NovoDynamics, Inc.
Ann Arbor, Michigan

Lynn F. Schneemeyer
Agere Systems/Bell Laboratories, retired
lschneemeyer@comcast.net

Amit Sehgal
Polymers Division
National Institute of Standards and Technology
Gaithersburg, Maryland
amit.sehgal@nist.gov

Minghu Song
Department of Chemistry
Rensselaer Polytechnic Institute
Troy, New York

N. Sukumar
Department of Chemistry
Rensselaer Polytechnic Institute
Troy, New York

Ted X. Sun
Intematix Inc.
Moraga, California
Tsun@intematix.com

Alan J. Vayda
NovoDynamics, Inc.
Ann Arbor, Michigan

Dorit Wolf
Degussa
Project House Catalysis
Frankfurt am Main, Germany
DWolf@crt.hoechst.com

Ronald Wroczynski
GE Global Research
Niskayuna, New York
wroczynski@crd.ge.com

PREFACE

Combinatorial and high throughput experimentation (CHTE) methods are changing the materials-development world in the first years of this century as thoroughly as they changed the drug-discovery world in the 1990s. Advances in chemistries, analytical techniques, preparation methods, software, and hardware have made this a tremendously exciting area in which to work. Suddenly materials scientists are thinking differently and attacking research areas as entire units rather than proceeding from experiment to experiment.

Many books are being written about combinatorial and high throughput techniques, but little has been written about the key intellectual step of planning the experiments. With the experimental horsepower now easily accessible, the experimenter really needs to map a route before careening across the landscape.

This book contains a selection of the best experimental design approaches currently being used or proposed for CHTE programs. However, as Cuthbert Daniel stated in the preface to his classic book, "It is impossible to make any very general statistical statement about industrial experiments. No claim is made here for the universal applicability of statistical methods. Rather, we proceed by examples and modest projections."* This book deals very largely with exploratory experiments. Most confirmatory and optimization experiments can be handled with conventional design-of-experiments methods.

The structure of the book is a progression from methods that are now standard, such as gradient arrays, to cutting-edge mathematical developments. I hope that the earlier chapters will be very useful to researchers entering the field and that the later chapters will inspire and challenge advanced practitioners. The book is intended for the entire array of materials scientists, from catalyst chemists to metallurgists to solid-state materials specialists, and for the statisticians and mathematicians who work with them.

I would like to thank GE Global Research and its management, particularly Eric Lifshin, Terry Leib, William Flanagan, Noreen Johnson, and Greg Chambers, for their support of combinatorial research, new experimental design methodology, and this book. Some of the work reported here was supported under the Department of Commerce, National Institute of Science and Technology (NIST), Advanced Tech-

*C. Daniel, *Applications of Statistics to Industrial Experimentation.* Wiley, New York, 1976.

nology Program Contract # 70NANB9H3038. NIST has been an important supporter of combinatorial materials development.

A large corps of reviewers, including many from GE Global Research, has contributed greatly to the clarity and focus of these chapters. They include Cheryl Bratu, Rich Beaupre, Steve Clarke, Jack Norton, Lihao Tang, Radislav Potyrailo, Eric Pressman, Bill Tucker, Tom Repoff, Ron Shaffer, Jay Spivack, Josef Schmee, David Bryant, Donald Whisenhunt, Dave Whalen, Jim Silva, Jack Reece, Rick Taylor, Mike Maclaury, Bob Mattheyses, John Hewes, Ron Wroczynski, Dan Hancu, and Mike Sutherland. The graphics in this volume have benefited greatly from the work of Richard J. Oudt of Talon Studios.

Finally, I would like to thank my wife, Marietta, and daughters, Lauren and Jeanne, for cheering me on through the bookmaking process.

JAMES N. CAWSE

Niskayuna, New York
October 2002

CHAPTER 1

THE COMBINATORIAL CHALLENGE

JAMES N. CAWSE
GE Global Research

1.1 INTRODUCTION

As the 21st century opens, the opportunities and challenges to materials scientists worldwide have never been greater. Market globalization, environmental performance, customer expectations, population expansion, and resource limitations present a need for materials with new or optimized properties. Although progress has been made in our ability to design or predict the properties of new materials, discovery of most new materials is still solidly based in experimentation.

As chemical and materials knowledge has accumulated over the centuries, the properties of simple materials have become better known and new developments have tended to occur in the regions of greater complexity. For example, "The properties of many functional solid-state materials arise from complex interactions involving the host structure, dopants, defects, and interfaces" [1]. At this level of complexity, the feared "combinatorial explosion" appears. If there are 76 useful, stable elements in the periodic table (Figure 1.1), there are 2850 binary, 70,300 ternary, 1,282,975 quaternary, and $>10^9$ heptanary combinations of these elements. In the case of organic structures, it has been estimated that there are 10^{63} "druglike" molecules synthesizable from the standard organic toolkit of elements [2]—and these are only a subset of the total number of molecules potentially of interest to materials scientists. Finally, the ultimate properties of useful materials will also be a function of relative compositions and of processing conditions. This further increases the number of possible combinations. Subsets of this enormous number of possibilities form the "experimental space" or "chemical space" of each individual program.

Combinatorial chemistry and high throughput experimentation have emerged during the last decade as a response to the challenges of materials development in these increasingly complex experimental spaces. Although its earliest antecedents can be traced to the beginning of this century, combinatorial experimentation really began to take off around 1990 in the pharmaceutical industry [3] and 1995 in mate-

Experimental Design for Combinatorial and High Throughput
Materials Development, Edited by James N. Cawse.
ISBN 0-471-20343-2 © 2003 John Wiley & Sons, Inc.

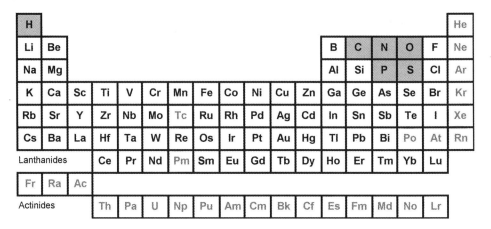

Figure 1.1 Elemental constituents for combinatorial chemistry. *Bold:* the 76 nonradioactive, nonnoble elemental possibilities. *Grey background:* the 6 elements in the standard organic toolkit.

rials development [1,4]. This resulted from a convergence of technologies in robots, semiconductor processing, computer software, and analytical capabilities, and from the simple realization that such productivity was possible. The rapid growth of this technology has led to a proliferation of terminology; we will try to use the definitions in Table 1.1. The general term *combinatorial and high throughput experimentation* (CHTE) is used in this book for the entire field of parallel, miniaturized, high speed experimentation.

The development of new experimental tools with awesome levels of productivity does not, however, exempt us from the need to plan our experiments well. Murphy's law continues to rule, and a poorly designed experiment will give us bad information with unprecedented speed and in outstanding quantities. In fact, these new capabilities will be used on harder problems, which require entirely new experimental designs and methods of data analysis and visualization. The chapters in this book report on the best current efforts in designing experiments in CHTE systems. The field is developing very rapidly and we expect to see many more such design tools during the next few years.

1.2 HISTORICAL BASIS

A discussion of experimental design for CHTE would be incomplete without at least a brief glance at the historical context. The art and science of experimentation has been evolving since the alchemists, but only in the last 75 years has systematic effort been made to include statistical concepts in the planning stage of experiments, rather than just the data analysis.

Table 1.1 Definitions of CHTE Terms

Experimental space	All possible combinations of formulation, and processing variables in the system under study
Chemical space	All possible molecules within a given set of constraints.
Property space	Multidimensional representation of a set of compounds in which the axes represent quantifiable properties, such as molecular weight and ionization potential, and individual compounds are represented by a vector or set of coordinates.
Combinatorial design	Experimental design whereby relationships between variables can be elucidated through the evaluation of their combinations.
Combinatorial library	A set of compounds prepared by combinatorial synthesis, or formulations prepared through a combinatorial design.
Combinatorial synthesis	Use of a combinatorial process (e.g., split/pool) to prepare sets of compounds from sets of building blocks.
Parallel synthesis	Synthetic strategy whereby sets of discrete compounds are prepared simultaneously in arrays of physically separate reaction vessels.
Split/pool synthesis	Synthetic strategy for assembly of a combinatorial library. A single compound is built onto each single support particle in a series of reactions using a process of splitting, reaction with synthons, pooling, mixing, and redividing.
High throughput experimentation	The use of miniaturization, robotics, and parallel techniques to increase the productivity of the research process.
High throughput screening/analysis	Process for rapid assessment of the activity of samples from a combinatorial screening/analysis library or other sample collection, often by running parallel assays.
Factory	Informal term for the entire combinatorial/high throughput process.
Hit	Library component whose activity exceeds a predefined, statistically relevant threshold.
Lead	A hit which has been substantiated by confirmatory experiments or secondary screening.
Descriptor	Numerical representation of a molecular property, including bulk properties (e.g., ionization potential, molecular weight), two-dimensional features (atom connectivities), or three-dimensional features (molecular shape).
Diversity	The "unrelatedness" of a set of, for example, building blocks or members of a combinatorial library, as measured by their properties (e.g., descriptors).

Source: Adapted from Maclean, D., et al., "Glossary of Terms used in Combinatorial Chemistry," *Pure and Applied Chemistry,* 1999, **71**(12): 2349–2365.

1.2.1   Edisonian Experimentation

Science and technology have no lack of examples of extensive—even heroic—experimentation. Table 1.2 gives a few historic examples of systematic and successful experimentation. Even combinatorial experimentation has precedents early in the century in the successful efforts to develop an ammonia synthesis catalyst. Mittasch reported the effects of 54 different metals (almost all the metals in the periodic table) on an iron catalyst, and found that "numerous combinations were found to possess catalytic activities." He quotes Willstatter that "multicomponent catalysts can be regarded, in their activities, as equivalent to new specific substances" [5].

"Edisonian" is a term sometimes used disparagingly in reference to CHTE, implying experimentation that lacks strong scientific underpinnings and is just near-random casting into uncharted waters. This is unfair to both Edison and modern CHTE practitioners. The intellectual effort and scientific rigor used by both to plan and carry out experiments are every bit as creditable as those used by more conventional experimenters.

1.2.2   Design of Experiments: Agricultural antecedents

Classic experimental design strategies grew up in the English agricultural research community in the 1920s, a period of slow, laborious, error-prone experimentation. The landmark designs developed by Fisher [6] were done when one experiment *per year* was the norm (Figure 1.2). The rationale for designed experiments in that period was principally one of determination of the *main effects* of the factors in the presence of enormous natural variation. Replication and blocking patterns were set to minimize and average out random factors such as soil type, wind, sun angle, cloud cover, and slope. Factorial designs and Latin squares were the preferred arrays.

1.2.3   Modern Industrial Design of Experiments

Classic industrial design of experiments (DOEs) [7,8] was introduced to the United States by Box after World War II and developed in the 1950s and 1960s. Experiments continued to be expensive, since they were typically still done one at a time

Table 1.2   Heroic Experimentation

Experimenter	Date	Goal	Approximate Number of Runs
Thomas Edison	1878–1880	Electric light	6000
Alwin Mittasch	1909–1912	Ammonia catalyst	20,000
Paul Ehrlich	1907–1910	Syphilis drug (Salvarsan)	900

Figure 1.2 Low throughput experimental design. A 5 × 5 Latin square planted about 1929. Photographed in 1945. Reproduced with the kind permission of the Forestry Commission, UK. from J. F. Box, *R. A. Fisher, A Life in Science,* Plate 6, copyright ©1978. John Wiley & Sons, New York.

by highly trained professionals. DOEs are usually attempts to determine the main effects and 2-factor interactions in a minimum number of runs. Factorial and fractional factorial designs were developed and heavily used as simple, effective tools that could be employed by nonstatisticians. The mathematical properties of 2-level factorials—from both the statistical standpoint of properties like orthogonality, and the computational properties useful in the precomputer era—led to them becoming the standard tool [9].

As statistical sophistication increased and computer use became the norm, more complex DOEs, such as central composite designs, mixture designs [10], and optimal designs, became common. These allow the experimenter to determine *curvature* as well as main effects and 2-factor interactions. All of these are now easily generated by standard software (e.g., Design-Expert [11], JMP [12], Minitab [13], and Echip [14]). The popularity of the Taguchi approach to experimentation [15] and the Six Sigma quality initiatives in many major companies [16] has made the use of DOEs increasingly common.

1.2.4 Design of Experiments for Combinatorial and High Throughput Experimentation

In screening for new materials, the emphasis is on the discovery of low-probability, high-value occurrences (hits) by searching extensive experimental spaces. These are typically the result of combinations or interactions of three or more important factors, and the system may have dozens of factors, each with multiple levels. The total number of combinations of all of these factors can easily be 10^6, 10^9, or much more. Experimental strategy and tactics must examine this space with a high probability of finding hits, and allow completion within the lifetime of the experimenter (or, more important, the resources of the project).

The basis for design of experiments in CHTE must include the following:

- Careful choice of the experimental space with a keen scientific understanding of the system.
- Selection of factors and levels to get a broadly diverse view of the possible chemical space that is being studied.
- Estimation of the level of complexity of the combinations and interactions that must be examined to have a reasonable chance of locating some hits.
- An understanding of the sources and amount of variation from the entire CHTE factory and the amount of "difference" that constitutes a hit.
- Recognition of the "procrustean bed"[1] of the (usually fixed) experimental array size and the limitations of the available technology (robots, processing steps, analytical tools).
- Use of the rapid turnaround of CHTE for effective sequential experimentation.

As so often occurs in science, few of the new methods examined here are strictly new. Inspired theft from other disciplines is a recurring theme. The other disciplines clearly include biology, biochemistry, and combinatorial mathematics [17], but also physics, geology, computer science, information theory, and mathematical geometry. One observation may be useful to people searching for other methods: **chemistry has now accelerated to approach the speed of the early computers.** Ideas and algorithms that were devised in the 1950s and 1960s to solve difficult problems on those machines can be adapted to the search problems chemists face now.

1.3 COMMERCIAL TARGETS

The National Institute for Science and Technology (NIST) has been a strong supporter of development of combinatorial methods, and its Vision 2020 Technology

---

[1]Near Eleusis, in Attica, there lurked a bandit named Damastes, called Procrustes, or "The Stretcher." He had an iron bed on which travelers who fell into his hands were compelled to spend the night. His humor was to stretch the ones who were too short until they died, or, if they were too tall, to cut off as much of their limbs as would make them short enough. (http://www.goines.net/Writing/procrustean_bed.html)

Roadmap for Combinatorial Methods [18,19] gave an overview of potential markets and key applications for CHTE. Table 1.3 is adapted from that document, and the following is a more detailed discussion of a few of the major commercial targets for CHTE.

1.3.1 Catalysts

Benefits After drugs, the highest potential value for combinatorial chemistry may be the discovery of new catalysts. Catalysts are ubiquitous in industrial chemical processing, and a composition that performs a new reaction or makes an old one more efficient can revolutionize whole areas of the industry. Catalyst-based manufacturing has been estimated to produce over 7000 compounds worth over $3 trillion globally [20]. Improved catalyst effectiveness, such as major yield and selectivity improvements, will reduce waste and energy consumption and minimize feedstock costs. New routes to currently high-cost materials will enable market entry of new feedstocks and raw materials.

Critical Technology Issues Industrial catalytic processes have highly important kinetic (time-varying) properties, and are often done in aggressive conditions of temperature, pressure, and chemical environment [21].

1.3.2 Polymers

Benefits A significant fraction of new plastic materials is developed through the process of blending known polymers, adding property-modifying chemicals, or performing surface modifications. These processes enhance desired properties such as weatherability or moldability. These blends have become more complex as additional outstanding properties are required to meet the needs of modern products. In addition, the cycle time of the products is decreasing; new computer hardware appears on cycles of 6 months or less. The market for these highly engineered materials in the computer industry alone exceeds $1 billion/year.

Critical Technology Issues Polymer properties are often highly dependent on the details of the mixing, extrusion, and molding processes, and the properties themselves are defined in macroscopic terms (e.g., breaking strength, surface hardness) [22].

1.3.3 Lamp Phosphors

Benefits Fluorescent lamp phosphors convert ultraviolet emission of a rare-gas/mercury discharge plasma into visible (white) light. General-purpose lighting consumes about one-quarter of all electricity produced in the United States. The incandescent lamp, known for its pleasing appearance to the human eye and its low purchase price, is a very inefficient light source. The replacement of incandescent lamps with compact fluorescent lamps (CFLs) could yield substantial savings in en-

Table 1.3 Markets and Current Key Applications for CHTS

Material Type	Market	Key Discovery Areas
Electro/Magnetic/Optical Materials		
Electronics	Appliances High-speed circuitry Miniaturization Fiber optics	Dielectric constant Multifunctional materials
Photonics	Optical computers Nonlinear optics Optical networking	Multifunctionality High speed High bandwidth
Magnetic materials	Transformers Power-conversion equipment MRI machines	New superconductors Magnetic materials
Scintillators	Medical imaging	New superconductors
Lighting	Fluorescent lamps Ceramic metal halide lamps Halogen lamps Light-emitting diodes	Phosphors Diode materials
Medical/Biotechnology		
	Lab-on-a-chip Patient-specific materials Biometric materials High throughput sampling Biodegradable materials	Proteomics Biomimetics Genetic-based drugs Diagnosis
Chemical		
Catalysis	Industrial chemicals Plastics Alternative fuels	Selectivity Yield Heterogeneous catalysis Homogeneous catalysis
Coatings, adhesives, lubricants	Automotive New adhesives Nonsolvent-based coatings	Multifunctionality
Chemical process/product design	Routes to commodity chemicals Third-world infrastructure Dream reactions Adsorbtion/separation processes New raw materials	Alternative pathways Process optimization Model refinement Kinetics
Energy technology	Energy conversion Energy storage	Fuel cell components Battery materials
Other Materials		
	Nonlead solder	Formulation/process interaction
	Cement Ceramics Hybrid products	Materials interaction Anisotropic materials Nanoscale materials Composites Smart materials

ergy and overall lamp-life cost with concurrent reduction in greenhouse gas emission from fossil-fuel power plants. CFLs use expensive rare-earth-based phosphors resulting in a high purchase price. Cheaper and/or higher performing phosphors will allow CFLs to find higher penetration in residential applications.

Critical Technology Issues   Phosphors consist of crystal lattices of complex elemental composition plus the presence of doping elements. Defining the constituting elements of a potential host lattice plus potential dopants creates a large number of material libraries from which the most efficient composition would be chosen for further evaluation.

1.3.4   Scintillators

Benefits   Scintillators absorb X rays and convert them to visible light for electronic detection. Scintillator based detectors are critical components in computed tomography (CT) and digital radiography (DR) medical imaging equipment, as well as X-ray-based systems for nondestructive industrial imaging systems. Computed tomography is a global $1.5B industry. Digital radiography is a new diagnostic imaging modality and is expected to substantially displace X-ray film, which currently is a large fraction of the $3.8B medical X-ray industry.

Critical Technology Issues   Scintillator quality depends on multiple interlinked properties, such as conversion efficiency, linearity, speed, radiation damage, and afterglow. These properties are affected by the scintillator host material and by chemical doping. This doping covers a very wide (5 ppm to 10 mole percent) range, with multiple dopants used to control several properties. Interactions between dopants are complex, and trade-offs between properties are frequently required.

1.3.5   Superconductors and Magnetic Materials

Benefits   A practical superconducting electrical transmission line must be high on the lists of the "holy grails" of materials development. Its benefits, from increased efficiency to improved use of real estate, are in the multi–100-billion-dollar range. Modest improvements in magnetic materials would lead to substantial improvements in transformers (already a $20–30B market for superconductors), power conversion equipment, and magnetic resonance imaging (MRI) machines.

Critical Technology Issues   The highest critical temperature of superconductors has tended to increase with the number of components in the formulation. The most complex of these materials are already five-component formulations; at six to eight components, the combinatoric explosion is formidable.

1.3.6   Coatings

Benefits   Most finished manufactured products have some sort of coating for protection, decoration, or other property enhancement. These processes are expensive, capital intensive (an automotive painting line may cost as much as $350M), and often environmentally offensive (U.S. regulatory costs are estimated at $50M per painting line). Combinatorial experiments in this area can be applied to systems as diverse as flexible flat-panel displays and weather-resistant clearcoats for auto body panels [23].

Critical Technology Issues The system is an experimental space of formulation/process interactions with complex application methodologies. Evaluation of the samples is primarily in terms of macrofunctionality.

1.4 CAPABILITIES OF COMBINATORIAL/HIGH THROUGHPUT EXPERIMENTAL EQUIPMENT

1.4.1 Overview

A wide array of equipment has been developed for CHTE programs during the past decade. Although the details of this hardware depend on the precise problem being studied, there are specific types, and they tend to fall into clearly defined size ranges (Figure 1.3). The limitations in array size, sample size, formulation accuracy, and resolution of analysis equipment all have a direct effect on the choices of experimental designs to use. There is a continual dialog between the hardware for producing sample arrays and the design of the array. This dialog will be apparent in many of the following chapters.

Since comparisons of materials are easiest (and of highest quality) when done within a single array, the hardware acts as a constraint and a goad to the experimentalist. This can sometimes result in a brilliant synthesis of array and experiment, as in the 1024-sample fractal array discussed by Sun in Chapter 3. More frequently, it results in a compromise of statistical ideal and practical needs.

1.4.2 Microreactors

The smallest array sizes tend to be groups of microreactors. These are constrained by the need to perform physical activities during the processing of the array—gas or liquid addition, mixing, sampling, or analytical measurements. They also tend to be somewhat larger than the following array types, often being in the milliliter rather than microliter scale. For this reason microreactors tend to be used for follow-up or

---

Figure 1.3 Combinatorial libraries. (a) A 6 × 4 array of microreactors. (Photo courtesy of Avantium Technologies BV.) (b) A 6 × 8 high-pressure vial reactor system. (Photo courtesy of GE Global Research.) (c) A 96-well plate of catalyst samples. (Photo courtesy of GE Global Research.) (d) Unfolded pentanary catalyst array. (Reprinted, with permission, from Mallouk, T.E., and Smotkin, E. *Fuel Cell Handbook* (forthcoming); copyright © 2002. John Wiley & Sons, New York.) (e) A 1024-sample fractal design array. (Reprinted, with permission, from Sun, T. X. *Biotechnology and Bioengineering* (*Combinatorial Chemistry*), 1999,

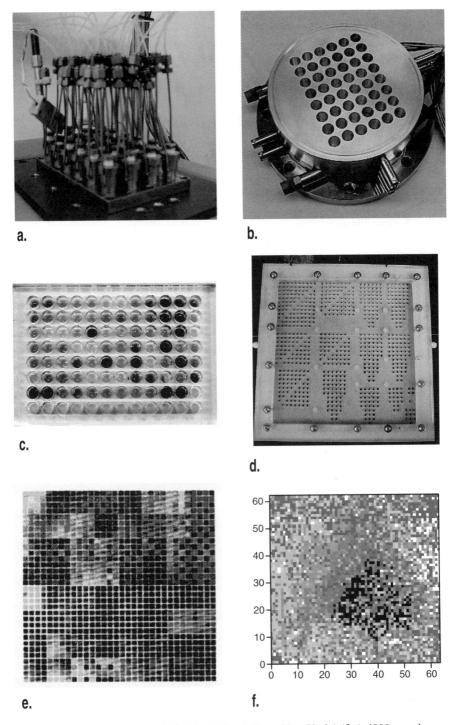

61(4), 193–201; copyright © 1999. John Wiley & Sons, New York.) (f) A 4000-sample continuous gradient. (Reprinted, with permission, from Takeuchi, I., van Dover, R.B., and Koinuma, H. MRS Bulletin, 2002, 27(4) 301–308; copyright © 2002 Materials Research Society, Warrendale, PA.) See color insert for color reproduction of this figure.

optimization rather than screening. Arrays of 6, 12, 24, and 48 reactors are typical, and the reaction and analysis times can be relatively long, so the throughput with these arrays is typically 10–100 samples/day.

1.4.3 Robot-Mixed Formulations

High-speed materials experimentation has had the advantage of access to a maturing technology of robotic materials handling equipment. This infrastructure, developed for the pharmaceutical industry, includes pipetting robots, handlers, and plate readers. Much of this was designed around the 96-well microtiter plate developed for biochemical screening. This makes the 96-well format a convenient one for materials development, although the robots are quite adaptable to other formats.

Well-based technologies are more flexible in the number and types of combinations possible than the thin-film technologies, but they are constrained by the need to physically move and mix components. This requires robotics. The most common robots in CHTE laboratories are pipettors that can accurately transfer 25–1000-microliter (μL) quantities using single or multiple pipet heads. The default design of these robots covers the 96-well plate, but most are reprogramable to other formats. Smaller quantities of material can be dispensed with very good accuracy using ink-jet-type devices. Given the speed of the robots and the need for some manual operations, the throughput of these systems is typically 1–5 well plates (96–480 samples) per day.

1.4.4 In Situ Mixing

This technology has primarily been used for the production of gradients or mixtures of thin-film materials. The earliest approach in the generation of thin-film libraries was the generation of continuous gradients of material on a substrate by sputtering several materials on a substrate at once. The sources are at an angle to the substrate, so the gradient is formed by the decrease of material with distance. This was first suggested by Kennedy et al. in 1965 [24] and Hanak [25] in 1970, but was not widely used for the next 20 years.

In its current form, sputtering has been joined by other methods for generating atomic or molecular vapors such as continuous vapor deposition (CVD) and evaporation. These technologies mix the materials on a molecular scale, but microdroplet deposition and mixing using ink-jet or spray methods is also possible. Continuous gradients can also be generated by physically spreading materials such as polymers on a surface [26]. In this approach the number of points is determined more by the resolution of the analysis tools than by the application technologies. Van Dover et al. have reported 4000 points on a 66 × 63 mm rectangle produced by sputtering three materials on a square plate [27].

A second use of in situ mixing has been the generation of thin-film libraries using the masking technologies pioneered by the semiconductor industry [28]. Deposition methods can include all the molecular-scale methods mentioned earlier. In this case, the materials are deposited in layers and annealed into a uniform film. The form factor has tended to follow a binary pattern. This began with the Xiang and Schultz 16-sample library, which quickly grew to 128 samples [4] and then to a 1024-sample library [29]. Addition of a shutter to provide a continuous gradient of material compositions has allowed preparation of ~25,000 distinct compositions on a single 25 × 25 mm substrate [30].

1.4.5 Organic Synthesis Technologies

Combinatorial organic synthesis is by far the most developed technology in the CHTE arena. Pharmaceutical development of methods and equipment has evolved extremely rapidly because of the large research budgets of the drug companies. The major method is solid-phase synthesis; either parallel synthesis using compounds affixed to pins or the like and "split and mix" synthesis with compounds attached to beads. The first methods typically produce hundreds to thousands of discrete compounds and the second from thousands to hundreds of thousands. Automated liquid-phase parallel synthesizers such as the Chemspeed (http://www.chemspeed.com) or the Myriad (http://www.mtmyriad.com/ps.htm) are more likely to be useful in materials development, where 10- to 1000-mg supplies of compounds are often needed. Liquid-phase synthesizers have 8–64 vessels and can be cycled once or twice a day. The rate-determining step in synthesis technologies is development of robust synthetic methods that can be applied to a wide range of compounds. This takes weeks to months, while the actual synthesis is done in a few days.

This area is thoroughly covered in books [31–34]; journals such as *Combinatorial Chemistry* (John Wiley & Sons) and the *Journal of Combinatorial Chemistry* (American Chemical Society); Web sites (www.combinatorial.com, www.combichemlab.com); and by companies such as Pharmacopeia (www.pcop.com), Arqule (www.arqule.com), Argonaut (www.argotech.com).

1.5 A SYSTEMATIC APPROACH TO PLANNING FOR A DESIGNED COMBINATORIAL EXPERIMENT

1.5.1 Background

Regardless of the excitement of using new and powerful technology for conducting an experiment, the "planning activities that precede the actual experiment are critical to successful solution of the experimenters' problem" [35]. If anything, planning must be even more careful, since we now have the opportunity of going in the wrong direction faster than ever. Montgomery and Coleman's classic paper, "Planning for a Designed Industrial Experiment" [35], contains a Predesign Master Guide Sheet (Table 1.4) that is still largely applicable for CHTE work. Their detailed discussion on each of the elements in the guide is also worthwhile reading. I will follow their outline, and then add additional elements unique to CHTE. This guide is useful both on the macroscale, in defining an entire program of experi-

Table 1.4   Predesign Master Guide Sheet

1. Name, organization, title
2. Objectives
3. Relevant background
4. Response variables
5. Control variables
6. Factors to be "held constant"
7. Nuisance factors
8. Interactions
9. Restrictions
10. Design preferences
11. Analysis and presentations techniques
12. Responsibility for coordination

Source: Reprinted, with permission, from Montgomery, D.C.; Coleman, D.E. *Technometrics* 1993, **35:** 1–12. Copyright © 1993 American Society for Quality.

ments that may cover months, and on the microscale, defining individual experiments.

1.5.2   Objectives

In a typical CHTE program, we know what we are looking for—catalyst activity, phosphor color or luminescence, polymer toughness, electronic properties, and so on. However, it is worthwhile to have an in-depth discussion with representatives of the key customers to ensure that all the critical needs are known, both qualitatively and quantitatively. The tools of Customer Needs Mapping and Quality Function Deployment can be very useful for this process [36]. This must be an ongoing process—target creep and mutation are very common. The objectives should also be "(a) unbiased, (b) specific, (c) measurable, and (d) of practical consequence" [35]. Once established, the objectives must be prioritized, because it is difficult to analyze for more than one or two critical objectives at maximum throughput. Those "most critical" objectives become the response variables to be measured at the high throughput stage, while less critical objectives become response variables during the secondary and tertiary stages.

1.5.3   Relevant Background

Many of the problems studied using CHTE methods have had years or even decades of study using conventional experimental methods. These will (at least) give some starting places for the study, but the team needs to be careful about being too limited by them. One common phenomenon is the moth-around-the-flame effect, in which experimentation consists largely of modest excursions from a known (pretty good) center, because bold experimentation is too expensive. The lively interplay between theory and experiment that marks a good scientific study should be,

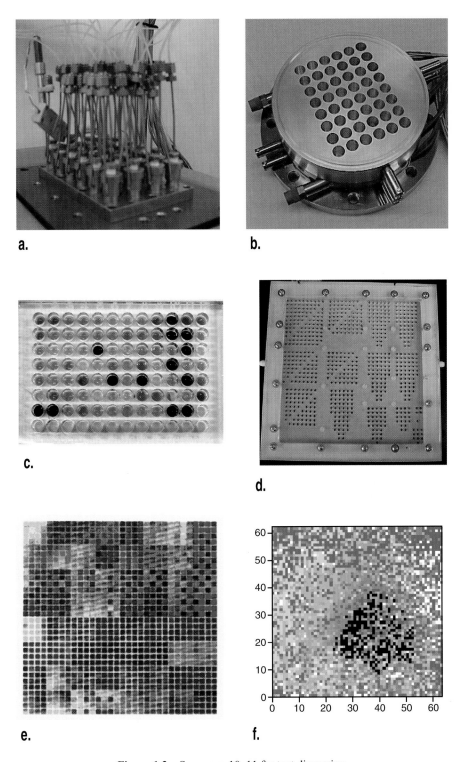

a.

b.

c.

d.

e.

f.

Figure 1.3 See pages 10–11 for text discussion.

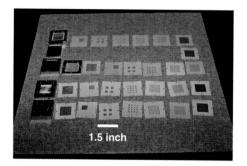

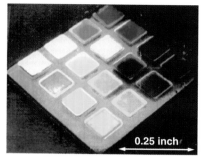

(a)

(b)

Figure 3.1 (a) and (b). See page 43 for text discussion.

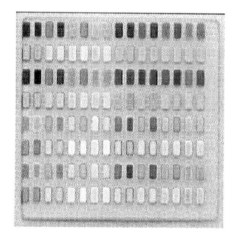

Figure 3.3 (a). See page 49 for text discussion.

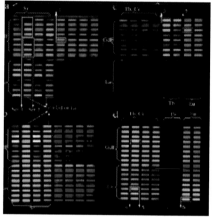

Figure 3.5 (a). See page 51 for text discussion.

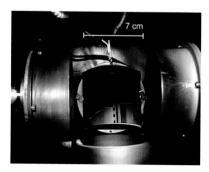

Figure 4.2. See page 61 for text discussion.

Figure 4.5. See page 65 for text discussion.

if anything, accentuated in CHTE because of the rapid feedback from experimentation.

A critical element of the background is the available resources. This is multivariate, with elements of personnel, stakeholders, money, machines, materials, and time.

1.5.4 Response Variables

A high-speed screening program can generally analyze for only one or two critical responses at a rate that matches the high throughput of sample preparation and processing. Definition of the method for performing the high-speed analysis is often the crucial step in setting up a new CHTE target. Often the analysis cannot measure the customer-critical property directly; instead an easily measurable property that correlates well with the critical property must be chosen. Sometimes a binary response (on or off, present or absent) is sufficient information for identification of a hit; in other cases, a quantitative measured response is necessary. This has a profound effect on all aspects of equipment design and experimental design. The remarkable progress that has been made in devising both parallel and rapid serial analytical methods for CHTE has been summarized in a recent review [37].

1.5.5 Control Variables

This part of the planning process is much more expansive than the factor definition in Montgomery's guidelines. Here we are defining an entire *experimental space,* which may contain:

- Qualitative formulation factors, such as specifically identified elements or compounds.
- Quantitative formulation factors, including both ranges and intervals.
- Process factors.
- Quantitative or qualitative levels of the process factors.
- Permutations of substituents on chemical structures.

The possible number of combinations of these factors is essentially unlimited. One of the critical intellectual tasks in the planning process is selection of a chemical space that is the right size. If the space is too restricted, the search will circle around known optima; if too expansive, it will contain too much sterile terrain.

1.5.6 Factors to Be Held Constant and Nuisance Factors

A CHTE "factory" usually has a number of process steps, each of which can introduce variation. Controlling this variation so that the output data from the factory has meaning requires intense attention to design and operation. The sources of variation in each process step can be considered as nuisance or noise factors in the statistical sense. A selection of the commoner nuisance factors is given in Table 1.5. We have found Six Sigma methodology to be a useful framework to design and maintain quality in the factory [36].

1.5.7 Interactions

CHTE is all about interactions, and usually high-order interactions. In many of the targets studied, the main effects and frequently the two-way interactions of the major process variables have been thoroughly studied. New and substantially enhanced properties will be found in high-order interactions, such as:

- A new phase in a three-way or higher mixture of electronically active ingredients [27,38]
- Synergistic ingredients in a catalyst formulation [39]

A sense of the degree and type of interactions to be searched is important in deciding on the experimental strategy and designs to be used. This can take the form of an estimate of the domain size of a new phase in a gradient search; the number of interactive components in a reduction/oxidation catalyst cycle; or the number of unique monomer combinations in a polymer.

1.5.8 Restrictions and Design Preferences

Every decision made during the engineering design of the CHTE factory will become a restriction on the factory operation. Therefore there must be an iterative process linking the customer requirements with the chemical analysis system and the reactor design specifications (Figure 1.4). Simply stated, both the analysis and

Table 1.5 Nuisance Factors in CHTE

Formulation Experiments	Gradient Experiments
Stock solution	Source
Chemical source	Purity
Amount	Uniformity
Sample	Sample
Amounts	Annealing
Total volume	Crystallization
Reactor	Processing
Temperature control	Thickness
Pressure control	System gases
Reaction time	
Analysis	Analysis
Component identification	Positional accuracy
Quantification	Quantification

the reaction system have to be sufficiently capable to allow identification of leads or hits above the composite noise of the systems. However, the ability to perform certain high throughput analyses is a function of the type of reaction and reactor employed, and similarly the reaction system can be configured to more easily allow rapid in-line or off-line analysis. Typical restrictions include:

- The standard size of the CHTE array, such as a 96-well plate.
- The sources, masks, and shutters of a system used for producing landscape libraries.
- The capabilities of the robot used for charging the ingredients, including:
  - Minimum and maximum aliquot size.
  - Accuracy of addition (as a function of aliquot size, viscosity, etc.)
  - Speed of completion of the array (which may affect sample stability, solvent evaporation, etc).
- Ramp up, ramp down, and uniformity of any processing steps.
- Speed, precision, and resolution of the analysis system.
- Capability of the analysis system to perform in situ measurements.

Because many CHTE systems operate with a fixed array size, fitting an optimized design to the array is often procrustean. Therefore issues of array-to-array consistency, blocking, and split-plot design must be included in the overall experimental plan. This is discussed in Chapter 8.

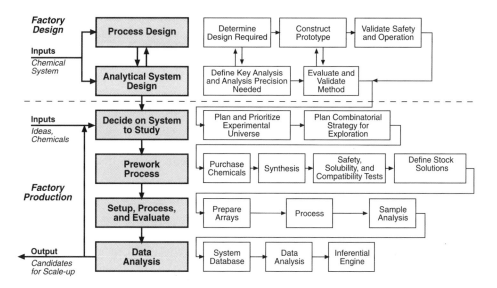

Figure 1.4 Process map. Interrelationships between design and construction, and operation of a CHTE system.

1.5.9 Analysis and Presentation Techniques

Analysis of the data from a CHTE study depends very much on the objectives and underlying design. In pure screening mode, a simple pick-the-winner strategy, with confirmation and optimization experiments following, may suffice. More complex strategies will require all the tools of statistical analysis and data mining. Visualization of multidimensional spaces is a major challenge; software such as Matlab [40] and Spotfire [41] is useful.

1.5.10 Responsibility for Coordination of the Experiment

The structure of a CHTE laboratory could be the subject of at least a full article if not a book. We have found that there are a number of roles that must be allocated:

- *Domain experts:* understanding of the chemistry and the commercial implications of the target.
- *Analytical team:* development of high throughput analyses for the critical parameters.
- *Synthesis team:* generation of noncommercial raw materials.
- *Engineering team:* development or modification of the reactor and robotic hardware.
- *Informatics team:* database, experimental design, data analysis, statistics, quality control, and visualization.

One effective structure for the overall laboratory is shown in Figure 1.5. Only the domain experts are completely dedicated to a single target. The other teams are set up to service all the targets. This improves personnel efficiency and improves cross-target communication and learning.

The intellectual effort in planning and analyzing the work in a CHTE system should not be underestimated. In our laboratory we have had instances where the opportunity to increase the throughput of the CHTE factory was rejected by the chemists because they could not generate new ideas and plans fast enough.

CHTE systems add areas of planning that were not addressed by Montgomery and Coleman. These include the handling and storage of the data from the experiment; multistage experimentation; and logistics.

1.5.11 Database and Data Handling

A CHTE program in full operation will generate anywhere from 200 to 200,000 experimental samples per week, and each sample will have at least 10, but frequently far more, data elements: factor names, factor settings, responses, and associated information. This deluge of data will swamp conventional methods of storage. We have found that a small CHTE program can operate reasonably well with the careful use of spreadsheet methods. Current spreadsheet technology (e.g., Microsoft

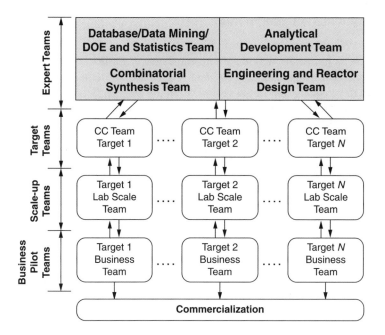

Figure 1.5 Organization of a CHTE laboratory.

EXCEL) can contain 230 columns × 65,536 (2^{16}) rows, so careful column assignment and the use of filtering functions will enable storage and retrieval of the data from a modest CHTE program. This works best when a single person with a good memory is the experimental planner and data analyst. Larger CHTE programs involving an operating team are better off with a full hierarchical database. Ideally this database should include a chemical structure search capability using such software as ISIS [42] or Chemdraw [43]. A database of this magnitude is not a small project and should not be underresourced.

One class of information that is often left out of database planning is the "metainformation" about the experiment. This includes:

- Connections to preceding experiments, literature, and experience.
- The geometry of the array design.
- Qualitative and anecdotal observations.
- Statistical analysis.
- Conclusions.
- Connections to follow-up experiments.

A well-designed data storage system should include a searchable capability for this type of information.

1.5.12 Multistage Experimentation

The high throughput screening process is only the first stage of a development project. It must be intimately tied to subsequent stages of further screening, optimization, lab-scale and pilot-scale development (Figure 1.6). After a lead is discovered, second-stage screening and initial optimization are frequently done using much of the same equipment as was used for the primary screen. Typically, the hit and its immediate neighbors are retested for confirmation and initial establishment of the active region. The optimization stage is a search for main effects, 2-factor interactions, and curvature in the region of the hit. The relatively low cost of high throughput runs allows for less parsimonious experimentation, so we frequently see full factorials or lightly fractionated designs in this stage. The confirmed and optimized lead is then scaled up to conventional laboratory equipment for detailed testing of all the customer requirements. Scale-up tends to become a bottleneck once the CHTE factory is at full production, so the scale-up teams should be in place and prepared for a rapid flow of candidates.

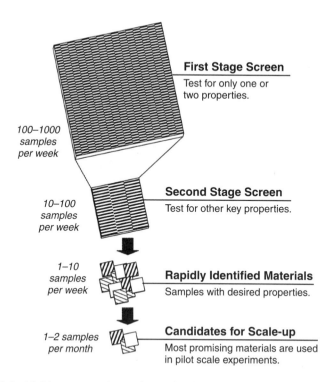

Figure 1.6 Multistage experimentation. The flow from screening to the pilot plant. (Reprinted, with permission, from Cawse, J. N. *Accounts of Chemical Research,* 2001, **34:** 213–221; copyright ©2001. American Chemical Society, Washington, DC.)

1.5.13 Logistics

Just like Henry Ford's assembly line, the CHTE factory must be a marvel of coordinated action. The Symyx mantra is "Analyze in a day what you make in a day" [44], and this applies to every other step in the factory as well. A full process map of the steps in the factory should be assembled and studied, both on a day-to-day basis and a longer-term prospect. Cross-training of personnel may be required to ensure that the project does not come to a stop when a team member is sick or on vacation. Raw material supply can be an issue; a project can come to a halt when it uses up every commercial variant of a critical component, if the synthesis team is not up to speed for generating new materials. Analysis of the data and recommendations for new experiments is often a slow step. Finally, a structure for communicating the progress of the experiment to the team members is essential. This is best tied in with the database.

1.6 STATISTICAL DESIGN ISSUES

1.6.1 Overview

The field of combinatorial and high throughput materials development is new enough that there has been relatively little examination of the statistical issues in the design of CHTE experiments. One concept that has reached general usage is the idea of *experimental space,* which in this work is the total number of combinations of all the factors and levels that may be varied in the system. It is clear from the previous discussion that this space can be unimaginably large. However, there has been little examination of the geometric and mathematical properties of this space that might help us define search strategies.

1.6.2 The NK Model

Fortunately, there have been important advances recently in a related area—the structure of the genome. The combinatorial issues there are at least as large; a mere decapeptide (with 20 amino acid choices per site) can have $20^{10} \approx 10^{13}$ possibilities, and a minor bacterium with 500 genes can have $2^{500} \approx 10^{150}$ potential genotypes. Kauffman [45,46] has formulated the concept of *correlated fitness landscapes,* which correspond very well with our ideas of experimental space. These landscapes have different degrees of "ruggedness," which bears directly on our ability to search the space and find new optima. Our concept of *factors* from traditional DOE (which can take on various levels and interact with each other) corresponds with *genes* (which similarly can take on various properties and interact with each other).

The essence of Kauffman's NK model (which is adapted from the physicist's spin-glass model) is the level of complexity that arises from K interactions of N genes (factors), where K can vary from 0 to $N - 1$. "The higher K is—the more interconnected the genes are—the more conflicting constraints exist, so the landscape becomes ever more rugged with ever more local peaks." If $K = 0$, the "landscape is a 'Fujiyama' landscape, with a single peak falling away along smooth gradual slopes" [46]. This is entirely analogous to a *main-effects* model (Figure 1.7a) in experimental design. Similarly, a $K = 1$ landscape is one of two-way interactions (Figure 1.7b) that can easily be handled by standard 2-level fractional factorial DOEs. I suspect that *the landscape of combinatorial chemistry is one where $K = 2$ or higher* (Figure 1.7c). In fact, I surmise that the landscape of problems that are approachable with our current tools is one where $K = 2$ to (at most) 4 or 5. There are fascinating opportunities for theoretical study in this area.

Some of the properties of these moderately rugged landscapes that may be applicable to our CHTE systems are (from [46]):

- "When K is low, peaks cluster near one another like the high peaks in the Alps . . . the landscape is *nonisotropic*."
- As K increases:
  - "Landscapes become increasingly rugged and multipeaked, while the peaks become lower."
  - "The high peaks spread apart from one another . . . the landscape is *isotropic*. . . . There is no point in searching far away on an isotropic landscape for regions of good peaks; such regions simply do not exist."
- "Landscapes with moderate degrees of ruggedness share a striking feature: it is the highest peaks that can be scaled from the greatest number of initial positions!"
- In searching a moderately rugged landscape by taking "long" jumps around it, the probability of finding a better region decreases exponentially.

This implies that a wise project strategy would therefore be to experiment boldly at the start, but switch to optimization of the best regions long before the project resources are exhausted. CHTE has the invaluable property of psychologically ener-

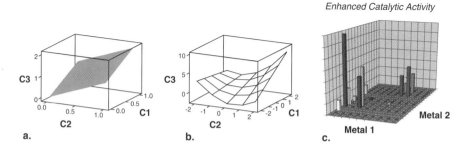

Figure 1.7   Landscapes. (a) Main effects only; (b) two-way interactions and curvature; (c) rugged correlated combinatorial landscape. (Figure 1.7c reprinted, by permission, from Cawse, J. N. *Accounts of Chemical Research,* 2001, **34,** 213–221, copyright ©2001 American Chemical Society, Washington, DC.)

gizing the research team to bold experimentation. By continually extracting useful scientific generalizations from the data we obtain in our screening, we can start to predict the large-scale features that will lead us to the optimum regions of our experimental space.

The landscape metaphor is a valuable tool in our development of methods for effective navigation of combinatorial space. It will be used extensively in Chapters 12 and 13.

1.6.3 Classic Statistical Issues

Regardless of the landscape we are exploring, the critical issues of statistically based experimental design remain: we are trying to make progress in a system containing experimental error while making efficient use of experimental resources.

Quality There has been relatively little examination of the quality issues in the generation of combinatorial arrays. Hortatory articles have appeared discussing high error rates in pharmaceutical high throughput efforts. In particular, ". . . some new users seem to be so mesmerized by the technology's power that they are forgetting basic principles of experimental design" [47]. In combinatorial and high throughput work, there are too many samples for each data point to be checked individually, and too many process steps for human oversight. Therefore all the steps of the process must be high quality. A defect minimization process such as Six Sigma must be in place during the design and operation of the factory.

Quality parameters that must be considered in planning the factory include:

- What kinds of improvements are we looking for? Are we looking for big jumps or incremental improvements in the output of our system? How big are the improvements needed relative to the current baseline?
- What risks are we willing to take? What are the consequences of false positives and false negatives? With what certainty do we want to declare a point a hit? Since multiple measurements are being made in parallel, care must be taken with simple statistical significance procedures inflating the rate of false positives. A false discovery rate procedure [48] has been proposed to improve the power of statistical tests in this regard.
- What is a lead? In the words of Gregory Bateson, what will "generate the difference which becomes information by making a difference" [49]?

From these parameters we can estimate an acceptable overall defect level for the process, and what level of replication may be required to deal with the actual defect level. They will also inform the search for the actual experimental designs with the statistical requirements required for successful experimentation.

Efficiency In the following chapters, a wide variety of approaches to exploration of a complex experimental space will be discussed. The efficiency and effectiveness of these approaches will depend on the detailed structure of the space to be examined and the tools available. However, basic statistical principles that apply to any investigation must be considered in the development of an experimental design. These include

- *Resolution.* In classic DOE, Resolution is the ability of a design to evaluate an effect or interaction [9]. In a Resolution III design, for instance, the main effects are confounded (mathematically combined so as to be indistinguishable) with 2-factor interactions. There is also the sense of pure physical resolution (as in the resolving power of a camera lens). This applies to gradient systems in which there is continuous variation of composition from point to point. How close can we physically place our sampling points? An understanding of the Resolution of our designs will help us understand what effects we can see (and what we cannot).

- *Point distribution.* Combinatorial space is vast and our resources are limited. We need to find effective ways to distribute our experimental points to sample the space. Fractional Factorial designs and the later D-Optimal designs were revolutionary improvements, allowing complex experiments to be performed with reasonable experimental resources. We are searching for the next generation of fractional and optimal designs.

- *Nesting.* In CHTE we are dealing with different sizes of experimental units [50]. Samples and arrays are formulated and processed in different ways. The error structure often contains sources of variation at the mixing, dosing, formulation, and processing level. If we wish to perform detailed statistical evaluations on these arrays, appropriate assignment of degrees of freedom and sums of squares must be made.

- *Signal/noise ratio.* An understanding of the overall quality of our system, and of the degree of improvement that constitutes a hit, will allow us to evaluate the signal/noise ratio, and consequently the probability of observing a hit correctly.

- *Blocking.* In our experience, the statistically preferred design generally does not fit in the fixed number of sample points required by the hardware. This will result in either statistical blocking (where the design is larger than the array) or a decision on the best use of the extra space. This space can be used for standards, replicates, comparison points, or minidesigns and "wild hair" experiments.

The following chapters detail a number of promising approaches to the issues of design for combinatorial and high throughput materials development. None are the last word on the subject.

REFERENCES

1. Jandeleit, B.; Schaefer, D. J.; Powsers, T. S.; Turner, H. W.; Weinberg, W. H. *Angew. Chem. Int. Ed.,* 1999, *38:* 2494–2532.
2. Bohacek, R. S.; McMartin, C.; Guida, W. C. *Med. Res. Rev.,* 1996, *16:* 3–50.

3. Borman, S. *Chemical and Engineering News,* February 24, 1997, 43–62.

4. Xiang, X.-D.; Sun, X.; Briceno, G.; Lou, Y.; Wang, K.-A.; Chang, H.; Wallace-Freedman, W. G.; Chen, S.-W.; Schultz, P. G. *Science,* 1995, *268:* 1738–1740.

5. Mittasch, A. In *Advances in Catalysis,* Vol. 2, Frankenburg, W. G.; Rideal, E. K.; Komarewsky, V. I., Eds., Academic Press, New York, 1950, 81–104.

6. Fisher, R. A. *The Design of Experiments,* Oliver and Boyd, Edinburgh, 1935.

7. Steinberg, D. M.; Hunter, W. G. *Technometrics,* 1984, *26:* 71–97.

8. Daniel, C. *Applications of Statistics to Industrial Experimentation,* Wiley, New York, 1976.

9. Box, G. E. P.; Hunter, W. G.; Hunter, J. S. *Statistics for Experimenters,* Wiley, New York, 1978.

10. Cornell, J. A. *Experiments with Mixtures: Designs, Models, and the Analysis of Mixture Data,* 2nd ed., Wiley, New York, 1990.

11. Stat-Ease, 6th ed. Stat-Ease, Inc., Minneapolis, MN, 2000.

12. JMP Statistical Discovery Software, SAS Inc., Cary, NC, 2001.

13. Minitab, 12th ed. Minitab Inc., State College, PA, 1999.

14. Echip, Inc., "Experimentation by Design" [last access April 4 2002]; available at http://www.echip.com.

15. NIST, *"Engineering Statistics Handbook"* [last access April 4, 2002]; available at http://www.itl.nist.gov/div898/handbook/.

16. Harry, M. J. *The Vision of Six Sigma: A Roadmap for Breakthrough,* Sigma, Phoenix, AZ, 1994.

17. Colbourn, C. J.; Dinitz, J. H. *The CRC Handbook of Combinatorial Designs,* CRC Press, New York, 1996.

18. NIST. "Technology Roadmap for Combinatorial Methods," National Institute of Standards and Technology, Gaithersburg, MD, 2001.

19. NIST. "Combinatorial Methods at NIST," National Institute of Standards and Technology, Gaithersburg, MD, 2001.

20. McCoy, M. *Chemical and Engineering News,* September 20, 1999, 17–25.

21. Department of Energy. "Vision 2020 Catalysis Report," Department of Energy, Washington, DC, 2001.

22. Department of Energy. "Materials Technology Roadmap," Department of Energy, Washington, DC, 2001.

23. NIST. "Combinatorial Methods for Coatings Development," Advanced Technology Program, National Institute of Standards and Technology, Gaithersburg, MD, 1999.

24. Kennedy, K.; Stefansky, T.; Davy, G.; Zackay, V. F.; Parker, E. R. *J. Appl. Phy.,* 1965, *36:* 3808–3810.

25. Hanak, J. J. *J. Mater. Sci.,* 1970, *5:* 964–971.

26. Meredith, J. C.; Karim, A.; Amis, E. J. *Macromolecules,* 2000, *33:* 5760–5762.

27. Van Dover, R. B.; Schneemeyer, L. F.; Fleming, R. M. *Nature (London),* 1998, *392:* 162–164.

28. Pirrung, M. C. *Chem. Rev.,* 1997, *97:* 473–488.

29. Wang, J.; Yoo, Y.; Gao, C.; Takeuchi, I. S., X.; Chang, H.; Xiang, X.-D.; Schultz, P. G. *Science,* 1998, *279:* 1712–1714.

30. Danielson, D.; Devenney, M.; Giaquinta, D. M.; Golden, J. H.; Haushalter, R. C.; Mc-

Farland, E. W.; Poojary, D. M.; Reaves, C. M.; Weinberg, H.; Wu, X. D. *Science,* 1998, *279:* 837–839.

31. Sucholeiki, I., Ed. *High-Throughput Synthesis,* Marcel Dekker, New York, 2001.

32. Jung, G., Ed. *Combinatorial Chemistry: Synthesis, Analysis, Screening,* Wiley, New York, 1999.

33. Ghose, A. K.; Viswanadhan, V. N. *Combinatorial Library Design and Evaluation,* Marcel Dekker, New York, 2001.

34. Gordon, E. M.; Kerwin, J. F., Eds. *Combinatorial Chemistry and Molecular Diversity in Drug Discovery,* Wiley-Liss, New York, 1998, 400.

35. Montgomery, D. C.; Coleman, D. E. *Technometrics,* 1993, *35:* 1–12.

36. "Method and Apparatus for Using DFSS to Manage a Research Project," U.S. Patent applied for. Available from World Intellectual Property Organization, WO 01/16785 A2, Cawse, J. N. (to General Electric Co., Inc.).

37. Potyrailo, R. A. In *Encyclopedia of Materials: Science and Technology,* Lifshin, E.; Cahn, R., Eds., Elsevier, New York, 2002.

38. Sun, T. X.; Xiang, X.-D. *Appl. Phy. Lett.,* 1998, *72:* 525–527.

39. Sun, Y.; Buck, H.; Mallouk, T. E. *Anal. Chem.,* 2001, *73:* 1599–1604.

40. Mathworks, "MatLab Website" [last access 2001]: available at http://www.mathworks.com/products/matlab/.

41. Spotfire, "Spotfire Website" [last access March 2002]: available at www.spotfire.com.

42. MDL Information Systems, "MDL Website" [last access August 2002]: available at http://www.mdli. com/products/isis.html.

43. CambridgeSoft, "ChemDraw Website" [last access December 2001]: available at http://www.camsoft.com/.

44. Cohan, P. In *Combi 2000* (*Knowledge Foundation*), San Diego, CA, 2000.

45. Kauffman, S. *The Origins of Order,* Oxford University Press, New York, 1993.

46. Kauffman, S. *At Home in the Universe,* Oxford University Press, New York, 1995.

47. Knight, J. *Nature,* 19 April 2001, 861–862

48. Benjamini, Y.; Hochberg, Y. *J. R. Stat. Soc. B,* 1995, *57:* 289–300.

49. Bateson, G. *MIND AND NATURE: A Necessary Unity,* Bantam Books, New York, 1980.

50. Milliken, G. A.; Johnson, D. E. *Analysis of Messy Data,* Van Nostrand Reinhold, New York, 1984.

CHAPTER 2

THE COMBINATORIAL DISCOVERY ARRAY: AN OVERVIEW OF THE CHAPTERS

JAMES N. CAWSE
GE Global Research

2.1 INTRODUCTION

Materials development is a highly varied field, so no one design strategy will fit all experimental situations. In fact, even in a single area of materials development a researcher will need a range of designs and methods to meet the challenges. However, an overall map (Figure 2.1) of the most common research areas will prove valuable in organizing our thinking and finding key commonalities.

In the figure, material types are the primary distinguishers that influence the choice of experimental hardware, strategy, and design. There are three broad, general material types now being examined using combinatorial methods:

1. *New phases*. These are typically inorganic materials in which the critical properties are controlled by the crystal structure and detailed compositions of the crystalline matrix. A simple example might be the perovskite structure $A_{n+1}B_nO_{3n+1}$, in which A = (La, Y, rare earth)$^{3+}$ partially substituted with (Ca, Sr, Ba, Pb, Cd)$^{2+}$ and B = (Mn, Co, Ni, Cr, Fe) [1].

2. *New formulations*. These can be wide-ranging combinations of inorganic, organic, and polymeric materials. There is no particular unifying principle in their preparation, and they are typically made by robotic mixing of premade stock solutions.

3. *New compounds*. This is the entire region of combinatorial organic chemistry, which has been and continues to be in rapid development by the pharmaceutical industry.

Experimental Design for Combinatorial and High Throughput Materials Development, Edited by James N. Cawse.
ISBN 0-471-20343-2 © 2003 John Wiley & Sons, Inc.

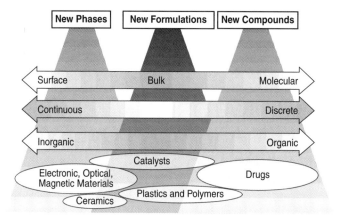

Figure 2.1   The combinatorial discovery array.

These general areas have some overlap, since both new phases and new compounds may be elements in new formulations. There are also three distinguishing tendencies as we move from phases to formulations to compounds:

1. *Geometry.* Phases tend to be studied as surface materials on a substrate; formulations are made and tested in bulk (which may be on a microliter scale); compounds are molecular, although they may be attached to polymeric supports.

2. *Quantitative vs. qualitative factors.* As we move across the array, the factors available for the design tend to shift from quantitative factors (such as compositions in a materials gradient) that can be measured on a continuous scale, to qualitative factors (such as the identity of elements in an perovskite structure), to discrete experimental units that each individual organic compound represents.

3. *Chemistry.* The general type of chemistry tends from inorganic to organic. Using the combinatorial discovery array as an organizing principle gives us some guide to the choices of experimental strategies and designs.

2.2   NEW PHASES

The ubiquity of electronic products based on planar geometries makes for a wide range of interest in inorganic film materials to meet a plethora of needs. Significant discoveries have been made in phosphors, scintillators, magnetic materials, superconductors, and other electronic materials (Table 2.1).

Finding a new material in this area requires search of the chemical universe at two levels. There is first the search for the just-right *combinations* of the elements; within each combination there is the search for the just-right *proportions* of those

Table 2.1 Materials Discoveries Using Thin-Film CHTE

Property	Material	Reference
Magnetoresistance	$La_x(Ba,Sr,Cs)_y-CoO_\delta$	[2]
Phosphor	$Gd_3Ga_5O_{12}/SiO_2$	[3]
Phosphor	Sr_2CeO_4	[4]
Dielectric	$Zr_{0.15}Sn_{0.3}Ti_{0.55}O_{2-\delta}$	[5]
Scintillator	$Sr-Ta-O/Tb^{3+}$	[6]
Organic light-emitting diode	$ITO/TPD/Alq_3$	[7]

elements. Engstrom and Weinberg called these *complexity* and *diversity,* and showed that, for instance, a search of the ternary combinations of 25 elements at a 5% interval in their proportions will require preparation of 399,025 compositions [8]. This can be multiplied further by the possible combinations of processing steps that may be required to form a functional material.

A fundamental concept used in generating libraries of materials in this area was formulated by Hanak [9]: "synthesizing, analysing, testing, and evaluating large parts of multicomponent systems in single steps." In this concept, gradients of materials are generated using methods that inherently produce such gradients, such as off-axis sputtering. Combinations of gradients of different materials generate a continuous composition gradient. This is typically two or three dimensional, with the gradients at 180° or 120° to each other. This method is an efficient tool for generating different *proportions* of active materials. The method languished, however, because it was ahead of its time in resolution of the evaluation tools and in computer technology. Schneemeyer and Van Dover discuss its revival in Chapter 4.

A fundamental limitation to this codeposition methodology is its restriction to systems of two or three formulation factors in a given area. In both the 2-factor and 3-factor case, the entire composition space can (in theory) be laid down on a plane. That 2-factor space can be represented on a two-dimensional surface is obvious; 3-factor space can also be represented on a two-dimensional surface because composition space is constrained. Three components must sum to 100%, so one degree of freedom is lost and a two-dimensional representation is possible (Figure 2.2). A system with four or more composition variables is therefore inherently ≥three-dimensional and cannot be projected on a plane.

In the mid-1990s, Xiang et al. added a technique from semiconductor manufacture: the use of masks to delineate distinct areas on the array [10]. These masks allowed introduction of different *combinations* of active materials to each element of the array. Their original 16-member binary library quickly grew to 128 members, then jumped to 1024 members with the introduction of the fractal quaternary mask system [3]. Superimposition of gradients on the masks by use of moving shutters allowed multielement gradient arrays in which *both* proportions and combinations were varied. These have been reported with arrays as large as 25,000 members, containing 27 2-factor gradients in block combinations with other elements [11]. A personal view of the development of this technology is given by Sun in Chapter 3.

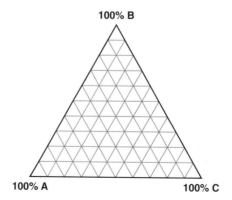

Figure 2.2   Two-dimensional representation of a ternary system. (Reprinted, by permission, from Cawse, J. N. *Accounts of Chemical Research,* 2001, 34: 213–221; copyright © 2001. American Chemical Society, Washington, DC.)

Scientists have been able to use this technology to build on the advances in miniaturization of both manufacture and measurement in the semiconductor industry, so the number of samples and measurements per array is quite high. The range of practical applications is 10^3 to $>10^4$ samples per array.

2.3   NEW FORMULATIONS

The most common targets of discovery efforts in the formulations area are new catalytic materials. Catalysts can be molecular species, new formulations, or new phases, but the largest amount of reported CHTE catalyst research is in the new formulations category. We will therefore discuss this category with catalysis as the primary example.

Once again, the challenge of discovering new formulations requires investigation of both complexity and diversity. Here, however, the complexity is not constrained by the restrictions of fitting elements in a crystal lattice. In most homogeneous and heterogeneous catalysts, each ingredient in a formulation can be selected from a large inventory of materials. A typical homogeneous catalyst formulation might include some or all of the materials given in Table 2.2.

It is easy to see that the number of choices available in this system can easily exceed 10^6, and the numbers of levels shown are probably modest.

Heterogeneous catalyst discovery must similarly search a vast chemical space. A typical heterogeneous catalyst experiment may include many choices of substrate; a mixture of 2 to 5 or more active metal species; and varied preparation and activation steps.

In either case, a practical commercial catalyst must be an incredibly robust system, since it must promote the formation of a low-energy transition state of the de-

Table 2.2 Constituents in a Homogeneous Catalyst Formulation

Factor	Type	Possible Choices (number of qualitative levels)	Possible Amounts (number of quantitative levels)
Primary catalyst	Group VIIIb metal compound (Pt, Pd, Rh, Ir, Ru)	1–6	1–3
Inorganic cocatalyst	Groups IVB, VB, VIB, VIIB metal compounds	12	2–4
Ligand	Organic compound	$10–10^3$	1–3
Counterion	Halides, many others	3–10	1–3
Solvent	Commercial material	5–20	1

sired reaction in the presence of a mixture of reactants, impurities, by-products, and products. It must further be resistant to deactivation mechanisms such as sintering, poisoning, coke formation, and oxidation.

Some of the design approaches to this general area are "true combinatorial design strategies," where combinatorial mathematics is used to calculate and select samples from the possible combinations [12]. In these strategies it is important to understand the degree of synergy or interaction that is being investigated. Figure 2.3 shows that the number of possible n-way combinations rises very rapidly as the degree of interaction rises. With a large number of choices for each factor, exhaustively studying combinations beyond three-way will require the very highest throughput systems.

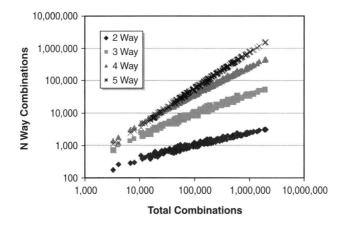

Figure 2.3 Total number of possible two-, three-, four-, and five-way combinations in a 5-factor experiment with 2–20 levels per factor. (Reprinted, by permission, from Cawse, J. N. *Accounts of Chemical Research,* 2001, 34: 213–221; copyright © 2001. American Chemical Society, Washington, DC.)

Fortunately, an experiment actually samples more than one combination of qualitative levels in each run. Thus, "in a five-factor combinatorial experiment, a single experimental run observes

- 10 two-way combinations (12,13,14,15,23,24,25,34,35,45)
- 10 three-way combinations (123,124,1125,134,135,145,234,235,245,345)
- 5 four-way combinations (1234,1235,1245,1345,2345)

Therefore the minimum required number of runs to observe all n-way combinations is much less than the total number of those combinations" [12].

Generating designs that observe all two-way combinations can be done with a standard Latin square or related designs. No such standard design exists for three-way and higher combinations, but they can be sampled using several strategies. These include:

- *Random runs.* These can be surprisingly effective in searching for three-way combinations. They also have the effect of removing the experimenter's biases and should be part of every CHTE arsenal.
- *Genetic algorithms.* These also have the advantage of being assumption free; they will work if there is any underlying structure to the experimental space. The critical design decisions in this methodology bear on the trade-off between the rate of convergence on the best material versus the certainty of convergence. Genetic algorithms are sometimes combined with neural nets. This is covered in detail in Chapters 9 and 10.
- *Exhaustive generation of all combinations of a given degree.* An algorithm developed for software testing can be used to generate these runs. This is available from a commercial Web site [13].
- *Uniform designs.* This type of design samples the system by design points that are uniformly scattered on the experimental domain. Uniform designs in high dimensions are quite difficult to generate (it is actually an NP-hard problem) [14]. Considerable effort has been made, particularly in China, to produce and tabulate a useful set of designs. They can be found at http://www.math.hkbu.edu.hk/UniformDesign.

Important limitations of these test plans are:

- They are highly dependent on the significant interaction effects being synergistic rather than antagonistic. Even a modest poisoning effect can obliterate a large portion of the design.
- They require that the desired high-order interaction effect be relatively large, while the main effects and low-order interactions remain small. Otherwise the desired observation will be drowned in the noise of the additive lower-order effects.
- The lack of redundancy requires that the quality of the experimental system be very high [15].

A second design type in the formulations region is the gradient design in which the gradient is sampled intermittently by robot-mixed formulations. This may be because the chemistry is inherently solution chemistry; the number of variable components in the gradient is greater than three; or because the sample evaluation requires relatively large amounts of material. In this type of problem, there may be a double combinatorial explosion: a large number of combinations of constituent materials compounded with large numbers of samples in each gradient. In this case, minimization of the number of points actually made and evaluated is important. Chapters 6, 7, and 13 detail current approaches to this type of design.

2.4 NEW COMPOUNDS

Searching for new compounds in the materials-development arena has distinct differences from the pharmaceutical. These include the following:

- The chemical space of a materials-development scenario is typically only part of the overall experimental space;
- The chemical space is far more constrained; frequently the key functional groups of the compound are defined;
- The resources, in time, money, and personnel, are far more limited.

For these reasons, a compound screen is most likely to begin with a search of commercially available chemicals with the right functionality. The Available Chemicals Database (www.mdli.com) can be effectively searched for such compounds. Once the screening of those compounds is complete, a synthetic effort will be required to supply new compounds. From practical experience, once the high throughput factory is up and running, it will screen through all the commercially available materials very quickly. If a synthetic effort is envisioned, it should be started early.

Searching compound space effectively requires some type of descriptors to move from a totally qualitative space to one with some level of quantititative character. This is the field of quantitative structure–activity relationships (QSAR), which has been active since the Hammet and Taft relationships were developed [16]. This is discussed in the drug-design context in Ghose and Viswanadhan [17]. For materials development we can consider four types of descriptors:

1. Measurable bulk or molecular properties of the compounds such as molecular weight, ionization potential, dipole moment, and partition coefficient;
2. Two-dimensional descriptors that express the topological properties of the molecule;
3. Three-dimensional descriptors that express the shape of the molecule;
4. Quantum-mechanical properties, such as highest occupied molecular orbital (HOMO) and lowest unoccupied molecular orbital (LUMO). Approaches to this area are presented in Chapter 11.

We do not expect these descriptors to directly correlate with the desired chemical behavior; however, these measured properties are related to the behavior because both "depend on the same *intrinsic molecular property*" [18].

There is a substantial literature on the approaches for combinatorial library design and evaluation in the pharmaceutical arena; a good starting place is the book by Ghose and Viswanadhan [17].

For materials development, a novel and promising approach would be to generalize the methods discussed by Carlson [18,19] for screening variations in organic synthesis. In his approach, a chemical space is defined generically by three axes: substrate, reagent, and solvent (Figure 2.4a). Sets of chemically reasonable descriptors and sets of candidate compounds are defined for each axis. Principal components analysis (PCA) is then used to determine the principal components for each axis, and a second-order Taylor series (interactions/quadratic) model is generated in all the principal components. Individual experimental runs are selected using a singular-value decomposition algorithm [20]. In generalizing this approach, the axes could be varied with the types of descriptors being used in the experiment (Figure 2.4b). A summary of the process is shown in Table 2.3.

2.5 PROCESS/FORMULATION DESIGNS

The discussion so far has focused exclusively on *formulation* factors, whether qualitative (the choices made for a given type of component) or quantitative (the proportion or amount of the component in the material). Real materials almost always go through some kind of process to express the desired property, and the final value of that property will be a function of both the formulation and the process. Therefore we must consider *process* factors, and the interactions between process factors and formulation factors. A few examples are given in Table 2.4

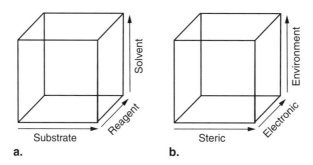

Figure 2.4 Representations of chemical spaces. (a) Organic reaction space, (b) generalized chemical space. (Figure 2.4a reprinted, with permission, from Carlson, R.; Carlson, J.; Grennberg, A. *J. Chemometrics,* 2001, 15: 455–474; copyright © 2001. John Wiley & Sons, New York.)

Table 2.3 Process Steps for Exploring Compound Space

Select a reaction theme
Key on an active center (e.g., nucleophile)
Enumerate major supportive centers (e.g., neighboring group participation)
Explore the standard variable possibilities
    Chain length
    Rings
    Aromatic/aliphatic
Test descriptors that could describe reaction parameters and rate-determining steps, e.g.,
    Size
    Dipole moment
    Ionization consant
    Partition coefficients
Define the virtual space (all combinations)
Define the available space (synthetic constraints)
Subset to a fractional design
Pick chemically reasonable descriptors
Use PCA to reduce dimensionality
Pick a subset and run

In some areas, methods have been developed to investigate process/formulation interactions within a single array. This is done by imposing a process gradient orthogonal to a formulation gradient. Karim et al. detail several approaches to this technique in Chapter 5.

It is more typical, however, for the CHTE process to be sequential, with arrays of premade formulations subjected to the process factors as a unit. This means that virtually all experiments of this type have a split-plot structure, with multiple sizes of experimental unit. The degrees of freedom and sources of error from formulation and from process must be carefully delineated [21]. If the experimental data are to be analyzed with anything beyond the simplest pick-the-winner approach, appropriate split-plot design and statistical techniques are required. Split-plot designs also have some very useful properties. They are often the most efficient design in com-

Table 2.4 Process Factors in Materials Discovery

Material Target	Process	Process Factors
Catalysts	Calcination	Temperature, time
Catalysts	Reaction	Temperature, pressure, gas composition, time
Polymers	Extrusion	Temperature, intensity, time
Coatings	Curing	Temperature, UV intensity, time
Electronic materials	Annealing, crystallization	Temperature, time

plex mixture/process situations, and they can be structured to enable accurate detection of high-level process/formulation interactions. Both the advantages and disadvantages are discussed in Chapter 8.

2.6 OPTIMIZATION STRATEGIES

Once a hit has been detected during the screening stage of an experiment, localization and optimization of the hit typically follows immediately. Localization in a gradient array can be done using the same strategy as screening, using a finer mesh. In a combinatorial array, the most common design strategy for optimization is simply a full factorial design with multiple levels for each factor. This becomes possible because the screening process typically reduces the numbers of factors and levels to an experimentally feasible set of combinations. This is aided by laboratory equipment that is inherently parallel in one dimension, such as multipipet robots or reactor arrays with orthogonal arrangements of gas pressure and temperature. At this stage, integration with the modeling tools that will facilitate scale-up becomes an important element of the experimental plant.

More sophisticated forms of optimization strategies are in the area of genetic algorithms (Chapter 9) and artificial neural networks (Chapter 10). These are tools for navigating highly complex, correlated, and nonintuitive experimental spaces.

2.7 VERY HIGH THROUGHPUT EXPERIMENTAL STRATEGIES

Most of the following chapters focus on the hardware, software, and strategic capabilities of the current generation of experimentation. A glance at our pharmaceutical neighbors makes it clear, however, that continued advances are likely. As pharmaceutical CHTE has progressed from 96-well to 384-well and 1536-well plates, we can expect materials CHTE to increase in throughput and complexity of problems to be addressed. Chapters 12 and 13 present two contrasting approaches for the next generation. In Chapter 12, the use of Monte Carlo methods is discussed, exploiting the virtues of randomness in finding high value regions in experimental space. Chapter 13 extols the exact opposite of randomness, finding extremely structured lattices that sample space with a high degree of efficiency. I expect that both methods will find effective application in the fascinating era of very high throughput experimentation that awaits us.

REFERENCES

1. Hsieh-Wilson, L. C.; Xiang, X.-D.; Schultz, P. G. *Acc. Chem. Res.* 1996, *29:* 164–170.
2. Briceno, G.; Chang, H.; Sun, X.; Schultz, P. G.; Xiang, X.-D. *Science,* 1995, *270:* 273–275.

3. Wang, J.; Yoo, Y.; Gao, C.; Takeuchi, I. S., X.; Chang, H.; Xiang, X.-D.; Schultz, P. G. *Science,* 1998, *279:* 1712–1714.

4. Sun, T. X.; Xiang, X.-D. *App. Phys. Lett.* 1998, *72:* 525–527.

5. Van Dover, R. B.; Schneemeyer, L. F.; Fleming, R. M. *Nature (London),* 1998, *392:* 162–164.

6. Sun, Y.; Buck, H.; Mallouk, T. E. *Anal. Chem.,*2001, *73:* 1599–1604.

7. Schmitz, C.; Posch, P.; Thelakkat, M.; Schmidt, H.-W. *Phys. Chem. Chem. Phys.,* 1999, *1:* 1777–1781.

8. Engstrom, J. R.; Weinberg, H. *AIChE J.,* 2000, *46;* 2–5.

9. Hanak, J. J. *J. of Mater. Science,* 1970, *5;* 964–971.

10. Xiang, X.-D.; Sun, X.; Briceno, G.; Lou, Y.; Wang, K.-A.; Chang, H.; Wallace-Freedman, W. G.; Chen, S.-W.; Schultz, P. G. *Science,* 1995, *268:* 1738–1740.

11. Danielson, D.; Devenney, M.; Giaquinta, D. M.; Golden, J. H.; Haushalter, R. C.; McFarland, E. W.; Poojary, D. M.; Reaves, C. M.; Weinberg, H.; Wu, X. D. *Science,* 1998, *279:* 837–839.

12. Cawse, J. N. *Acc. Chem. Res.,* 2001, *34:* 213–221.

13. Telecordia Technologies, "AETG Web" [last access April 3 2001], available at http://aetgweb.argreenhouse.com/.

14. Fang, K.-T.; Lin, D. K. J. "Uniform Experimental Designs and Their Applications in Industry," Hong Kong Baptist University, Hong Kong 2001.

15. Cawse, J. N. U.S. Patent Application 09/681222, 2001 (to General Electric Co., Inc.).

16. Gould, E. S. *Mechanism and Structure in Organic Chemistry,* Holt, Rinehart & Winston, New York, 1959.

17. Ghose, A. K.; Viswanadhan, V. N. *Combinatorial Library Design and Evaluation,* Marcel Dekker, New York, 2001.

18. Carlson, R.; Lundstedt, T. *Acta Chem. Scand. B,* 1987, *41:* 164–173.

19. Carlson, R.; Carlson, J.; Grennberg, A. *J. Chemometrics,* 2001, *15:* 455–474.

20. Press, W. H. *Numerical Recipes in C: The Art of Scientific Computing,* Cambridge University Press, Cambridge, UK, 1992.

21. Milliken, G. A.; Johnson, D. E. *Analysis of Messy Data,* Van Nostrand Reinhold, New York, 1984.

CHAPTER 3

FRACTIONAL MASKING METHODS IN COMBINATORIAL SYNTHESIS OF FUNCTIONAL MATERIALS

TED X. SUN

Intematix Inc.

3.1 INTRODUCTION

3.1.1 Conventional Material Research Approach and Combinatorial Principles

Materials chemistry is the oldest chemical science, beginning perhaps with the firing of clay pots in prehistoric eras or certainly with the smelting of ores in the Bronze Age. Centuries ago, alchemists developed a method to synthesize solid-state materials [1]: multiple raw materials were mixed together, thermally fused in a high-temperature furnace, and quenched for alloys. Although modern techniques allow searching for materials in a more rational way, the current material synthesis approach still inherits such "shake and bake" methodologies, and materials are usually synthesized and studied one at a time, which typically takes days of work in a laboratory.

At the current level of material and functional complexity, there is little definitive predictive power for composition, structure, and reaction pathways leading to an advanced material with desirable properties. "It is self-evident that one cannot by any rational process make a material for a particular desired application, almost all the highly important materials are found as a result of the largely empirical, partly serendipitous (chance plus informed response) process" [2]. The universe of possible new compounds remains largely unexplored due to the inefficiency of the experimental approach and lack of theoretical guidance, which usually needs more experimental data to support or develop for itself [3]. Up to 1989, only 24,000 inorganic phases were known, in which 8000 are ternary compounds, and 16,000 binary compounds [3]. Little work was done on solid-state phases with more than three metallic components.

Experimental Design for Combinatorial and High Throughput Materials Development, Edited by James N. Cawse. ISBN 0-471-20343-2 © 2003 John Wiley & Sons, Inc.

On the other hand, there is a single broad trend in materials research: the degree of complexity is increasing [4]. This is reflected not only in the structures and compositions of materials but also in the corresponding material synthesis and characterization techniques, and in the required material functions, which could be multiple. The drive to study complex materials and structures is that some novel material properties, which are hard to realize in simple material systems, may exist in complex material compositions. For example, a historic view on the increase of superconductor transition temperature (T_c) indicated that higher T_c results from an increase of compositional complexity for superconductors [3]. According to this trend, a room temperature superconductor exists only for material compositions with more than seven different elements. For about 60 tangible and nonradioactive elements from the periodic table, there are approximately two trillion different 7-element compounds with a single composition ratio, not considering possible variation of elemental composition ratios in each 7-element compound series.

Therefore, the question arises: Is there a more efficient experimental approach to search for advanced materials without reliance on predictive power? We can learn from the approach that Mother Nature uses in making and selecting her compounds or species, and she solved this problem with a combinatorial approach. For example, to select the species to survive in an environment, multiple species with different deviations from the previous generations were created and tested against the environment, with the fittest surviving and being selected [5]. Another good example is the human immune system, which can generate and screen up to 10^{12} different antibody molecules to identify the one that specifically recognizes and binds a foreign invading pathogen [6]. This process has to take place in a short enough time to avoid the disease, and it is essential to generate and screen large collections of molecules efficiently because of the complexity and unpredictability of the external world. Such a combinatorial approach, first introduced in the pharmaceutical industry, revolutionized the drug-discovery process [7].

Combinatorial chemistry is in essence based on the principle of parallelism [8], which allows chemists to prepare many more substances than the number of chemical operation steps by generating a large number of diverse molecules in a template and studying them simultaneously in a library.

3.1.2   History of Developing Discrete, Spatially Addressable Material Libraries

Prior to 1993, J. Hanak of RCA Laboratories made a limited effort to make a continuous compositional gradient of material systems [9]. An RF cosputtering technique of two metal targets was used to generate a continuous thin-film phase gradient of the two elements. There was no boundary between different samples, and it was applied to study the phase diagram of some binary superconductor alloys. A 30 fold increase in the rate of finding new materials were claimed. This elegant concept is the first step toward a combinatorial material process in material research. However, this continuous gradient method was not highly successful in discovering new materials, partially due to the following technical issues:

1. The number of elements in this approach is intrinsically limited. The continuous phase gradient method typically can be used to study compositional variation of only two or three elements, corresponding, respectively, to either a binary or ternary phase diagram.
2. Different samples are in contact with one another, increasing the chance of cross-contamination.
3. Thin-film libraries can only be made using a vapor-deposition approach. There are many materials used in powder forms that cannot be studied with this approach.

In 1993, a Berkeley group, led by Prof. Peter Schultz and Dr. X.-D. Xiang, started investigating combinatorial synthesis of materials. A series of combinatorial methods was developed to synthesize discrete material libraries in both thin-film and powder forms, using vapor-deposition and ink-jet dispensing techniques [13]. The material libraries made were then screened for various advanced physical properties, including superconducting [10], magnetoresistance [11], and fluorescence [12]. Most of our inventions and original works were published [14–17].

One of our inventions for making a discrete material library is the application of fractional masking methods to make thin-film libraries. To reach the goal of making various spatially separated material blocks on a substrate, we applied a dual-masking scheme, with a primary mask consisting of an array of holes under a series of secondary masks with interchangeable patterns of openings. By combining different patterns of such masks with the deposition of various amounts of different elements through the mask patterns, a large number of elemental compositions can be generated in parallel on a substrate.

We further expanded the scope of masking strategies for making discrete material libraries by developing both binary masking and a novel rotary quaternary-masking scheme. The details on these masking strategies are discussed in the next section. We also applied both contact photolithography masks and physical shadow masks to make material libraries. These masking methods were combined with various gaseous or liquid vapor-deposition apparatuses to make high-density discrete material libraries. Since different masking strategies need to be used in the context of various vapor-deposition methods, in the following section I will also discuss some vapor-deposition methods commonly used in our lab and their corresponding optimum masking methods. I applied these fractional masking methods and made various functional materials libraries at the Lawrence Berkeley National Lab (LBNL), including superconductors, magnetoresistance material, and luminescent materials. They are covered in the last section of this chapter.

3.2 FRACTIONAL MASKING METHODS AND VAPOR-DEPOSITION SYSTEMS IN MAKING THIN FILM MATERIAL LIBRARIES

Various high-vacuum thin film deposition apparatuses (sputtering, laser ablation, thermal and e-beam evaporation, ion-beam deposition, etc.) have been used in making thin film libraries. The key to the art of making a thin-film library using these vacuum-deposition systems is the masking strategy, which includes the choices of masking forms as well as masking schemes. So far three types of mask forms have been successfully used in making libraries: shadow mask [10,11,12], lithographic mask [18], and movable shutter mask [18,19]. The first two masks are better used for a wide-ranging screen of chemical space for elemental components with the desired properties; a shutter mask is more useful for optimizing the composition in a n identified system of materials. In addition, the natural thickness profiles generated from vapor deposition methods (e.g., sputtering) were applied to make continuous gradient material libraries without using masks [20].

3.2.1 Physical Masking Method (or Shadow Masking)

In this method, two masks serving different functions are overlaid and aligned and then placed on the top surface of a substrate. The primary mask consists of a metal grid, which spatially separates different samples on the substrate. The mask can be made from dual thin metal foils or a single stainless-steel sheet (Figure 3.1a). On top of the primary mask, a sequence of secondary masks can be overlaid, through which a controlled quantity of a thin-film precursor can be deposited. The sequence and pattern of the secondary masks determine the final stoichiometry of materials in the library. Figure 3.1b is the photo of the first discrete compositional material (superconductor) library ever made. I designed, deposited, and synthesized this library in early 1994 by combining sputtering deposition with a dual-layer masking strategy. This secondary masking pattern is binary masking, with each different mask covering one-half of the total primary mask.

The shadow masking method allows a fast, but low-density (100–1000 compositions /inch^2) synthesis of a library. The design of a thin-film library is, to a large extent, the choice of secondary masking schemes. Although there are virtually infinite formats of masking schemes, those that were developed and used most frequently at LBNL include binary masking (Figure 3.2a), quaternary masking (Figure 3.2b), and the movable shutter mask scheme.

3.2.2 Binary-Masking Scheme

In binary masking, one-half of the total primary masking area will be covered for each elemental deposition step, and none of these binary masks has an identical pattern of coverage. The number of different materials synthesized is 2^n, where n is the total number of masking and deposition steps. Binary masking can generate a large number of different compositions with a finite number of deposition steps, and it corresponds to very high masking efficiency in generating a discrete material library.

With the binary-masking method, every combination of the elemental components can be generated in the library. There is no grouping or discrimination of elements according to their elemental properties or periodic table position. This can generate mixtures of elements that are known not to form an alloy or compound. Therefore, some samples on the binary library may not have meaningful composi-

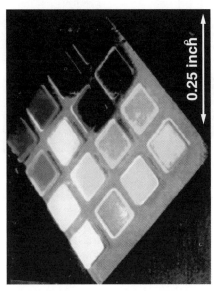

(a)

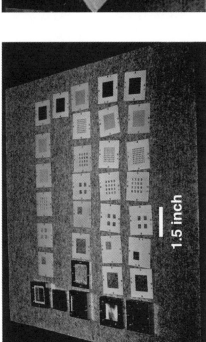

0.25 inch

(b)

Figure 3.1 (a) Some shadow masks that were used in LBNL for making spatially addressable material libraries. (b) A photo of the first discrete material library, composed of Bi, Sr, Ca, Cu elements in different compositions and spatially separated. (Unpublished). See color insert for color representation of this figure.

43

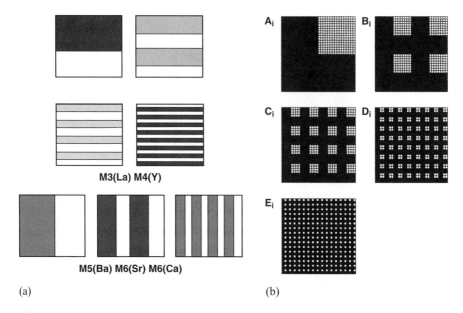

(a)   (b)

Figure 3.2 (a) A binary-masking scheme for making a 128-member material library. (Reprinted, with permission, from Xiang, X. D., et al. *Science,* 1995, 279: 1738–1740. copyright © 1995. American Association for the Advancement of Science, Washington, DC.) (b) The quarternary-masking scheme for making a 1024 member material library. (Reprinted, with permission, from Wang, J., et al. *Science,* 1998, 279: 1712–1714; copyright © 1998. American Association for the Advancement of Science, Washington, DC.)

tions (e.g., it is known that they cannot form a compound at all, or there are so many elements that it is impossible to form a single phase). In this case, other masking patterns were devised to increase the efficiency of the materials search with libraries, using input knowledge of materials and chemical science, thus sacrificing masking efficiency. It is a good practice to separate immiscible elements spatially instead of crossing them with binary masks and generating some useless mixtures. Basically, the more knowledge input into the design of the materials library, the more effective the library is in searching for materials with the desired properties, and the smaller the library has to be to discover interesting materials. Combinatorial methods do not mean an ignorance of material science; instead, they increase the speed that materials are developed with the aid of material knowledge.

3.2.3 Quaternary-Masking Scheme

A quaternary-masking scheme was devised to increase the selectivity and effectiveness in the design of libraries. A set of quaternary fractal masks was used, wherein a group of up to four different elements can be inscribed in each mask, and whose rotation can generate up to four different masking patterns. This combinatorial masking scheme involves a series of n different masks, which successively subdivide the substrate into a series of self-similar patterns of quadrants (Figure 3.2b). The rth ($1 \leq r \leq n$) mask contains 4^{r-1} windows where each window exposes one-quarter of the area where the preceding mask is deposited. Within each window there are an array of 4^{n-r} gridded sample sites. Each mask is used in up to four sequential depositions, each time the mask is rotated by 90°. This process produces 4^n different compositions with $4n$ deposition steps. The advantage of the quaternary mask is that elements in the same mask group will not spatially overlap, which provides selectivity in the design of the library. Only five different mask patterns are necessary to make 1024 different compositions. The disadvantage, though, is that it takes at least 15 steps (with one quarter for each mask not deposited, compared to 10 steps for a primary mask) to get 1024 different materials. Also, the constant change and physical rotation of the mask can introduce some misalignment during thin-film deposition. This will be more of a problem as the library sample size gets smaller.

3.2.4 Movable-Shutter Mask

An X-Y shutter mask allows for in situ changes in the mask patterns and the stepwise generation of a compositional gradient library. The two independent halves of the shutter mask are driven by two respective step motors to control the exposure gap between the vapor and substrate. The shutter should be close to the substrate to minimize the shadowing effect. A larger gap between substrate and mask will result in vapor diffusion between them and reduce the spatial resolution and library sample density. The current system is equipped with a pulsed laser deposition apparatus, although it should be compatible with any other physical vapor-deposition apparatus. It is a secondary mask, and a primary mask grid is also needed to separate the different samples to make a discrete materials library. Compared with the other masking strategies, this is the most inefficient way to generate a materials library, because it usually makes only one deposition line at each step, and only one library can be made each time. Its advantage, though, is that it can generate a library without ever breaking the vacuum to change the mask, which makes it a convenient way to make a gradient composition library.

Another advantage of using the X-Y shutter masking method in making a phase gradient library is that, by controlling the exposure time of the substrate to a constant flux of chemical deposit, the composition of every sample is well controlled. If the thin film thickness profile from vapor deposition is well known or can be characterized, a discrete compositional gradient thickness library can be made with accurate compositional profiles by simply applying a grid mask to the top surface of the substrate during the fabrication of a continuous-phase gradient library.

These three physical masking methods have their own merits and problems. Table 3.1 lists the efficiency for each masking strategy based on 1024 different compounds. It is apparent from the table that the number of materials produced in a library using any masking scheme is far more than the number of operating steps required. Indeed, the number of compounds increases exponentially as a function of the number of operating steps. Therefore, the real power of the combinatorial method is more dramatic with higher-complexity materials, where more deposition steps are needed. However, in materials science it is not common to get a single-phase material con-

Table 3.1 Efficiencies and Selectivity of Various Masking Strategies to Make 1024-Member Libraries of Different Discrete Compounds

	Binary Masking	Quaternary Masking (0–5 components)	Quaternary Masking (5 components)	X-Y Shutter Masking
Number of compounds	1024	1024	1024	1024
Number of steps	10	15	20	64
Efficiency (relative)	100	67	50	16
Selectivity	Bad	Good	Good	Fair

sisting of, for example, ten different elements. Therefore, not every elemental combination will result in a single-phase compound. A compromise has to be made between the efficiency of combinatorial methods and the selectivity in elemental combination. For example, binary masking has no selectivity in elemental combination, while in the quaternary-masking scheme elements can be grouped to cross combine only intergroup elements in the library. Therefore, if a library designer wants to keep certain elements in the library apart to avoid generating meaningless combinations from a material science point of view or due to incompatibility, a quaternary-masking scheme where those elements can be put in the same group can be used. This results in lower masking efficiency, but greater library effectiveness.

3.2.5 Lithographic-Masking Method

The patterns of lithographic masking can be identical to those of physical masking, the only difference being in the way that mask is generated and used. Compared to the physical-mask method, the apparent advantage of contact lithographic masking is the high spatial resolution and the lack of any shadowing, due to the optical alignment of markers for different masking pattern-generation steps. Indeed, it is a routine process to get 10-micron multilayer thin films without any misalignment, as compared to the physical masking method. The disadvantage of this method is the time spent in preparing a polymer mask pattern, since each masking pattern needs to be generated and dissolved after deposition. In addition, some films experience compatibility problems when the lithographic chemical process is used, which limits its application. The melting temperature of a photoresist mask is normally below 200°C, which, given adequate cooling, is compatible with the surface temperature in all the physical vapor deposition methods cited so far.

3.2.6 Vapor-Deposition Methods for Making Discrete Material Libraries

Three general methods for doing thin/thick film deposition have been applied to make discrete material libraries:

1. Chemical vapor deposition, and chemical vapor transport;

2. Chemical solution deposition, including spin casting, spray coating, dip coating;

3. Physical vapor deposition, including various sputtering methods, pulse-laser deposition methods, ion-beam sputtering methods, and thermal or electron evaporation methods.

Chemical vapor deposition (CVD) is widely used in semiconductor processing (e.g., growth of SiO_2, Si_3N_4, Polysilicon, GaAs). Elemental precursors are prepared in volatile form, mixed, and reacted on a hot substrate. After thermal decomposition, oxides or other films can form. One way to use CVD in the generation of materials libraries is to use refractory inorganic contact masks, due to the high substrate temperature during deposition. CVD shows some limited success in making thin-film libraries [21]. Other chemical-solution deposition methods, including spin coating, dip coating, and spray coating, have little control of the stoichiometry, and are incapable of making high-quality thin films. The precursor solutions also need chemical modification for adequate rheological characteristics. Applying spraying and spinning methods with masking strategies to make material libraries was tried in early 1994 and made some discrete superconductor libraries. The materials in the libraries looked like a thick powder coating.

Sputtering deposition, which, for many laboratories, is a low-cost and readily accessible physical vapor-deposition approach, can make very smooth and dense thin films. It also has other such advantages as good thin-film spatial uniformity, good maintenance of film stoichiometry with target, and long target operational lifetime. Its deposition speed is fairly slow, however, and the target has to be coupled to the power-supply system, which makes it more difficult to switch the target in situ during library fabrication. Both laser ablation and ion beam sputtering decouple the power supply from the target, which means an easy switch of target using a target carousel system. The problems are the high cost of an excimer laser and the damage it makes to a target, plus the very high vacuum requirements of an ion-beam gun during operation.

3.2.7 Thin-Film Synthesis With the Multilayer-Elemental Precursor Approach

Most of the as-deposited thin-film libraries are in the form of multiple elemental layers. It is a challenge to develop a generic synthetic approach for various classes of complex multilayered material libraries. Most of the masking strategies discussed previously would not be useful if we could not convert the as-deposited multilayer precursor library into a high-quality uniform material library. A generic synthetic method was developed by adapting the method that Johnson used in making intermetallic phases from diffusion couples [22]. The trick is to decouple the diffusion process from the sintering process. Uniform amorphous phases can usually be formed by extensive low-temperature (usually a few days) annealing of multilayered precursor libraries. High-temperature sintering can then usually make high-quality thin-film samples from the uniform amorphous precursors. The $Yba_2Cu_3O_x$

film superconductor can only be made from the multilayered elemental precursors using this approach. Its effectiveness was also confirmed later by experiments in making phosphors [18] and discrete dielectric libraries [23].

3.3   APPLICATIONS OF THE FRACTIONAL MASKING METHOD IN COMBINATORIAL SEARCH OF FUNCTIONAL MATERIALS

3.3.1   Application of the Binary Masking Method in Making Superconductor Libraries

In 1994, the binary masking scheme (shown in Figure 3.2a) was used to design and synthesize the first 128-member discrete superconductor library, which comprised seven elements: Bi, Sr, Ca, Cu, Y, Ba, and Pb. This library targeted two series of well-known high Tc superconductors: the BSCCO and YBCO systems. It contains every combination of these seven elements, and demonstrated both the feasibility of making superconductors using such discrete libraries, and of studying the unexplored elemental combinations among these seven elements.

A few months were required to discover the synthetic route for superconductor thin films from multilayer elemental precursors, as described in the previous section. Once the generic synthetic approach to make a homogeneous phase from multilayer elemental precursors was found, synthesizing this first 128-membered superconductor library became straightforward. The natural reflected-light appearance of the library after the synthesis is shown in Figure 3.3a [10]. A few sites on the library, which include YBCO, BSCCO, BSCCO-Pb systems, were found to be superconducting at their well-known transition temperatures (Figure 3.3b). It was also found that different deposition sequences for the same superconductor compositions (e.g., BSCCO) made different superconductors with different normal states (Figure 3.3b).

3.3.2   Application of a Combination of Binary and Quaternary Masking Methods in Making Libraries of New Magnetoresistance Ceramics

It is suggested in the literature that a ceramic material should possess the following two characters to become a good candidate with magnetoresistance (MR) properties: it should be ferromagnetic below a certain temperature (curie temperature), and electric conductive (or semiconducting). The scope of the search for advanced MR materials from such a rational approach was narrowed to start with cobaltites.

The first MR ceramic library was designed, fabricated, and synthesized based on elemental series: $La(Y)Ba(Sr,Ca,Pb)CoO_x$. To design the selectivity into the library, the quaternary-masking scheme (eight of them, labeled (Q1)–(Q8) in Figure 3.4a) was blended with the binary-masking (four of them, labeled (B1)–(B4) in Figure 3.4a) scheme for the most efficient library design. Figure 3.4a shows the layout of the library; note that the number of masks increased to 12, but there is no mean-

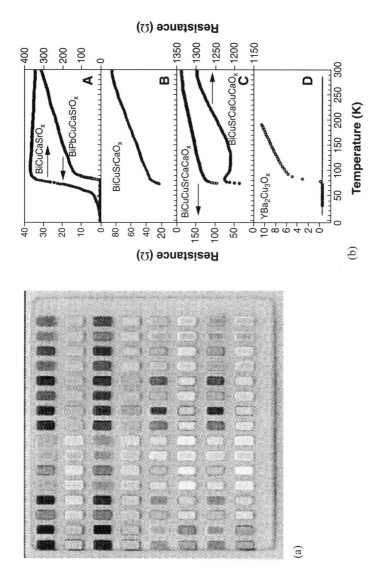

(a)

(b)

Figure 3.3 (a) Photo of a 128-member superconductor material library, after thermal processing. (b) The resistance vs. temperature (kelvin) of selective samples in the library. (Both reprinted, with permission, from Xiang, X. D., et al. *Science*, 1995, 268: 1738–1740; copyright © 1995. American Association for the Advancement of Science, Washington, DC.) See color insert for color representation of figure (a).

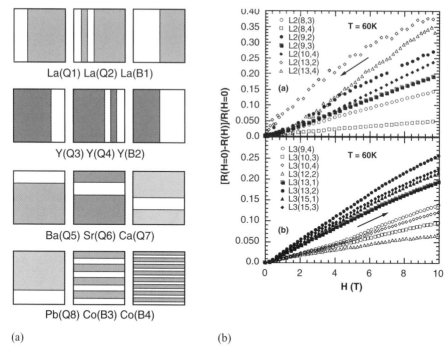

(a)   (b)

Figure 3.4 (a) Photoluminescent image of four discrete phosphor libraries made by binary masking schemes under UV illumination. (b) Photoluminescent image of two discrete 1024-member phosphor libraries made by binary masking schemes under UV illumination. (Both reprinted, with permission, from Sun, T. X. *Biotechnology and Bioengineering* (*Combinatorial Chemistry*), 1999, 61(4): 193–201; copyright © 1999. John Wiley & Sons, New York.)

ingless cross of some elements in the sample, such as La with Y, or any duals from Ba, Sr, Ca, Pb. They were divided into different regions of the substrate and physically separated.

Three categories of compounds that exhibit a large MR effect have been identified from this library: $La_x M_y CoO_\delta$, where M = Ca, Sr, Ba. The normalized MR of representative samples as a function of the magnetic field at a fixed temperature (60 K) are shown in Figure 3.5b. The largest MR ratio measured in this library was 72%, obtained for the sample with library coordinate (15, 2) at $T = 7$ K and $H = 10$ T, corresponding to $La_{0.8}Ba_{0.4}CoO_x$. This value is comparable to those measured for films generated in a similar fashion in a Mn-based library [24]. In contrast to the behavior of Mn-oxide MR materials [24], the MR effect increases as the size of the alkaline earth ion increases. Optimization of the composition, stoichiometry, substrate, and synthetic conditions can lead to increases in the MR ratio. The corresponding Y-(Ba, Sr, Ca)-Co compounds show much smaller (<5%) MR effects.

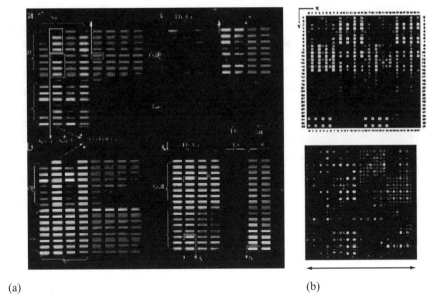

(a)   (b)

Figure 3.5   (a) Hybrid binary/quaternary masking schemes that were used to make the MR ceramic material library. (b) Change of field-induced resistance vs. magnetic-field measurement on selective samples in the library, and a few new MR ceramic materials were identified. (Reprinted, with permission, from Briceno, G., et al. *Science,* 1995, 270: 273–275; copyright © 1995. American Association for the Advancement of Science, Washington, DC.) See color insert for color representation of figure (a).

3.3.3   Application of a Combination of Binary and Quaternary Masking Methods in Making Libraries of New Luminescent Materials

Both binary and quaternary masking schemes were applied to make and search for advanced luminescent materials. The first series of phosphors is based on the formula series $La(Gd,Y)_m(Sr)_nAlO_x:Eu^{2+}$, Eu^{3+} (or Tb^{3+}, Ce^{3+}). These series of phosphors were studied using binary masking schemes. After high-temperature synthesis, the photoluminescent emission color and intensity can be directly observed by using a photographic approach in the dark room to subject the phosphor library to a uniform field of UV exposure. Figure 3.4a alone contains a tremendous amount of information on the composition/processing and luminescent properties for this series of phosphors. Lead compositions with desirable colors and intensity can be readily picked out from such photography. Indeed, several single-phase compositions, such as $(Gd_{0.6}Sr_{0.4})Al_{1.6}O_xF_{1.8}:Eu^{2+}_{0.08}$ (green), $(Gd_{2-x}Zn_x)O_{3-\delta}:Eu_y^{3+}$ (for $x > 0.18$) (red) were found to be efficient [quantum efficiency (QE) about 90%] as compared to standard phosphors of known QE. The emission color of the red phosphor is centered at 621 nm, thus providing better red color properties than the standard $Y_2O_3:Eu^{3+}$ red phosphor [25].

A 1024-member phosphor library was made on 1 inch^2 Si(100) substrate by using quaternary masks with either the lithographic process (Figure 3.4b upper) or shadow masks (Figure 3.4b lower). The five quaternary masks, A, B, C, D, and E (illustrated in Figure 3.2a), were used to generate high-density phosphor libraries. Elements were grouped into five families, with each family having no more than four elements. The five rotating masks were applied to the substrate, with various amounts of different elements deposited through the fractional openings. A total of 15 steps (each mask rotating three times) made 1024 different combinations among the elements of the five groups. Some interesting efficient blue composite phosphors were discovered from these libraries [18].

SUMMARY

In summary, various masking-method schemes were discussed and compared for making thin-film combinatorial materials libraries. The number of different samples in a library can increase exponentially with the number of deposition steps, for example, up to 1024 (2^n) different material compositions can be generated by only 10 steps of deposition when binary masking is used. Other deposition methods, such as quaternary masking, offered better selectivity in combining elements, although at a loss of masking efficiency. These masking schemes have been successfully applied to a variety of material problems, such as superconductors, magnetoresistance, and luminescent materials.

ACKNOWLEDGEMENT

I would like to thank Dr. X.-D. Xiang for discussion and help in codeveloping the majority content of this contribution. I would also like to thank Dr. Peter Schultz for his support of my project when I worked in Lawrence Berkeley National Lab.

REFERENCES

1. DiSalvo, F. J. *Science,* 1990, 247: 649–655.
2. Rustum, R. *Solid State Ionics,* 1989, 32/33: 3–22.
3. Phillips, J. C. *Physics of High-T$_c$ Superconductors,* Academic Press, New York, 1989.
4. Praveen, C. *Phys. Today,* October 1992, 22–23.
5. Darwin, C. *On the Origin of Species by Means of Natural Selection,* J. Murray, London, 1859.
6. Nisonoff, A.; Hopper, J. E.; Spring, S. B. *The Antibody Molecule,* Academic Press, New York, 1975.
7. GlaxoWellcome. *Nature,* 1996, 384(Suppl.): 1–5.

8. Pirrung, M. C. *Chem. Rev.,* 1997, 97: 473–488.

9. Hanak, J. J. *J. Mater. Sci.,* 1970, 5: 964–971.

10. Xiang, X.-D.; Sun, X.; Briceño, G.; Lou, Y.; Wang, K.-A.; Chang, H.; Wallace-Freedman, W. G.; Chen, S.-W.; Schultz, P. G. *Science,* 1995, 268: 1738–1740.

11. Briceno, G.; Chang, H.; Sun, X.; Schultz, P. G.; Xiang, X.-D. *Science,* 1995, 270: 273–275.

12. Sun, X. D.; Gao, C.; Wang, J.; Xiang, X.-D. *Appl. Phys. Lett.,* 1997, 70: 3353–3355.

13. Sun, X.; Wang, K. A.; Yoo, Y.; Wallace-Freedman, W. G.; Gao, C.; Xiang, X. D.; Schultz, P. G. *Adv. Mater.,* 1998, 9: 1046.

14. Schultz, P. G.; Xiang, X.-D.; Goldwasser; I. U.S. Patent 5,776,359, 1998 (to The Regents of the University of California and Symyx Technologies).

15. Schultz, P. G.; Xiang, X.-D.; Goldwasser; I. U.S. Patent 5,985,356, 1999 (to The Regents of the University of California and Symyx Technologies).

16. Schultz, P. G.; Xiang, X.-D.; Goldwasser; I. U.S. Patent 6,004,617, 1999 (to The Regents of the University of California and Symyx Technologies).

17. Schultz, P. G.; Sun, X.; Xiang, X.-D. U.S. Patent 6,048,469, 2000 (to The Regents of the University of California).

18. Wang, J.; Yoo, Y.; Gao, C.; Takeuchi, I.; Sun, X.; Chang, H.; Xiang, X. D.; Schultz, P.G. *Science,* 1998, 279: 1712–1714.

19. Danielson, E.; Golden, J. H.; Mcfarland, E. W.; Reaves, C. M.; Weinberg, W. H.; Wu, X.-D. *Nature,* 1997, 389: 944–948.

20. Van Dover, R. B.; Schneemeyer, L. D.; Fleming, R. M. *Nature,* 1998, 392: 162–164.

21. Wang, Q. "Combinatorial HWCVD" Presented at Combi–2000, Knowledge Foundation, Jan. 23, 2000, San Diego, CA.

22. Fister, L.; Novet, T.; Grant, C.; Johnson, D. *Adv. Synth. React. Solids,* 1994, 2: 155–234.

23. Chang, H; Gao, C; Takeuchi, I; Yoo, Y; Wang, J.; Schultz, P. G.; Xiang, X.-D.; Sharma, R. P.; Downes, M.; Venkatesan, T. *Appl. Phys. Lett.,* 1998, 72: 2185–2187.

24. Jin, S.; O'Bryan, H. M.; Tiefel, T. H.; McCormack, M.; Rhodes, W. W. *Appl. Phys. Lett.,* 1995, 66: 382.

25. Sun, X.-D.; Xiang, X.-D. *Appl. Phys. Lett.,* 1998, 72: 525–527.

CHAPTER 4

HIGH THROUGHPUT SYNTHETIC APPROACHES FOR THE INVESTIGATION OF INORGANIC PHASE SPACE

LYNN F. SCHNEEMEYER AND R. BRUCE VAN DOVER

Agere Systems

4.1 INTRODUCTION

High throughput synthesis and screening approaches offer materials research sophisticated new tools that have demonstrated speed and effectiveness in a few recent cases. Perhaps even more significantly, these new, more-efficient methodologies prod researchers to rethink traditional materials research strategies in order to maximize the productivity and impact of their work.

Recent interest in high throughput synthesis and screening approaches applied to studies of inorganic materials grew out of the success of combinatorial chemistry in the drug-discovery process. As rapid evaluation of biological activity using rapid screening techniques found widespread use, the preparation and evaluation of large numbers of biologically interesting organic molecules were enabled.

In contrast with organic molecules, where each different chemical composition and molecular arrangement is a unique chemical entity, phase space in a given material system tends to contain a limited number of unique phases. Intermediate compositions are mixtures of a few neighboring phases. There are cases in which the property of interest is a function only of the composition of a material. More often, the form of the material and factors such as the defect structure and microstructure play significant roles. High throughput experimentation (HTE) is being used to identify promising lead compounds. These experimental methods are of significant interest for applications such as catalysis and polymers, and are expected to increase dramatically the pace of materials discovery and development.

In place of the extensive chemical and structural diversity available in molecular organic compounds, many properties of solid-state chemical systems are strongly dependent on processing details. Such factors as cooling rates or the distribution of

Experimental Design for Combinatorial and High Throughput
Materials Development, Edited by James N. Cawse.
ISBN 0-471-20343-2 © 2003 John Wiley & Sons, Inc.

impurities can affect gross microstructures and crystalline defect structures. Thus, processing variables become an important consideration in designing a combinatorial-type materials study.

In general, there are two different philosophies for carrying out a combinatorial study. In the first, the scope of the problem is constrained to limit the number of possible chemical compositions to be studied. The second is a broad-brush approach, in which a large amount of phase space is explored coarsely; those data are used to home in on interesting materials systems. Thus, a typical combinatorial-type study comprises:

- Experimental design: selection of regions of inorganic phase space (often referred to as libraries), and compounds of interest;
- Synthesis and characterization of compounds: compound libraries or phase spreads;
- Testing materials for properties of interest;
- Developing models, where possible, to guide additional investigations.

4.2 CHALLENGES IN INORGANIC MATERIALS RESEARCH

Materials research is a very broad field that includes polymers and composites as well as inorganic materials. Combinatorial approaches, in general, provide research tools that enable efficient searches for new materials with specific properties of interest. In addition, such techniques can be used to tailor the properties of a given material for some particular application. In all cases, properly utilized highly parallel materials search tools increase the probability that a material with desired properties will be found. The goal of this section is to discuss some of the active research areas in inorganic materials that can be explored using the combinatorial-type approaches discussed in this chapter.

Inorganic materials can be parsed in a number of different ways. Here, we distinguish materials classes based on the degree of long-range order; that is, whether they are crystalline or amorphous. We also discuss metallic alloys separately from inorganic compounds, such as oxides or nitrides.

Typically, the research challenge is defined by the intended application for a material. Table 4.1 lists typical materials used in a number of common applications and whether the material is typically used in bulk or thin film form or as nanophase materials. As noted earlier, searches are most likely to be successful if the synthetic approach used matches the form of the material in the application.

4.3 OPPORTUNITIES FOR PARALLEL SYNTHESIS AND HIGH THROUGHPUT EXPERIMENTATION IN MATERIALS

The three main classes of materials being investigated using HTE are catalysts, electronic and optical materials, and polymer materials. Inorganic electronic and

Table 4.1 Inorganic Materials Matched Against Relevant Search Approaches

Material	Form			Application Examples	Search Approach		
	Thin Film	Thick Film	Bulk		Phase Spread	Discrete	Other
Complex Oxide			Single crystal	Bulk optics, Scientific studies, Phase ID			Not developed
Oxides Nitrides	Amorphous			Electronic devices, Optical devices	✓	✓ (in some cases)	
Oxides Nitrides	Polycrystalline			Optical devices, Electronic devices, e.g., DRAM, FRAM	✓	✓	
Oxides Nitrides	Epitaxial			Scientific studies, Superconductors, Compound semiconductor devices	✓	✓	
Oxides		Polycrystalline		Multilayer capacitors, Phosphors, Catalysts	✓	✓	Sol-gel, Solution precursors
Oxides Nitrides			Polycrystalline	Substrates, RF devices, Bulk optics			Solution precursors, New approaches
Metals	✓	✓	✓	Magnetic Shape memory Corrosion studies, Catalysts	✓		Reduction of solution precursors

optical materials are used in the production of electronic devices such as micro-processors, and optical components and devices. High rates of growth and a rapid pace of technological change have driven the fields of optical and electronic materi-als. In particular, the technical challenges being confronted as the electronics indus-try nears the limits of miniaturization, will apparently in many cases require the use of new materials that may be best discovered using high throughput approaches.

4.4 STRATEGIES FOR EXPLORING INORGANIC PHASE SPACE

4.4.1 Approaches

There are two common approaches to the exploration of inorganic phase space. First, though, it is important to recognize that there is an inherent tension between working too dumb (trying to do everything) and working too smart (taking a known material and modifying it slightly). The problem with working dumb is that it is simply impossible to make and examine everything. The number of pos-sible combinations of 100 elements is too vast. For example, 10 elements taken 3 at a time is 720 combinations. If more than one ratio of elements is considered for each combination of those three elements, the number of samples rises rapidly. Therefore, it is necessary to constrain the number of materials to be examined us-ing some criteria.

In the case of research aimed at optimizing a known material using combinatori-al-type approaches, it is tempting to conduct studies similar to those carried out us-ing traditional handcrafted sample approaches. By working too smart in this case, researchers eliminate the possibility of a serendipitous discovery in an unexpected compositional domain.

The two common approaches to the exploration of inorganic phase space are dis-crete combinatorial approaches and phase spread approaches, which are discussed in more detail below as well as elsewhere in this book. The strength of the discrete combinatorial approach is the ability to explore wide swatches of phase space. However, it is also possible for important compositions to be missed (fall through the grid). In contrast, the phase-spread approaches can more thoroughly explore phase space, but are less well suited to comprehensive searches. Two considera-tions can guide the researcher in a choice between these approaches. First is the equipment that the researcher has available. Also, the researcher needs to consider the constraints on the problem to be able to decide how many candidate elements must be considered.

4.4.2 Discrete Compositions

Arguably the most flexible approach in terms of compositions that can be explored is the discrete combinatorial approach first discussed by Xiang et al. [1–4] and re-viewed in Chapter 3.

The most serious limitation of this approach is that the compositions examined in the initial lead identification may fail to sample the properties exhibited by various compounds made up of the constituent elements. For example, in exploring the Ba-Y-Cu-O pseudoternary phase diagram for high temperature superconductors, it is necessary for a composition to lie within the $Ba_2YCu_3O_7$ phase domain in order for any of that phase to be present. Yet, a very generous estimate for the $Ba_2YCu_3O_7$ phase domain shows that it represents only about a quarter of the Ba-Y-Cu-O phase space [5]. Only the $Ba_2YCu_3O_7$ phase will provide indications of superconductivity in that compositional space. Other higher T_c cuprates have much smaller phase fields. In addition, investigators must be aware of other processing variables that can influence their search. Again, considering the example of the Ba-Y-Cu-O pseudoternary phase diagram [5], the synthesis atmosphere plays a crucial role in the preparation of the superconducting phase [6]. Samples heated under a nitrogen atmosphere will not display superconductivity, because the oxygen stoichiometry of the $Ba_2YCu_3O_{7-x}$ material will be too low. In general, the influence of processing variables is an issue in all combinatorial-type investigations.

4.4.3 Continuous Compositional Spreads

Wedges This technique for preparing composition spreads uses automated shutters that pass over the substrate with a fixed (or programmable) speed during the depositions, creating a gradient thickness of the deposited material. By creating gradients in various directions for each deposition species, an overall composition spread can be prepared (e.g., a standard ternary spread is implemented by arranging three gradient vectors at 120° intervals). As with the discrete masked approach, a high-temperature homogenizing anneal is necessary to obtain interdiffusion of the layers, yielding a uniform composition through the thickness of the film. For materials that can only be formed at a high temperature, this is usually not a severe limitation. However, it should be kept in mind that reactions between adjacent layers might form binary precursors that do not then react in the same way as a film with three or more intimately mixed species.

Sputter deposition is particularly suited for this approach, since the sputter rate is generally very constant with time, though the technique can also be implemented fairly straightforwardly using feedback-controlled evaporation. Because the composition gradient can be varied directly, the gradient-composition technique is particularly useful for investigations on a coarse or fine scale, where the natural composition spread approach runs into practical limitations. For example, a complete ternary phase diagram can be prepared on a single chip. This has proven to be a powerful technique when appropriate substrates can only be obtained in a fairly small size.

Chang and Xiang have prepared two-component linear spreads of doped rare-earth manganites epitaxially on 1-inch-square single-crystal substrates using this technique. As another example, a gradient spread was fabricated that mapped the $(Ba,Sr,Ca)TiO_3$ pseudoternary system on a triangular chip of $LaAlO_3$ (100) [7]. Three gradient depositions were performed in three different directions for three precursors: $BaCO_3$, $CaCO_3$, and $SrCO_3$. A large fraction of this phase diagram had never been produced in either bulk or thin films prior to this study. Figure 4.1

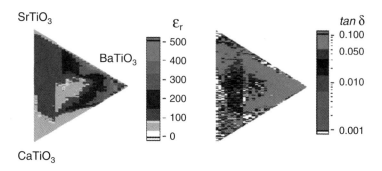

Figure 4.1 Two-component linear spreads of doped rare-earth manganites epitaxially deposited on 1-inch-square single-crystal substrates by Xiang et al. using the wedge technique. (Reprinted, with permission, from Xiang, X.-D., *Biotechnology and Bioengineering, 1999,* 61(4): 227–241; copyright © 1999. John Wiley, New York.)

shows the results of scanning evanescent microwave microscopy (discussed below) performed on the chip. The left image is a plot of the dielectric constant (i.e., the real part of the complex dielectric constant), ε_r, and the right image is a plot of the dissipation factor, expressed as $\tan \delta = \text{Im}(\varepsilon)/\text{Re}(\varepsilon)$. From this study, a new low-microwave-loss region, $Ba_{0.12-0.25}Sr_{0.35-0.47}Ca_{0.32-0.53}TiO_3$, was successfully identified.

The controlled-gradient technique is closely related to the use of so-called "wedge" structures for investigating the effect of thickness (typically of a single component) on some property or behavior. This technique dates at least as far back as 1961, when Middelhoek [8] used a wedge-geometry thin film of Ni-Fe (Permalloy) to demonstrate the transition in magnetic-domain wall structures from Bloch walls (where the film was thick) to Néel walls (where the film was thin). In 1981, the technique was used in a much more sophisticated experiment to investigate extension of the A15 phase boundary in homoepitaxial $Nb_{100-x}Al_x$ [9]. In that work, Kwo and Geballe used tunneling measurements to determine that layers with a composition $x = 24$ could be stabilized in the A15 phase if grown on (equilibrium) material with $x = 20$, but only for thicknesses less than 200 Å, neatly resolving a puzzle that had been noted earlier [10]. More recently, the technique was used by Ungaris et al. to great advantage in studying the effect of metal interlayer thickness on exchange coupling between magnetic thin films [11,12]. This group used scanning electron microscopy with polarization analysis (SEMPA) (a technique that is sensitive to the magnetic orientation of the imaged film) to image an Fe-Cr-Fe trilayer with varying Cr thickness. Results showed a quasi-periodic reversal of the sign of the exchange coupling illustrating the change in the interactions between the electrons in the metallic conductor. This result is what is expected for the RKKY interaction [13]. This wedge technique also has been employed in a two-dimensional geometry, e.g., with variation of Cr thickness along one direction and variation of the Fe overlayer thickness in the orthogonal direction.

Codeposition   Oxide or metal phase spreads can easily be produced using off-axis cosputtering to produce binary or ternary composition spreads, as illustrated in Figure 4.2. In off-axis deposition onto a fixed substrate, the thickness of the deposited film decreases approximately exponentially with distance from the gun. The application of a radio frequency (rf) field to the growing film during deposition has proven to be a crucial factor [14]. This bias causes ion bombardment of the growing film that probably enhances the surface mobilities of the deposited species and results in the preparation of a denser film. Note that these phase spreads have inherent thickness variations of approximately a factor of 2, a limitation of this method. The film in Figure 4.2 shows interference fringes that are a combination of the thickness gradients and index gradients present in the phase spread that has been deposited.

This system, shown schematically in Figure 4.3, allows the codeposition of up to three metals. Thus, a large portion of a pseudoternary oxide phase diagram can be deposited in a single run. Relatively high oxygen partial pressures, 10–40%, can routinely be used to ensure fully oxidized films. Metal films can also be made using this system. However, the base pressure of such a chamber is likely to be too high to prevent oxidation of more reactive metals. A cryotrap inside the vacuum chamber, surrounding the deposition sources and substrate (a Meissner trap), can expand the usefulness of such a chamber for studies of metal alloy systems. Such a system greatly reduces oxidation of the metals during deposition so that clean metals can be obtained. Nitrides and oxynitrides can also be explored in a system like this by introducing nitrogen gas into the sputtering atmosphere. Indeed, oxynitride can be expected unless efforts are made to limit oxidation by using a system such as the cryotrap just mentioned.

Phase-spread techniques being employed today are descendants of related approaches to the parallel preparation of thin-film materials with inherent composi-

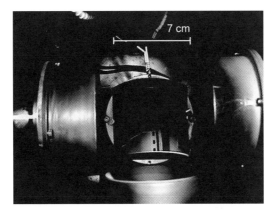

Figure 4.2   Off-axis codeposition system employing three independently controlled rf sputtering guns and a substrate holder with an independent rf power supply. See color insert for color representation.

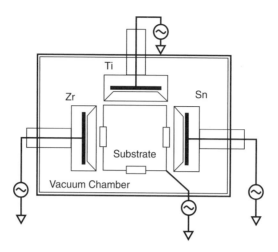

Figure 4.3 Off-axis codeposition system employing three independently controlled rf sputtering guns and a substrate holder with an independent rf power supply.

tional gradients that have been employed for more than 30 years. In 1965, Kennedy and coworkers described a rapid method for determining ternary metal alloys that involved simultaneous e-beam evaporation of three elements onto a heated substrate to form phase-spread films [15]. A series of rf-sputtered gadolinium iron garnet films with varying defect concentrations were prepared in 1969 by Sawatzky and Kay [16], working at IBM. A two-target sputtering system was employed for their combinatorial-type approach. Other attempts were also reported in the late

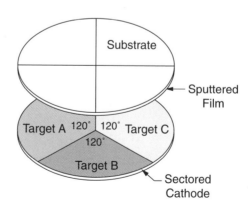

Figure 4.4 Sectored sputtering target used by Hanak to codeposit samples with inherent compositional gradients. (Reproduced, with permission, from Hanak, J. J. J. *Materials Science,* 1970, 5: 964–971; copyright © 1970. Kluwer Academic Publishers, Dordrecht, The Netherlands.)

1960s aimed at achieving more efficient syntheses by coevaporating or cosputtering elements from independent sources onto a suitable substrate. A "multisample approach in materials science" [17] was championed by Hanak, working at RCA Labs. He investigated the synthesis, compositions, and preparations of entire binary and ternary systems through the use of a one cathode, multiple-target, rf cosputtered method of synthesis. A simple compositional analysis method for these cosputtered films was based on film thickness measurements. Samples were sputtered from a composite target, as indicated in Figure 4.4, producing cosputtered, and thus intimately mixed, samples having compositional spreads.

4.5 TYPES OF MATERIALS: MATCHING THE SYNTHETIC APPROACH TO THE APPLICATION

4.5.1 Overview

A combinatorial-type search is most likely to be successful if the materials that are investigated are made in the same form, for example, as thin films or as bulk materials, as they will be used in the application of interest. Fortunately, a variety of preparation approaches is available either for discrete compositions or for phase spreads. Table 4.1 outlines the search approaches suitable for various types of materials as well as application examples for those materials. The discussion of high throughput synthesis approaches for inorganic materials is divided into thin-film deposition techniques and bulk preparation methods. While films can be either organic or inorganic, the focus of this section is on techniques suited to the deposition of inorganic materials.

4.5.2 Thin-Film-Deposition Techniques

Thin films are layers of a material that are a few atoms thick to a few thousand atoms thick (angstroms thick). By way of scale, a human hair is 10–100 μm or 100,000 to 1,000,000 angstroms thick. A thin film is deposited on a substrate, typically silicon wafers, sapphire wafers, or glass wafers. Structurally, thin films can be amorphous, polycrystalline with randomly oriented crystallites, polycrystalline with texture, or single crystal. The most important factors controlling film structure are the nature of the deposited material, the temperature, and the substrate used. Obtaining single-crystal films usually requires that the substrate have a close lattice match to induce epixaty. There are a number of standard texts devoted to thin films. Two recommended discourses are *Thin Film Processes,* by Vossen and Kern [18], and the *Handbook of Thin Film Technology,* by Maissel and Glang [19].

While thin films can be conveniently deposited using a variety of techniques, the use of thin films for combinatorial-type materials investigations has both advantages and limitations. In general, the impact of a search for new materials is likely to be greatest where the unique properties provided by a particular material are required. This is the case in photonics where dense-wavelength division multiplexing

(DWDM) and increasing data rates are driving the need for innovations in materials properties [20]. Also, in integrated circuits the drive to ever-smaller feature size is likely to necessitate the use of more exotic materials [21]. Certain materials properties are easily measured in a thin film. For example, some electrical measurements such as resistivity are more easily measured on a thin film than on a bulk sample. Measurements of certain optical properties, such as fluorescence intensity and lifetime, are also straightforward.

However, there are also disadvantages to the study of thin films. In general, determination of the composition of a film is difficult. Most analytical techniques give the composition with about a 5% uncertainty. Rutherford backscattering spectroscopy, (RBS) and energy dispersion spectroscopy (EDS) are often used to obtain compositional information about thin-film samples. Detailed structural data are also hard to obtain using thin-film samples. A thin-film sample may simply not have enough material along certain crystallographic directions to give adequate scattering intensity. Thus, the X-ray diffraction data from a thin film may not properly sample all of reciprocal space for a particular structural arrangement. In the case of textured films, this is an even more significant problem.

Where a material is intended for use as a thin film, the study of combinatorial-type thin-film samples is an appropriate choice. For materials that will be used in bulk form, any process-dependent properties may not be accurately reflected in studies of thin-film samples. Other issues that must be examined in the case of thin films include interactions between the substrate and the growing film that lead to unexpected compositions. Also, films often grow under either tensile or compressive stress. Certain physical properties are strongly affected by film stress.

4.5.3 Physical-Deposition Approaches

Various physical deposition approaches are commonly used for both research and commercial film growth. Many of the physical-deposition techniques fall under the general heading of sputtering. [18,19] A target made of the material to be deposited is bombarded by ions that ablate, or sputter, atoms, or more complicated clusters of atoms, from the target surface. These are then deposited on a substrate, which is held on a substrate holder, facing the target. The substrate can be heated or cooled, and grounded, biased, or floated electrically. The bombarding ions can be created using a glow discharge. More commonly, rf magnetron arrangements are used because of their high efficiencies. Alternatively, vapor species can be created by thermal evaporation using resistance, electron beam-, rf-, or laser heated sources.

Phase spreads, compositionally inhomogeneous samples that are created deliberately, can be created easily by using multiple sources. An example is the off-axis sputtering arrangement shown in Figure 4.4 and discussed before. It is important that the sources have reasonably stable deposition rates. While heroic efforts can be made to produce somewhat stable e-beam evaporation deposition rates, magnetron sputtering has inherently more stable deposition rates. Indeed, for a given deposition system, the sputtering rate can be expected to be reproducible from day-to-day for the same power setting and same target material.

4.5.4 Pulsed-Laser Deposition

In pulsed laser deposition (PLD), a target material is congruently vaporized by a pulsed UV laser beam from an eximer laser. An important advantage of PLD films is that the stoichiometry of the target is preserved in the laser-generated plume. A PLD system modified to allow for the simultaneous deposition of thin films from two targets by use of a dual-beam–dual-target configuration has been described by Schenck and Kaiser of NIST [22] and shown in Figure 4.5. The dual-beam PLD system includes an adjustable beam diverter, variable attenuator, and independent focus control of both beams. These modifications allow control over the deposition rate from each target. The targets are physically separated in space, thus the composition of the film varies with position on the substrate. Films produced using this system can have substantial variations of composition, and therefore properties, over a small substrate. Note, however, that these films may also have thickness variations.

4.5.5 Molecular-Beam Epitaxy

Molecular beam epitaxy (MBE) [18] is used to produce very high quality thin films of a wide variety of materials, including metals, semiconductors, and insulators. MBE is most commonly a research rather than a production technique. In MBE, atoms of an element or compound are delivered to a substrate through an ultrapure, ultrahigh vacuum (UHV), normally less than ~10–11 torr, atmosphere. The UHV atmosphere in the MBE chamber allows the atoms to arrive on the substrate without colliding with other atoms or molecules, keeping the growing film free of contaminants. Films are usually grown on a heated substrate that allows the arriving atoms to distribute themselves evenly across the surface to form an almost perfect crystal structure. Normally the substrate is placed in an UHV chamber with a direct line of

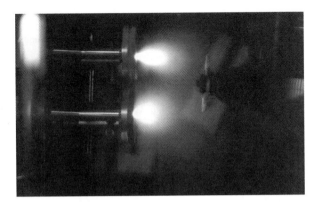

Figure 4.5 Plumes from two targets being illuminated simultaneously by the pulsed eximer laser beam of a PLD system. (Photo courtesy National Institutes of Science and Technology, http://www.ceramics.nist.gov/programs/thinfilms/PLD.html.) See color insert for color representation.

sight to several elemental species, each of which is slowly evaporated from an evaporation furnace known as an effusion cell, as shown schematically in Figure 4.6. MBE provides precise control of deposition parameters and a variety of in situ characterization techniques.

By arranging the distances to the sources so that the arriving beams of atoms are not uniform, inhomogeneous, phase-spread-type samples can be produced. MBE techniques applied to combinatorial-type synthesis and characterization of novel materials has been discussed by a number of researchers. Among the early proponents of the use of combinatorial approaches using MBE deposition is Professor Hideomi Koinuma of the Tokyo Institute of Technology, who has used the approach for studies of phosphors, nonlinear optical materials, and other materials-discovery problems [24]. The experimental setup used by the Koinuma group is illustrated in Figure 4.7.

4.5.6 Chemical Vapor Deposition Thin-Film Techniques

Chemical vapor deposition (CVD), is a common commercial process for producing thin films. As is clear from the name, volatile chemicals are the source of the vapor species that are deposited as the thin film. Here, we will consider metal-organic chemical vapor deposition (MOCVD) to be simply a variant of CVD. There are a number of other chemical methods for the deposition of films, including electrochemical methods such as electroplating and anodization. While metal multilayers can be produced using electrodeposition techniques, controlled production of phase-spread samples by these techniques has not been demonstrated yet.

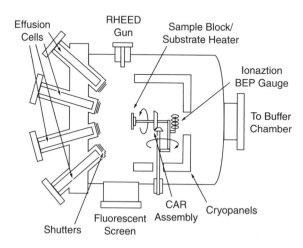

Figure 4.6 Schematic drawing of an MBE chamber showing multiple effusion cells with shutters and diagnostic tools such as reflection high-energy electron diffraction (RHEED). (Reproduced, with permission, from Streetman, B. G., http://www.ece.utexas.edu/projects/ece/mrc/groups/street_mbe/mbechapter.html, copyright © University of Texas, Austin.)

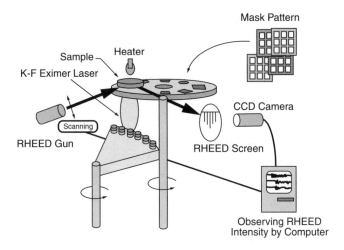

Figure 4.7 Schematic drawing of a combinatorial MBE apparatus. (Reproduced, with permission, from H. Koinuma, http://oxide.rlem.titech.ac.jp/koinuma/comb/comb1.html; copyright © H. Koinuma.)

CVD has a number of advantages as a thin-film growth approach. Conformal coverage of complex shapes is possible. In addition, highly uniform large-area films can be deposited. In a CVD reactor, the gaseous precursors are typically introduced via a "showerhead" arrangement that helps the uniformity of the source material, as shown in Figure 4.8. Gladfelter and coworkers have recently shown that the use of multiple showerheads diffusing different precursor chemicals can produce compositionally nonuniform, phase-spread, samples [25].

4.5.7 Bulk Materials Approaches

Many materials, such as catalysts, are typically used in bulk form. Therefore, approaches to parallel synthesis that produce bulk samples are needed. Several techniques can be automated to deliver accurate amounts of starting materials to reaction wells, including ink-jet and other liquid delivery systems, as well as automated powder delivery systems. The aim of these approaches is to produce bulk materials in small quantities, but with wide compositional and structural diversity. The special challenges associated with hydrothermal synthesis, which is important for zeolite catalysts, are also discussed.

4.5.8 Solution Deposition Techniques

The use of liquid precursors has a number of advantages in the synthesis of bulk ceramic materials. In general, liquid precursors allow for lower reaction temperatures and provide more uniform mixing. The use of metal halides or, alternatively, metal

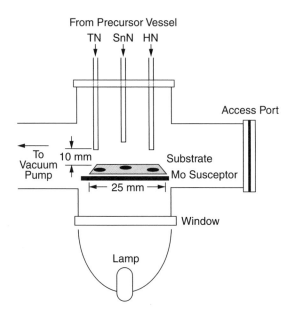

Figure 4.8 Schematic drawing of a MOCVD reactor showing shower heads for producing phase-spread samples. (Reproduced, with permission, from Smith, R. C.; Hoilien, N.; Roberts, J.; Campbell, S. A.; Gladfelter, W. L. *Chem. Mater.,* 2002, 14: 474–476.

alkoxides as precursors has a number advantages for combinatorial-type synthesis routes. They provide a great deal of synthetic flexibility by covering a wide range of chemistry. Specifically, the chlorides, bromides, or iodides of nearly all metals are highly soluble. Alkoxide-based sol-gels using the common solvent, 2-methoxyethanol, also allow access to essentially the entire periodic table [4]. The hydrolytic characteristics of the systems can be influenced through the addition of specific additives that are ligands such as hexanoates or alcoholamines. Citrates are another commonly used group of precursors. The high solubilities of these compounds are attractive for use in the synthesis of a broad array of materials.

Importantly, both the halide and alkoxide precursors are compatible with robots. Precursors that are too reactive or corrosive would rapidly damage or clog autopipetters, automated robotic syringes, and other automated delivery tools. Such automated tools are commonly adapted from parallel organic synthesis use, and thus work with organic solvents, but not with hydrolytically unstable materials.

The inorganic community has borrowed autopipetters and the standard 96-well and 384-well plates used in combinatorial organic synthesis. The plates, however, are usually either borosilicate glass or plastic, and thus are unsuitable for the high-temperature reaction step that inorganics, especially oxides, often need. Some researchers have drilled fused silica or alumina plates to make custom multiwell high-temperature reaction plates. At this time, however, these are not commercially available. In the synthesis of small quantities of a material, any reaction with the well material would drastically alter the composition of the product.

Delivery of controlled amounts of solution species using ink jets has been reported. Mallouk and coworkers made a library of ternary and quaternary metal catalysts based on platinum and palladium using ink-jet delivery of solution precursors [26]. An important contribution of their study was a highly parallel optical screening approach that allowed them to identify the best electrocatalysts. A new quaternary alloy showed significantly improved activity. It is unlikely that such a complex material would have been discovered using conventional metallurgical approaches.

McGinn and coworkers have examined ink-jet-based liquid citrate precursor syntheses to identify new catalysts for a variety of applications [27]. Infrared thermal imaging is used to screen the libraries for those compounds showing the highest levels of exothermic activity. For example, perovskite compounds are being studied for use in CO oxidation applications. More traditional routes are then employed for additional studies of the most interesting materials.

While automated instruments can be essential in laboratories dedicated to high throughput synthetic approaches to materials development, much cheaper and readily available equipment can play a useful role in preliminary or smaller-scale studies. For example, Eppendorf pipettes used in conjunction with selected oxylate, citrate, or halide precursors, representing a few of the many possible precursor compounds, and homemade multiwell reaction plates can easily create families of interesting oxide materials for further characterization.

4.5.9 Parallel Hydrothermal Synthesis

Zeolites and periodic mesoporous hosts offer important properties such as molecularly selective sorption, ion exchange, diverse catalytic activity, and applications in nanotechnology. The Bien group at the Univerity of Munich has described a high throughput approach to zeolite synthesis, and has developed automated, parallel methodologies that allow them to explore a wide range of synthetic parameters. The group's aim is to access the multiparameter, multicomponent space for the exploration and optimization of synthesis conditions, and to facilitate development of new materials with interesting properties. Their instrumental approach employs a totally inert, miniaturized environment for synthesis over wide pH and temperature ranges (up to at least 180°C). A typical synthesis experiment consists of the following steps: (1) the reagents are directly dispensed into multiclaves, followed by homogenization, (2) after crystallization at an elevated temperature, the samples in the multiclaves are repeatedly washed and recovered in library format using centrifugation techniques without individual manipulation of any library element; (3) the resulting sample libraries are then measured directly using automated X-ray diffraction. The synthesis fields within the $K_2O-Na_2O-Al_2O_3-SiO_2-(TEA)_2O-H_2O$ model system have been studied in detail [28]. The study emphasized the effect of dilution, aging, temperature, and postsynthesis treatment on the phase distribution within in the system. Successful investigation of zeolite systems using combinatorial approaches will require careful characterization of morphology, particle size, and product purity.

4.6 SUMMARY

A number of different approaches are now demonstrated for the rapid synthesis of large numbers of materials. Essential to successful application of combinatorial-type materials investigations is first the researcher's understanding of the research goal. Also required are a rapid synthesis approach that produces samples in a form similar to that of the intended use, and a meaningful measurement and evaluation protocol.

REFERENCES

1. Xiang, X.-D, Sun, X; Briceno, G.; Lou, Y.; Wang, K. A.; Chang, H.; Wallace-Freedman, W. G.; Chen, S.-W.; Schultz, P. G. *Science,* 1995, 268: 1738–1740.
2. Sun, X.; Gao, C.; Wang, J.; Xiang, X.-D. *Appl. Phys. Lett.,* 1997, 70: 3353–3356.
3. Danielson, E.; Devenney, M.; Giaquinta, D. M.; Golden, J. H.; Haushalter, R. C.; McFarland, E. W.; Poojary, D. M.; Reaves, C. M.; Weinberg W. H.; Wu, X. D. *Science,* 1998, 279: 837–839.
4. Giaquinta, D.; Devenney, M.; Danielson, E. "Synthesis of Inorganic Materials by Combinatorial Solution Techniques," American Ceramics Society National Meeting, Indianapolis, IN, April 25, 2001.
5. Roth, R. S.; Davis K. L.; Dennis, J. R. *Adv. Ceram. Mater.,* 1987, 2: 303–308.
6. Gallagher, P. K. *Adv. Ceram. Mater.,* 1987, 2: 632–636.
7. Chang, H.; Xiang X.-D. *Ferroelectrics,* 2000, 29: 113–112.
8. Middelhoek, S. In *Ferromagnetic Domains in Thin Ni-Fe Films,* Drukkerij Wed. G. van Soest N.V., Amsterdam, 1961, 20.
9. Kwo, J.; Geballe, T. H.; *Phys. Rev.,* 1981, B23: 3230–3239.
10. Dayem, A. H.; Geballe, T H.; Zubeck, R. B.; Hallak, A. B.; Hull, Jr., G. W. *J. Phys. Chem. Solids,* 1978, 39: 529–536.
11. Unguris, J.; Celotta, R. J.; Pierce, D. T. *Phys. Rev. Lett.,* 1991, 67: 140–144.
12. Celotta, R. J.; Pierce, D. T.; Unguris, J. *MRS Bull.,* 1995, 20: 30–33.
13. Van Dover, R. B.; Schneemeyer, L. F.; Fleming, R. M.; Huggins, H. A.; *Biotechnol. Bioeng. (Comb. Chem.),* 1998/1999, 61: 217–223.
14. Kittel, C. *Introduction to Solid State Physics,* Wiley, New York, 1996.
15. Kennedy, K.; Stefansky, T.; Davy, G.; Zackay, V. F.; Parker, E. R. *J. Appl. Phys.,* 1965, 16: 3808–3811.
16. Sawatzky E.; Kay, E. *IBM J. Res. Dev.,* 1969, 13: 696–702.
17. Hanak, J. J. *J. Mater. Sci.,* 1970, 5: 964–971.
18. Vossen, J. L.; Kern, W. *Thin Film Processes,* Academic Press, New York, 1978.

19. Maissel L. I.; Glang, R., eds., *Handbook of Thin Film Technology,* McGraw-Hill, New York, 1970.

20. White, A. E. *Optics & Photonics News,* March 2000, 27–30.

21. Chaudhari, P. *Bull. Mater. Res. Soc.,* 2000, 25: 55–56.

22. Kaiser D. L.; Schenck, P. K. http://www.ceramics.nist.gov/programs/thinfilms/PLD.html

23. Cho, A. Y., Ed. *Molecular Beam Epitaxy,* Springer-Verlag, New York, 1994.

24. Zhengwu J.; Fukumura, T.; Kawasaki, M.; Ando K.; Saito, H.; Sekiguchi, T.; Yoo, Y. Z.; Murakami, M.; Matsumoto, Y.; Hasegawa, T.; Koinuma, H. *Appl. Phys. Lett.,* 2001, 78: 3824–3827.

25. Smith, R. C.; Hoilien, N.; Roberts, J.; Campbell, S. A.; Gladfelter, W. L. *Chem. Mater.,* 2002, *14,* 474–476.

26. Reddington, E.; Sapienza, A.; Gurau, B.; Viswanathan, R.; Sarangapani, S.; Smotkin E. S.; Mallouk, T. E. *Science,* 1998, 280: 1735–1737.

27. Reichenbach, H. M.; McGinn, P. J. *J. Mater. Res.,* 2001, 16: 967–974.

28. Choi, K.; Gardner, D.; Hilbrandt, N.; Bein, T. *Angew. Chem., Int. Ed.,* 1999, 38: 2891–2894.

CHAPTER 5

COMBINATORIAL MAPPING OF POLYMER BLENDS PHASE BEHAVIOR

ALAMGIR KARIM, AMIT SEHGAL, AND ERIC J. AMIS
National Institute of Standards and Technology

J. CARSON MEREDITH
Georgia Institute of Technology

5.1 INTRODUCTION

Combinatorial methods (CM) and high-throughput measurements of relevant chemical and physical properties, when combined with the informatics approaches of data mining and automated analysis, allow for efficient development of structure–processing–property relationships. The benefits include efficient characterization of novel regimes of thermodynamic and kinetic behavior (knowledge discovery) and accelerated development of functional materials (materials synthesis and discovery). Although historically applied to pharmaceutical research, there is increasing interest in applying CM to materials science, as indicated by recent reports of combinatorial methodologies for a wide range of organic/polymeric materials [1–17].

Interface directed phase segregation and phase-separated microstructure are important properties of thin (1 to 1000 nm) polymer blend films. A precise understanding of phase separation phenomena in confined or reduced dimensions is crucial for the preparation of nanometer scale functional materials from multicomponent polymer blends. However, phase behavior in thin films is inherently complex due to its dependence on a large number of parameters, including molecular mass, polymer-polymer and polymer-surface interactions, temperature (T), composition (ϕ), and film thickness (h) [1–3,18–25]. In addition, these multiple variables interact in a complex fashion by phenomena that include surface segregation of one blend component, shift of the phase boundary with film thickness [18,21], substrate interactions, film viscosity increases or decreases relative to the bulk [4,5], and coupling between phase separation and surface deformation modes [23]. A simple estimate of possible variable combinations based on realistic ranges

Experimental Design for Combinatorial and High Throughput
Materials Development, Edited by James N. Cawse.
ISBN 0-471-20343-2 © 2003 John Wiley & Sons, Inc.

of temperature ($15°C < T < 180°C$, $\Delta T = 2°C$), mass fraction ($0 < \phi < 1$, $\Delta\phi = 0.01$), and thickness (1 nm $< h < 1000$ nm, $\Delta h = 15$ nm) yields nearly 1.5 million unique T–ϕ–h data points for a single polymer blend pair. The additional dependence on time, variations in molecular mass and chemistry, copolymerization, and extension to multicomponent mixtures causes the magnitude of unexplored variable space to become overwhelming. Faced with the large number of variable combinations and interactive phenomena, unraveling the T–ϕ–h dependence of the kinetics and thermodynamics of thin-film phase behavior is indeed a challenging task using the conventional one-sample for one-measurement approach.

Early efforts in combinatorial materials science used sputtering methods to prepare composition-gradient libraries for measuring the phase behavior of ternary metal alloys [25] and other inorganic materials [26]. However, limitations in computational capability and robotics for instrument automation have severely limited the benefits of combinatorial materials characterization until recently [10–17,25]. The primary limitation to characterizing polymer thin films with combinatorial methods has been a shortage of techniques for preparing libraries with systematically varied composition ϕ, thickness h, and processing conditions, e.g., T. In a previous study, we reported a T-gradient heating stage and a velocity-gradient coating procedure that was used to prepare T-h gradient film libraries to investigate dewetting and block copolymer segregation in nanoscale polymer films [7–9]. Here, we apply the gradient technique to investigate the thickness dependence of binary polymer-blends phase separation in nanoscale thin films.

We briefly review recent advances in applying CM to library design and characterization of polymer-blends phase behavior and characterization. We present applications of several novel CM developed by the authors for the preparation of T, ϕ, h, and surface energy (γ_{so}) continuous polymer film libraries. We focus in particular on the novel library preparation and high throughput screening steps, since these have been the principal limiting factors in CM development for polymers. The use of continuous-gradient libraries in the measurement of fundamental properties is described for polymer-blends phase behavior.

5.2   PREPARATION OF POLYMER THIN-FILM LIBRARIES

5.2.1   Overview

In investigations of polymer films and coatings libraries with variations in ϕ, h, T, and, γ_{so}, we have found that the deposition of films with continuous gradients of these properties is a convenient and practical alternative to the deposition of libraries containing discrete regimes. However, this experimental design has the associated caveat that the scale of the continuous gradients in these libraries should accommodate point measurements and prevent cross talk between the measurement elements. Of course, the introduction of chemical, thickness, and thermal gradients drives nonequilibrium transport processes that will eliminate the gradients over very long periods of time. The timescale and lengthscale over which gradient library measurements are valid are determined in part by the magnitude of these

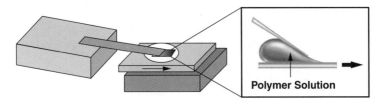

Figure 5.1 Schematic of the combinatorial experimental method for preparation of thickness gradients. The substrate is accelerated across the knife-edge, creating a gradient in film thickness.

transport fluxes. In most cases high molecular mass[1] ($M_w > 10000$ g/mol) polymers have relatively low transport coefficients, e.g., diffusivity, and high viscosity. Thus the mass transport, flow length, and time scales are often several orders of magnitude less than those of the measurements, allowing properties to be measured near equilibrium.

5.2.2 Thickness-Gradient Libraries

A velocity-gradient knife-edge flow coater [14–17], depicted in Figure 5.1, was developed to prepare coatings and thin films with a continuous thickness gradient. A 50 μL drop of polymer solution (mass fraction 2% to 5%) was placed under a knife-edge with a 2.5-cm-wide stainless-steel blade, positioned at a height of 300 μm and at a 5° angle with respect to the substrate. A computer-controlled motion stage (Parker Daedal) moves the substrate under the knife-edge at a constant acceleration, usually 0.5 to 1 mm/s². This causes the stage velocity to gradually increase from zero to a maximum value of 5 to 10 mm/s. The increasing substrate velocity results in a progressive increase in the residual fluid volume passing under the knife edge (inertial effect) giving films with controllable thickness gradients. Figure 5.2 shows an h-gradient obtained immediately after flow coating for polystyrene (PS) as a function of solute concentration on a 4-inch Si substrate. The immediately evident variation in shade shows the systematic variation of film thickness from approximately 20 nm to 100 nm across the substrate. Thin film thickness-dependent phenomena can be investigated from nanometers to micrometers employing several h-gradient films with overlapping gradient ranges. The relatively weak thickness and temperature gradients do not induce appreciable flow in the polymer film over the experimental timescale. [14,15] A unidirectional Navier–Stokes model for flow over a flat plate estimates lateral flow at a characteristic velocity of 1 μm/h at $T = 135°C$, in response to gravitational action on the thickness gradient [27]. This small flow is several orders of magnitude slower than the flow induced by the physical phenomena that these libraries are designed to investigate, such as dewetting [15]

---

[1]According to ISO 31–8, the term "molecular weight" has been replaced by "relative molecular mass," symbol Mr. The conventional notation, rather than the ISO notation, has been employed for this publication.

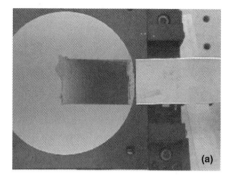
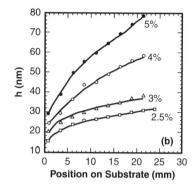

Figure 5.2 (a) A gradient thickness film on a 4-inch Si substrate immediately after flow coating. The knife-edge and the translation stage are shown. The variation in color of the film corresponds to thickness changes from 20 nm to 100 nm, approximately. (b) Thickness, h (nm) vs. x position on substrate (mm), for various h-gradient film libraries composed of polystyrene (M_w = 1800 g/mol) on Si as a function of mass fraction PS in the toluene coating solution. Standard uncertainty in thickness is ±3 nm. (Courtesy of the National Institute of Standards and Technology, Gaithersburg, MD.)

and phase separation [14]. To check for flow, we examined thickness-gradient libraries before and after annealing at $T > T_g$, as shown in Figure 5.3 for a PS film (M_w = 1800) on Si/SiO$_x$. The difference of thickness gradients across the 2-cm × 3-cm library area before and after annealing was within a standard uncertainty of ±1.5 nm [15].

5.2.3 Composition-Gradient Libraries

Three steps are involved in preparing composition gradient films: gradient mixing (Figure 5.4a), gradient deposition (Figure 5.4b), and film spreading (Figure 5.4c). Gradient mixing utilizes two syringe pumps (Harvard PHD2000)[2] that introduce and withdraw polymer solutions (of mass fraction $x_A = x_B = 0.05$ to 0.10) to and from a small mixing vial at rates I and W, respectively. Pump W was used to load the vial with an initial mass M_o of solution B ($M_o \approx 1$ g). The infusion and withdrawal syringe pumps were started simultaneously under vigorous stirring of the vial solution, and a third syringe, S, was used to manually extract ≈ 50 μL of solution from the vial into the syringe needle at the rate of $S = (30$ to $50)$ μL/min. At the end of the sampling process, the sample syringe contained a solution of polymers A and B with a gradient in composition, ∇x_A, along the length of the syringe needle.

---

[2]Certain equipment and instruments or materials are identified in the chapter in order to adequately specify the experimental details. Such identification does not imply recommendation by the National Institute of Standards and Technology, nor does it imply the materials are necessarily the best available for the purpose.

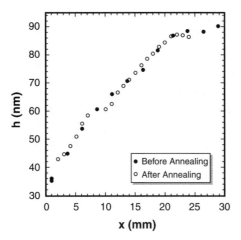

Figure 5.3 Thickness gradients for PS (M_w = 1800) on Si/SiO$_x$ before and after heating at 135°C for 2 h. There is no observable change in the gradient due to flow during annealing.

The relative rates of I and W were used to control the steepness of the composition gradient, e.g., dx_A/dt. The sample time, t_s, determines the endpoint composition of the gradient. The gradient produced by a particular combination of I, W, S, M_o, and t_s values was modeled by a mass balance of the transient mixing process, given elsewhere [21]. This balance predicts that the composition gradient will be linear only if $I = (W + S)/2$, a prediction supported by Fourier transform infrared (FTIR) measurements of composition. An 18-gauge needle long enough to contain the sample volume ensured that the gradient solution did not enter the syringe itself. This

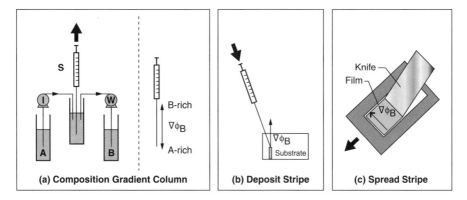

Figure 5.4 Schematic of the composition-gradient deposition process involving (a) gradient mixing, (b) deposition of stripe, and (c) film spreading. (Figure courtesy of the National Institute of Standards and Technology, Gaithersburg, MD.)

prevented turbulent mixing that might occur upon expansion of the solution from the needle into the larger-diameter syringe.

Under the influence of the gradient in the syringe needle, ∇x_A, molecular diffusion will homogenize the composition. However, the timescale for molecular diffusion is many orders of magnitude larger than the sampling time. For example, consider gradient solutions of PS (M_w = 96.4 kg/mol, M_w/M_n = 1.01, Tosoh Inc.) and poly(vinylmethyl ether) (PVME) (M_w = 119 kg/mol, M_w/M_n = 2.5) in toluene, a system used to characterize the ϕ-gradient-deposition procedure [14,28]. For a typical ϕ-gradient with $\Delta\phi \approx 0.025$ mm^{-1}, ϕ_{PS} and ϕ_{PVME} change negligibly by 0.004% and 0.001% in the 5 min required for film deposition.[3] Fluid flow in the sample syringe remains in the laminar regime, preventing turbulence and convective mixing, as discussed elsewhere [14].

The next library preparation step (Figure 5.4b) is to deposit the gradient solution from the sample syringe as a thin stripe, usually 1 to 2 mm wide, on the substrate. This gradient stripe was then spread as a film (Figure 5.4c) orthogonal to the composition gradient using the knife-edge coater described earlier. After a few seconds most of the solvent evaporated, leaving behind a thin film with a gradient of polymer composition. The remaining solvent was removed under vacuum during annealing, described in the next section (T-gradient annealing). Because polymer melt diffusion coefficients, D, are typically of order 10^{-12} cm^2/s, diffusion in the cast film can be neglected if the lengthscale resolved in measurements is significantly larger than the diffusion length, $\sqrt{Dt}$. Composition-gradient films of blends of PS/PVME [21] and poly(D,L-lactide) (PDLA; Alkermes, Medisorb 100DL; M_w = 127,000 g/mol, M_w/M_n = 1.56)/poly(ε-caprolactone) (PCL, Aldrich, M_w = 114,000 g/mol, M_w/M_n = 1.43) [29] were used to test the ϕ-gradient procedure. The composition variation was typically verified by FTIR spectra measured with a Nicolet Magna 550 and were averaged 128 times at 4 cm^{-1} resolution. The beam diameter, 500 μm (approximate), was significantly larger than the diffusion length of 3 μm (approximate) for the experimental timescale. Films 0.3 to 1 μm thick were coated on a sapphire substrate and a translation stage was used to obtain spectra at various positions on the continuous ϕ-gradient. Figure 5.5a shows typical FTIR spectra for a ϕ-gradient film of PS/PVME. As position is scanned along the film, a monotonic increase in PVME absorbances, and a corresponding decrease in PS absorbances, is observed. For the PS/PVME blend, compositions were measured based upon a direct calibration of the $\nu = 2820$ cm^{-1} peak using known mixtures, yielding $\varepsilon(2820$ cm$^{-1}) = 226 \pm 3$ A/hc, where A = absorbance for this peak, h is the film thickness measured in micrometers, and c is the molar density of PVME in moles per liter. For PDLA/PCL system, $\varepsilon(\nu)$ values for pure PDLA and PCL were determined over the $C\text{–}H$ stretch regime of (2700 to 3100) cm^{-1}, based upon $\varepsilon_i(\nu) = A_i(\nu)/(ch)$, where A_i is the absorbance for each peak. Unknown PDLA/PCL mass fractions

---

[3]The diffusive flow rate of PS and PVME were calculated as $J = L\pi r^2 D_i\rho(d\phi_i/dx)_{max}$, where ρ is the solution density, $r = 2.3$ mm is the syringe diameter, and $L = 4.2$ mm is the length of the fluid column in the syringe. We estimate $\Delta\phi_i$ as $(Jt)/(x_pL\pi r^2\rho)$, where $x_p = 0.08$ is the total polymer mass fraction in solution.

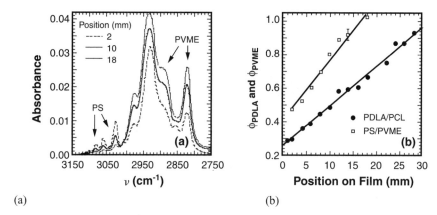

(a)   (b)

Figure 5.5 (a) FTIR spectra at various positions x (mm) along a ϕ-gradient PS/PVME library, as described in the text. PS absorptions decrease and PVME absorptions increase, monotonically, as one samples spectra across the film. (b) Mass fractions ϕ_{PVME} and ϕ_{PCL} vs. position, x (mm), for typical PCL/PDLA and PS/PVME ϕ-gradient libraries. Composition of PS/PVME blends is calculated by calibration of the $\nu = 2820$ cm^{-1} PVME absorption. Coating parameters were: PS/PVME ($I = 0.51$ mL/min, $W = 1.0$ mL/min, $S = 20$ μL/min, $M_o =$ 1.57 mL, sample time = 94 s) and PDLA/PCL ($I = 0.76$ mL/min, $W = 1.5$ mL/min, $S = 26$ μL/min, $M_o = 1.5$ mL, sample time = 95 s) Unless otherwise indicated by error bars, standard uncertainty is represented by the symbol size. (Figure 5.5b courtesy of the National Institute of Standards and Technology, Gaithersburg, MD.)

were determined to within a standard uncertainty of 4% by assuming the observed spectra were linear combinations of pure PDLA and PCL spectra, e.g., $A_{mix} = h(\alpha\varepsilon_{PDLA}c_{PDLA} + (1 - \alpha)\varepsilon_{PCL}c_{PCL})$ and α is related to the mass fraction PDLA. Figure 5.5b shows typical composition gradients for PDLA/PCL blends coated from CHCl$_3$ and PS/PVME blends coated from toluene. Essentially linear gradients were obtained and the endpoints and slope agree with those predicted from mass balance [14]. It is possible to create gradient films with wider composition ranges than those shown in Figure 5.5, by sampling the mixing vial, Figure 5.4a, for longer times.

5.2.4 Temperature-Gradient Libraries

To explore a large temperature (T) range, the h- or ϕ-gradient films are annealed on a T-gradient heating stage, with the T-gradient *orthogonal* to the h- or ϕ-gradient. This custom aluminum T-gradient stage, shown in Figure 5.6, uses a heat source and a heat sink to produce a linear gradient ranging between adjustable endpoint temperatures. Endpoint temperatures typically range from $160 \pm 0.5°C$ to $70.0 \pm 0.2°C$ over 40 mm, but are adjustable within the limits of the heater, cooler, and maximum heat flow through the aluminum plate. To minimize oxidation and convective heat transfer from the substrate, the stage is sealed with an O-ring, glass plate, and vacuum pump. Each two-dimensional T–h or T–ϕ parallel library contained about 1800 or 3900 state points, respectively, where a "state point" is defined

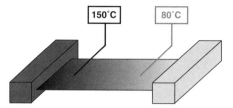

Figure 5.6 Schematic of the temperature-gradient stage, highlighting a typical temperature range investigated.

by the T, h, and ϕ variation over the area of a 200× optical microscope image: $\Delta T = 0.5°C$, $\Delta h = 3$ nm, and $\Delta \phi = 0.02$. These libraries allow T, h, and ϕ-dependent phenomena, e.g., dewetting, order–disorder, and phase transitions, to be observed in situ or post-annealing with relevant microscopic and spectroscopic tools.

5.2.5 Surface-Energy Gradients

In many polymer-coating and thin film systems, there is considerable interest in studying the film stability, dewetting, and phase behavior on substrates with surface energies varying between hydrophilic and hydrophobic extremes. Therefore, a gradient-etching procedure has been developed in order to produce substrate libraries with surface energy, γ_{so}, continuously varied from hydrophilic to hydrophobic values [30]. The gradient-etching procedure involves immersion of a passivated Si-H/Si substrate (Polishing Corporation of America) into an 80°C Piranha solution [31] at a constant immersion rate. The Piranha bath etches the Si-H surface and grows an oxide layer, $SiO_x/SiOH$, at a rate dependent on T and the volume fraction H_2SO_4 [31]. A gradient in the conversion to hydrophilic $SiO_x/SiOH$ results from one end of the wafer with increasing time of exposure to the Piranha solution along the length. After immersion, the wafer is withdrawn rapidly (≈ 10 mm/s), rinsed with deionized water, and blow-dried with nitrogen.

We have also developed a simple method for chemical modification of chlorosilane self-assembled monolayers (SAMs) on Si surfaces by exposure to a gradient in UV-ozone radiation to create stable surface energy gradient substrates. Typical deionized water contact angles are shown in Figure 5.7. By preparing several gradient substrates covering overlapping ranges of hydrophilicity, it is possible to screen a large range in surface energy, from hydrophilic ($\theta w \approx 0°$) to hydrophobic ($\theta w \approx 90°$) values of the water contact angle. In another procedure for varying substrate energy, mixed SAMs of alkanethiolates are deposited with a composition gradient [32]. In this procedure, ω-substituted alkanethiolates with different terminal groups, e.g., —CH_3 vs. —COOH, cross-diffuse from the opposite ends of a polysaccharide matrix deposited on top of a gold substrate. Diffusion provides for the formation of a SAM with a concentration gradient between the two thiolate species from one end of the substrate to another, resulting in controllable substrate energy gradients. The

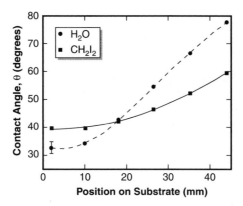

Figure 5.7   Deionized water and dioido methane contact angle versus spatial position (mm) for a gradient UV oxidized SAM substrate. The method allows a systematic variation of contact angles and corresponding surface energy across the substrate.

polysaccharide matrix is removed after a period of time, halting the diffusion process. These gradient SAM substrates were subsequently used to investigate the effect of surface energy on phase separation of immiscible polymer blends [33].

5.3   UNCERTAINTY AND STATISTICAL CONSIDERATIONS OF LIBRARY MEASUREMENTS

A potential drawback of the method is that the polymer-surface libraries are not composed of distinct sample areas, but rather have continuous gradients in ϕ and annealing T. These gradients induce variance in observed properties, which is not an issue with uniform samples. Many properties measured with combinatorial libraries are obtained from microscope images [optical, fluorescent, atomic force (AFM),] or spectroscopy (UV, FTIR). Thus it is important to understand how uncertainties associated with the gradients and the lateral resolution affects properties measured on the libraries. Figure 5.8 demonstrates how a combinatorial library is divided into a grid of "virtual" measurement sites (e.g., microscope images) of lengthscale L. The T and ϕ for each measurement site is taken as the average $<T>$ and $<\phi>$ over the length L. Because gradients are present on the library, each measurement site has systematic variances $\Delta\phi$ and ΔT that increase as the measurement lengthscale L increases. Hence lower measurement resolution (lower L) results in lower $\Delta\phi$ and ΔT. A typical 500-μm $\times$ 500-μm image would have reasonable variances of $\Delta\phi = 0.01$ and $\Delta T = 0.3°C$. The number of measured features decreases as L decreases, causing an increase in measurement uncertainty.

A key question to be answered is how to select the optimum measurement scale L, reflecting a balance between counting statistics, $\Delta\phi$ and ΔT. The effects of these

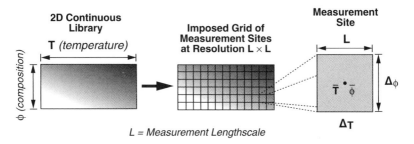

Figure 5.8 Distribution of discrete measurement sites of resolution $L \times L$ over a continuous gradient library. Measurement sites have average T and ϕ, with gradient variance ΔT and $\Delta\phi$.

contributions on the variance about the mean of any property $<p>$ within a measurement site is accounted for using a standard uncertainty propagation,

$$\Delta<p> = (\partial<p>/\partial N)\Delta N + (\partial<p>/\partial T)\Delta T + (\partial<p>/\partial\phi)\Delta\phi. \tag{5.1}$$

Here $<p>$ is a function of T, ϕ, and the number of observations made in the measurement site, $N \sim L^2$. It is assumed that the number of features (microstructures, cells, etc.) can be counted exactly, so that $\Delta N = 0$. The partial derivatives can be estimated from finite difference approximations of the measured data, e.g., $\partial<p>/\partial T = (\Sigma p(T_{i+1}, \phi_i) - \Sigma p(T_i, \phi_i))/[N(T_{i+1} - T_i)]$. The values of $\Delta\phi$ and ΔT are $\Delta\phi = m_\phi L$ and $\Delta T = m_T L$, where m_T and m_ϕ are the slopes of the linear gradients, known from the library preparation procedure. Making these substitutions shows that the error propagation for property p scales as

$$\Delta<p> \sim (m_T + m_\phi)/L. \tag{5.2}$$

Constants have been removed in order to reveal only the dependence on gradients and the measurement scale, L. Equation (5.1) demonstrates that the uncertainty of any property measured on the libraries at a given ϕ and T will decrease if the measurement scale L is increased (because more features are counted) and if the magnitude of the gradients is decreased (reducing ϕ and T uncertainty). Thus the following guidelines will be followed during experimental design and data analysis: (1) L will be made as large as possible while still being able to resolve features of interest, and (2) the gradient slopes will then be adjusted to attain an acceptable uncertainty ($<1\%$) in the measured property. The preceding analysis considers only uncertainty contributions from the library gradients. Additional sources of uncertainty, treated later, arising from the culturing and assay steps themselves, are also present for uniform conventional samples.

5.4   FUNDAMENTAL PHASE BEHAVIOR AND PROPERTIES OF POLYMER-BLEND FILM LIBRARIES

5.4.1   Composition Gradients for Phase Boundary

Figure 5.9 presents a photographic image of a typical temperature-composition library of the PS/PVME blend (just discussed in the composition-gradient preparation section) for 16-h annealing time. As Figure 5.9 indicates, the lower critical solution temperature (LCST) cloud-point curve can be seen with the unaided eye as a diffuse boundary separating one-phase and two-phase regions. Cloud points measured on bulk samples with conventional light scattering are shown as discrete data points and agree well with the cloud-point curve observed on the library [14,34]. The diffuse nature of the cloud-point curve reflects the natural dependence of the microstructure evolution rate on temperature and composition. Near the LCST boundary the microstructure size gradually approaches optical resolution limits (1 μm), giving the curve its diffuse appearance. Based upon a bulk diffusion coefficient of $D \approx 10^{-17}$ m^2/s, the diffusion length ($\sqrt{Dt}$) for a 2-h anneal is 270 nm. In Figure 5.9 each pixel covers about 30 μm, which is over 100 times the diffusion length, and ϕ-gradient-induced diffusion has a negligible effect on the observed LCST cloud-point curve. The combinatorial technique employing T–ϕ polymer-

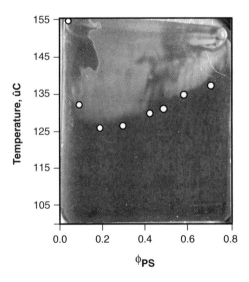

Figure 5.9   Digital optical photographs of a PS/PVME T-ϕ library after 91 min of annealing, showing the LCST cloud point curve visible to the unaided eye. The library wafer dimension is 31 mm × 35 mm and the film thickness varies from approximately 400 to 600 nm from low to high ϕ$_{PS}$ values. White circles light-scattering cloud points measured on separate uniform samples. (Figure courtesy of the National Institute of Standards and Technology, Gaithersburg, MD.)

blend libraries represents a powerful means for rapid and efficient characterization of polymer-blend phase behavior in orders of magnitude less time than with conventional light- or neutron-scattering techniques. A large database of phase behavior of a host of binary polymeric mixtures, and the influence of additives, fillers, or other processing variables can be thus generated with tremendous implications for application of multicomponent-blend films.

5.4.2 Influence of Substrate Surface-Energy

The presence of an interface in an ultrathin polymer-blend film guides or even significantly alters the phase-separation process. Confinement between the air and substrate interfaces and the preferential wetting of the components at the walls determines the in-plane and the surface-directed compositional distribution and the spatial scales. Genzer et al. have shown a wetting reversal transition from a symmetric three layer to an asymmetric two-layer structure with a change in substrate surface energy using compositional depth profiling with forward-recoil spectrometry (FRES) [33]. The segregation of a particular component to the polymer–substrate interface is governed by the interfacial energy resulting in a change in the compositional distribution and blend-phase microstructure. Other work points at the substrate dependence in the time evolution of the phase-separated morphology for Si, Au, and Co interfaces [35]. Gradient substrates with systematic variation of surface energy were used to probe the influence of surface energetics of the phase behavior of a model LCST PS/PVME blend at the critical composition of $\phi_{PS} = 0.2$. Figure 5.10 shows AFM images (30 μm × 30 μm) of a PS/PVME film having a fixed thickness $h = 57$ nm cast on a substrate with a gradient in surface energy. The surface-energy gradient was obtained by gradient UV ozonolysis of an octyl-

h = 57 nm

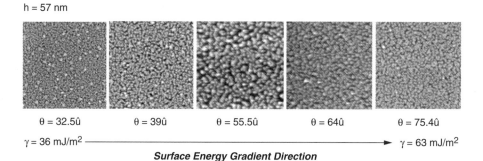

$\theta = 32.5\hat{u}$ $\quad\quad\theta = 39\hat{u}$ $\quad\quad\theta = 55.5\hat{u}$ $\quad\quad\theta = 64\hat{u}$ $\quad\quad\theta = 75.4\hat{u}$

$\gamma = 36$ mJ/m^2 ⎯⎯⎯⎯⎯⎯⎯⎯⎯⎯⎯⎯⎯⎯⎯⎯⎯⎯⎯⟶ $\gamma = 63$ mJ/m^2

Surface Energy Gradient Direction

Figure 5.10 Blend-phase morphology dependence on surface energy. AFM images of PS/PVME blend film ($\phi_{PS} = 0.2$) at a constant thickness $h = 57$ nm on a SAM gradient energy substrate. A nonmonotonic change in the lateral scale of phase separation is evident as we span surface-energy space. Maximal scale seen for $\theta = 55°$ corresponding to a surface energy of 48 mJ/ m^2. (Figure courtesy of the National Institute of Standards and Technology, Gaithersburg, MD.)

dimethylchlorosilane SAM ($32.5° < \theta_{H_2O} < 75.4°$; 36 mJ m^{-2} < γ_{so} < 63 mJ m^{-2}). The phase-separation process has developed to long times ($T = 145°C$) where "pinning" occurs. It is evident from Figure 5.10 that the pattern scale varies nonmonotonically with the surface energy of the substrate. The surface structures of the annealed films were mapped using AFM, preprogrammed to uninterruptedly map out a total of approximately 100 scans over the phase-separated film regions of interest.

Automated image analysis includes radially averaging the isotropic fast Fourier transform (FFTs) of the AFM height-topography images to give the in-plane scale of phase separation, λ. We observe that the pattern scale (λ) is maximum for an intermediate characteristic value of $\gamma^* = 48$ mJ m^{-2} ($\theta = 55°$) that is nearly independent of h. We tentatively assign this value of $\gamma^* = 48$ mJ m^{-2} to the surface energy where both polymer components of the blend have a similar polymer–substrate interaction and the effective film thickness is maximal with the depletion of the substrate–polymer boundary layer. Measurement of polymer–substrate interactions is key to understanding the stability and phase behavior of polymer with blend films on surfaces. The continuous-gradient approach provides this information in an efficient manner without involving numerous substrates with creating incremental variations of surface energy and characterizing the surface energy of each substrate.

5.4.3 Film-Thickness Gradients

Figure 5.11 shows the progressive change of scale of the late stage of phase separation morphology with increasing film thickness for the PS/PVME film at fixed surface energy, $\gamma_{so} = 63$ mJ m^{-2}. This h-gradient range is low enough (<200 nm) that surface-directed spinodal decomposition is suppressed, and phase separation occurs laterally [24]. In addition, $h > 2R_g$, where $R_g = 8$ nm, is the estimated radius of gyration of the polymers. Hence the film is not confined completely to two dimensions,

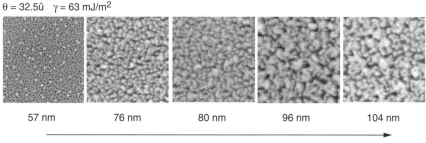

$\theta = 32.5û$ $\gamma = 63$ mJ/m^2

| 57 nm | 76 nm | 80 nm | 96 nm | 104 nm |

Film Thickness Gradient Direction

Figure 5.11 Blend-phase morphology dependence on film thickness. AFM images of PS/PVME ($\phi_{PS} = 0.2$) blend-phase morphology at a constant surface energy $\gamma_{so} = 63$ mJ/ m^2 for a gradient thickness film. A systematic increase in the lateral scale of phase separation is evident with film thickness. (Figure courtesy of the National Institute of Standards and Technology, Gaithersburg, MD.)

but is in a transition regime between a three-dimensional (bulk) and two-dimensional geometry. Figure 5.11 shows that thickness has a profound effect on microstructure size, film topography, and roughness. A bicontinuous lateral microstructure superimposed upon a topography of peaks and valleys is observed at all h-values. This bicontinuous structure is consistent with phase separation in the spinodal regime, and the peak and valley topography has been attributed to surface-tension differences between PS-rich and PVME-rich regions. The pattern scale λ is determined by FFTs of the AFM height images of the patterns. No statistically significant variation of λ was determined by changing scan dimensions. We observe a linear increase of λ with increasing film thickness, h. There have been some recent studies to indicate this behavior in phase separation in thin films [36–38]. However, other work has also suggested nonlinear trends depending on the blend chemistry and experimental conditions [36]. Care must be taken to generalize results, since the behavior can change significantly with substrate surface energy or in other film-thickness regimes where changes in phase-separation kinetics are anticipated. The nonmonotonic variation of lateral scale with substrate surface energy suggests a possible explanation for the variance in observed experimental trends for different chemistries. The combinatorial approach allows us to rapidly scan for both substrate surface-energy variations and film-thickness changes in this case [39].

5.5 CONCLUSIONS

To allow more efficient measurement of phase behavior over large ranges of T–ϕ–h parameter space, we report novel combinatorial library preparation and high throughput screening methods for nanoscale polymer-blend thin films. The benefits include efficient characterization of large regimes of thermodynamic and kinetic behavior (knowledge discovery) and accelerated development of functional materials (materials discovery). This is accomplished by coupling sample libraries containing hundreds to thousands of variable combinations with high throughput measurements. Although historically applied to pharmaceutical research, there is an ongoing movement to apply this methodology in materials characterization and development, as indicated by recent reports for inorganic and organic or polymeric materials. In this chapter we have presented recent advances in which combinatorial methodologies have been used for efficient measurement of chemical and physical properties of polymer blends over large regimes of variable space. Methodologies for phase-behavior characterization allow for the discovery of new models for structure–processing–property relationships in polymer blends. Several recent developments were presented in the combinatorial characterization of polymers blends phase behavior and properties using high throughput libraries of films and coatings using continuous-gradient polymer libraries with controlled variations in temperature, composition, thickness, and substrate surface energy. The use of these new library techniques facilitates characterization of polymer-blend phase behavior, and more generally multicomponent polymeric materials.

More recently, the continuous-gradient high throughput approach has been used for mapping phase behavior of olefinic [40], as well as nanocomposite polymer blends [41], as representative blend systems of commercial importance. These studies demonstrate the utility of the continuous-composition-gradient methods for investigating industrially important and relevant problems in phase-separated and multicomponent polymeric materials, an approach that is likely to be adopted widely by the plastics and materials manufacturing industries in the future.

REFERENCES

1. Jandeleit, B.; Schaefer, D. J.; Powers, T. S.; Turner, H. W.; Weinberg, W. H. *Angew. Chem. Int. Ed.,* 1999, 38: 2494–2532.

2. Gravert, D. J.; Datta, A.; Wentworth, P.; Janda, K. D. *J. Am. Chem. Soc.,* 1998, 120: 9481–9495.

3. Brocchini, S.; James, K.; Tangpasuthadol, V.; Kohn, J. *J. Am. Chem. Soc.,* 1997, 119: 4553–4554.

4. Brocchini, S.; James, K.; Tangpasuthadol, V.; Kohn, J. *J. Biomed. Mater. Res.,* 1998, 42: 66–75.

5. Dickinson, T. A.; Walt, D. R.; White, J.; Kauer, J. S. *Anal. Chem.,* 1997, 69: 3413–3418.

6. Reynolds, C. H. *J. Comb. Chem.,* 1999, 1: 297–306.

7. Schmitz, C.; Posch, P.; Thelakkat, M.; Schmidt, H. W. *Phys. Chem. Chem. Phys.,* 1999, 1: 1777–1782.

8. Schmitz, C.; Thelakkat, M.; Schmidt, H. W. *Adv. Mater.,* 1999, 11: 821.

9. Schmitz, C.; Posch, P.; Thelakkat, M.; Schmidt, H. W. *Macromol. Symp.,* 2000, 154: 209–222.

10. Gross, M.; Muller, D. C.; Nothofer, H. G.; Sherf, U.; Neher, D.; Brauchle, C.; Meerholz, K. *Nature,* 2000, 405: 661.

11. Takeuchi, T.; Fukuma, D.; Matsui, J. *Anal. Chem.,* 1999, 71: 285–290.

12. Terrett, N. K. *Combinatorial Chemistry,* Oxford University Press, Oxford, 1998.

13. Kennedy, K.; Stefansky, T.; Davy, G.; Zackay, V. F.; Parker, E. R. *J. Appl. Phys.,* 1965, 36: 3808–3810.

14. Meredith, J. C.; Karim, A.; Amis, E. J. *Macromolecules,* 2000, 33: 5760–5762.

15. Meredith, J. C.; Smith, A. P.; Karim, A.; Amis, E. J. *Macromolecules,* 2000, 33: 9747–9756.

16. Smith, A. P.; Meredith, J. C.; Douglas, J. F.; Amis, E. J.; Karim, A. *Phys. Rev. Lett.,* 2001, 87: 015503.

17. Smith, A. P.; Douglas, J. F.; Meredith, J. C.; Amis, E. J.; Karim A. *J. Polym. Sci. B: Polym. Phys.,* 2001, 39: 2141–2158.

18. Reddington, E.; Sapienza, A.; Gurau, B.; Viswanathan, R.; Sarangapani, S.; Smotkin, E.; Mallouk, T. *Science,* 1998, 280: 1735–1737.

19. Xiang, X.-D.; Sun, X.; Briceno, G.; Lou, Y.; Wang, K.-A.; Chang, H.; Wallace-Freedman, W. G.; Chen, S.-W.; Schultz, P. G. *Science,* 1995, 268: 1738–1740.

20. Wang, J.; Yoo, Y.; Gao, C.; Takeuchi, I.; Sun, X.; Chang, H.; Xiang, X.-D.; Schultz, P. G. *Science,* 1998, 279: 1712–1714.

21. Sun, X.-D.; Xiang, X.-D. *Appl. Phys. Lett.,* 1998, 72: 525–527.

22. Danielson, E.; Golden, J. H.; McFarland, E. W.; Reaves, C. M.; Weinberg, W. H.; Wu, X. D. *Nature,* 1997, 389: 944–948.

23. Danielson, E.; Devenney, M.; Giaquinta, D. M.; Golden, J. H.; Haushalter, R. C.; McFarland, E. W.; Poojary, D. M.; Reaves, C. M.; Wenberg, W. H.; Wu, X. D. *Science,* 1998, 279: 837–839.

24. Klein, J.; Lehmann, C. W.; Schmidt, H.-W.; Maier, W. F. *Angew. Chem. Int. Ed.,* 1998, 37: 3369–3372.

25. Hanak, J. J. *J. Mater. Sci.,* 1970, 5: 964–971.

26. Newkome, G. R.; Weis, C. D.; Moorefield, C. N.; Baker, G. R.; Childs, B. J.; Epperson, J. *Angew. Chem. Int. Ed.,* 1998, 37: 307–310.

27. Leal, L. G. *Laminar Flow and Convective Transport Processes,* Butterworth-Heinemann, Boston, 1992.

28. Daivis, P. J.; Pinder, D. N.; Callaghan, P. T. *Macromolecules,* 1992, 25: 170–178.

29. Meredith, J. C., Amis, E. J. *Macromol. Chem. Phys.,* 2000, 201: 733–739.

30. Ashley, K.; Meredith, J. C.; Karim, A.; Raghavan, D. *Polym. Int.,* 2002, in print.

31. Kern, W., Ed. *Handbook of Semiconductor Wafer Cleaning Technology,* Noyes, Park Ridge, NJ, 1993.

32. Liedberg, B.; Tengvall, P. *Langmuir,* 1995, 11: 3821–3827.

33. Genzer, J.; Kramer, E. J. *Europhys. Lett.,* 1998, 44: 180–185.

34. Meredith, J. C.; Karim, A.; Amis, E. J. Manuscript in preparation.

35. Winesett, D. A.; Ade, H.; Rafailovich, M.; Zhu, S. *Polym. Int.,* 2000, 49: 458–462.

36. Müller- Buschbaum, Stamm, M. *Colloid Polym. Sci.,* 2001, 279: 376–381.

37. Walheim, S.; Böltau, M.; Mlynek, J.; Krausch, G.; Steiner, U. *Macromolecules,* 1997, 30: 4995–5003.

38. Dalonki- Veress, K.; Forrest, J. A.; Dutcher, J. R. *Phys. Rev. E,* 1998, 57(8): 5811–5817.

39. Sehgal, A.; Douglas, J.F.; Amis, E. J.; Karim, A. Manuscript in preparation.

40. Wang, H.; Shimizu, K.; Hobbie, E. K.; Wang, Z.-G.; Meredith, J. C.; Karim, A.; Amis, E. J.; Hsiao, B. S.; Hsieh, E. T.; Han, C.C. *Macromolecules,* 2002, 35: 1072–1078.

41. Karim, A.; Yurekli, K.; Meredith, C.; Amis, E. J., Krishnamoorti, R. *Polym. Eng. Sci.,* In Press.

CHAPTER 6

COMBINATORIAL EXPERIMENTAL DESIGN USING THE OPTIMAL-COVERAGE ALGORITHM

DAVID S. BEM AND RALPH D. GILLESPIE
UOP LLC

ERIK J. ERLANDSON, LAUREL A. HARMON, STEVEN G. SCHLOSSER, AND ALAN J. VAYDA
NovoDynamics, Inc.

6.1 INTRODUCTION AND BACKGROUND

Combinatorial approaches have opened exciting new methods to investigate novel materials and processes. These methods are being applied to a broad array of materials, including polymers and biomaterials, homogeneous and heterogeneous catalysts, phosphors, and coatings [1–6]. The development of high throughput and parallel laboratory equipment and techniques has introduced significant challenges for designing experiments that effectively leverage these new capabilities.

Particular challenges to experimental design in the materials domain include:

- *High-dimensional problems.* Combinatorial materials discovery operates in high-dimensional experimental spaces and must address not only composition and structural variables but also synthesis and process parameters. These spaces can never be searched exhaustively, due to the combinatorial explosion of variables [7], leading to a reconceptualization of experiment design.

- *Nonlinear responses.* Material properties of interest are often highly nonlinear or even discontinuous within parameter ranges of experimental importance. Such behaviors violate the underlying assumptions of traditional experiment design and corresponding data analysis approaches. As a consequence, experiment design is critically important in this new field, yet well-established methods are no longer adequate.

Experimental Design for Combinatorial and High Throughput Materials Development, Edited by James N. Cawse.
ISBN 0-471-20343-2 © 2003 John Wiley & Sons, Inc.

Combinatorial discovery is often conceptualized in two modes or phases [8]:

1. *Screening.* The search over a broad region in chemical space for either discrete material candidates (leads) or ranges of materials with desired properties.

2. *Optimization or focused screening.* The refinement of molecular or material properties in the vicinity of a lead or within a small range of parameters identified during screening.

Each phase imposes its own requirements on experimental design, and strategies are needed to address both, as well as to support the transition from screening into optimization.

This chapter presents a particular approach to combinatorial materials experiment design originated by NovoDynamics, Inc., and termed *optimal coverage.* The approach is suitable for either screening or optimization experiments. Prior knowledge, novel opportunities and practical constraints must all be leveraged during the design of combinatorial experiments. The optimal-coverage approach is specifically intended for iterative experiments in which an arbitrarily complex chemical space of interest may be defined based on prior knowledge and previous experimental points. When coupled with appropriate regression or other data-modeling methods, the method can guide the discovery process to materials meeting the experimental objectives. The method allows the experiment design to be mapped to any configuration of laboratory equipment, with user-specified replicate and reference samples. Application of the method is illustrated in combinatorial discovery for new heterogeneous catalysts.

6.1.1   Experiment Design in Combinatorial Discovery

In either screening or optimization experiments, the role of combinatorial experiment design is to specify sets of experimental data points that will steer the discovery process toward its goals, while maximizing the information obtained from each experiment and making effective use of available resources.

New methods are being developed and applied that explicitly take an iterative, exploratory approach to discovery rather than attempting to systematically quantify all relationships within a single experiment, as in classic statistical DOE [9,10]. These methods treat the discovery process as a search in which information is developed through a succession of experiments. In contrast to traditional methods, the iterative approach requires not only a design for the currently planned experiment but also a well-defined method for guiding the next design based on current and previous experimental results. It is clear from the behaviors of heterogeneous catalysts and materials described later that any method for designing experiments in this domain must probe beyond local maxima, allow the identification of discontinuities, and ultimately direct experiments toward global maxima in the desired properties. The optimal coverage method is one approach that has been developed for and successfully applied in combinatorial heterogeneous catalyst discovery.

6.1.2 Prior Knowledge and Laboratory Constraints in Experiment Design

Despite breakthroughs in automation and throughput, the number of experiments that can be pursued is necessarily bounded by limitations in both resources and time. The space to be considered in combinatorial experiments can be reduced by leveraging the hundreds of years worth of understanding in conjunction with new experimental design techniques and combinatorial methods. Chemical and physical information is available to constrain and focus experiment designs, reflecting current understanding of the chemistry under investigation. Prior research can be used to identify promising areas for combinatorial research as well as to eliminate areas unlikely to produce results. At the same time, combinatorial approaches offer the unique opportunity to investigate outside the scope of current chemical knowledge and to challenge existing assumptions. Effective experiment-design tools must facilitate the incorporation of prior knowledge and chemical intuition while enabling exploration of novel materials and processes.

The day-to-day efficiency of combinatorial experiments can be improved by taking into account constraints in the laboratory environment. Laboratory constraints may be imposed by equipment design, such as the number of wells in a synthesis plate or the number of reactors in a parallel testing system. Designs which do not align with equipment configurations waste experimental resources and time. Laboratory-imposed constraints can also stem from the need to include replicates or reference samples for quality control. Downstream economic factors can also impose real constraints on the experiment space to be considered, such as metallurgy, which can limit processing temperatures, or raw material prices, which can limit the practical concentrations of certain components.

6.1.3 Limitations of Statistical Design of Experiments

While highly successful in many applications, statistical experiment-design approaches have significant limitations for combinatorial materials research. These limitations stem from differences between the characteristics of the combinatorial materials discovery problem domain and the domains for which traditional methods have been developed.

Traditional methods typically assume continuous or even approximately linear responses to independent variables. In the materials domain, particularly for complex materials like heterogeneous catalysts, this assumption breaks down due to the intrinsic properties of the materials of interest. Under certain conditions, such as optimization experiments, however, responses can be approximated as continuous or linear, and statistical DOE may be very effective.

Traditional statistical DOE is designed to explore a regular hypercube in chemical space. As described earlier, physical, chemical, and knowledge-based constraints can render parts of the experimental hypercube uninteresting or even impossible to access experimentally, leading to smaller but highly irregular regions of real experimental interest. Dropping or moving points from a traditional design leaves the experiment unbalanced and the ability to fully leverage the data is lost. Furthermore, traditional methods require a specific number of points to be tested in order to achieve statistical significance. For serial experimentation, the only practical implication is the amount of time required to complete the experiment. In parallel experiments, equipment configurations and capabilities must be taken into account. For example, a two-level factorial design with 6 factors requires 64 points. If available equipment handles 48 points at a time, there are two choices: make two runs, wasting the capacity for 32 experimental points, or drop 16 points, reducing the power of the design. Of course, parallel laboratory equipment designed to accommodate a variety of experimental configurations can be mapped to a correspondingly wider range of experiment designs.

Traditional methods do not take into account necessary laboratory protocols, such as referencing and replication for quality control. Reference samples may be required to monitor consistency over multiple runs. These points are not part of the experiment design, but they must be run on the equipment along with the experimental points. Also, to measure consistency within a run, it may be necessary to include replicates of experimental points. Planning for references and replicates requires specific knowledge of equipment configurations and laboratory procedures. Some traditional methods incorporate replicates and user-defined reference points, but without taking into account details of equipment configuration and layout.

Finally, traditional methods assume that each design starts with a clean sheet of paper. There is no capability to build a new design around previous data points, as is desirable when examining an area of the experimental space in more detail or when exploring beyond the boundaries of previous experiments. For example, if an experiment determines that a particular area is of interest but it was sampled only by a small number of points in a prior experiment, a new design within that area will place its points without consideration of the points that have already been tested. This may result in unnecessary duplication of data and suboptimal sampling of the region of interest.

6.2 OPTIMAL-COVERAGE APPROACH

The optimal-coverage approach has been developed to design a series of experiments that explore complex and high-dimensional experiment spaces. It enables the type of iterative experimentation that arises when it is not possible to assume that a single experiment will provide all of the desired information. Other experimental strategies, such as genetic algorithms [11] and Monte Carlo methods [12,13], are also being brought to this domain, as described elsewhere in this volume.

6.2.1 Overview

In the optimal-coverage approach, chemical, physical, and knowledge-based constraints are applied to establish and limit the region of experimental interest before the specification of experimental points. The result is an arbitrarily complex geometric subset of the nominal chemical space of interest. Once defined, the region of interest is "covered" with individual design points of maximum diversity. Laboratory constraints are taken into account before the final experiment points are computed and mapped to available equipment.

The overall approach comprises the following steps:

- Determine the region of initial experimental interest;
- Design a set of experiment points to explore the region using the optimal coverage algorithm;
- Perform an experiment that supplies response values for these experimental points;
- Identify a new region of interest based on the results of this and all previous experiments;
- Design the next set of experiment points using the optimal-coverage algorithm in the new region of interest.

This process is repeated as many times as needed.

The optimal-coverage algorithm begins with a functional description of the region to be covered in the experiment. Here, the term *coverage* is defined with respect to the experimental space itself [14] rather than to a library or set of samples, as it has sometimes been defined in combinatorial drug discovery [15]. The desired number of experiment points is distributed within the region to maximize diversity among them, where diversity is based on a geometric distance in the space of the independent variables. In effect, the optimal coverage algorithm minimizes the maximum distance from any point in the region of interest to the nearest point in the experiment design. This notion of diversity is similar to that used in combinatorial small-molecule libraries [16,17]. In pharmaceutical applications, however, diversity is computed from finite (though perhaps very large) sets of existing or virtual molecules, typically using geometric distances in the space of selected molecular descriptors [18]. In the application addressed here, experiment points (and ultimately samples) are selected from a continuum of possibilities in order to maximize the diversity function and thus cover the region of interest.

The proprietary optimal coverage algorithm, as currently implemented, assumes that each independent variable represents a continuum rather than a discrete set of values. The method can be extended in a straightforward manner to handle discretized variables, as long as an appropriate distance measure can be defined. When optimizing diversity or coverage, the algorithm can take into account previous experiment points as well as fixed points specified by the researcher.

6.2.2 Illustration of Optimal-Coverage Experiment Design

A simple example is helpful to demonstrate how the optimal coverage approach is used to iteratively explore an experimental space. A three-component mixture problem can be visualized with a simplex where shades of gray are proportional to the

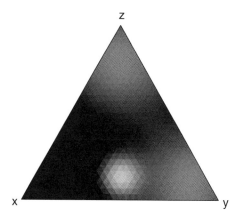

Figure 6.1 Synthetic three-dimensional response surface with three maxima.

height of the response, with black representing zero and white representing the maximum value. Figure 6.1 shows a synthetic response surface with three maxima. We present here a series of 24-point experiments directed at finding the global maximum in this response. In this illustration, a function returns the value of the underlying response surface for each data point as a surrogate for experimental measurement.

Figure 6.2a shows the first experiment where the 24 points, marked with outline boxes, are spread over the entire experimental space by the optimal coverage algorithm. Figure 6.2b shows all of the experimental points that will be used to determine the region of interest. For this example, we are using a simple "greedy" algorithm that defines the region of interest as the volume within a defined distance from the five highest experimental points. This region may be noncontiguous. The distance used to define the region of interest is proportional to the average spacing between all of the experimental points. Thus, as more experimental points are

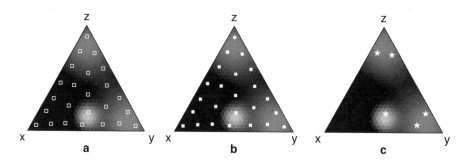

Figure 6.2 Experiment 1: (a) experimental points; (b) all points; (c) top five points.

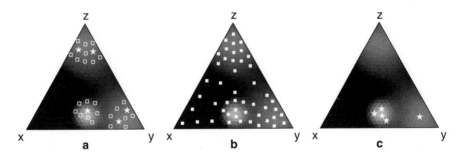

Figure 6.3 Experiment 2: (a) experimental points; (b) all points; (c) top five points.

added in each iteration, the boundary for the next iteration becomes tighter. Figure 6.2c shows the five top points, marked with stars, completed in the first experiment. Notice that there is one point near the highest peak and two points near each of the other two peaks.

Figure 6.3a shows the 24 experimental points for the next experiment clustered around the top five points. These points have been placed in the region of interest by the optimal coverage algorithm, taking into account the location of previous points. Figure 6.3b shows all of the experimental points from Experiments 1 and 2, and Figure 6.3c shows the top five points. Notice that there are four points near the highest peak and one point near one of the other peaks. Figures 6.4 and 6.5 show the corresponding data for Experiments 3 and 4. After Experiment 3, the focus is entirely in the region around the highest peak.

This approach works the same way in higher-dimensional spaces, but it is more difficult to visualize. One visualization method is to extract three-dimensional slices from an N-dimensional space where each slice represents a single value of the other $N–3$ dimensions. A set of slices that vary in only one dimension can show how the response varies with respect to that dimension and the three dimensions of

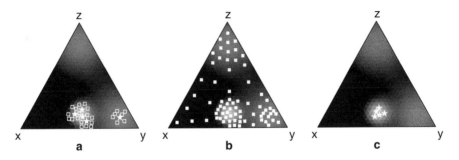

Figure 6.4 Experiment 3: (a) experimental points; (b) all points; (c) top five points.

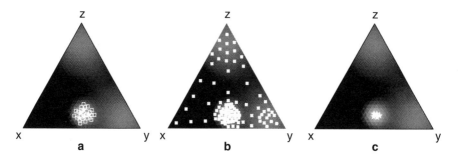

Figure 6.5 Experiment 4: (a) experimental points, (b) all points, (c) top five points.

the slices [19]. By selecting different sets of four dimensions from the N-dimensional experimental space, more effects can be studied visually.

The number of points used for each experiment iteration depends on many factors, including the dimensionality of the space and laboratory equipment constraints. More complex algorithms are available for selection of the region of interest, some of which also take into account the estimated error of each experimental point. For a more complex space with a not-necessarily well-behaved response surface, which frequently arises when experiment points correspond to materials in different phases, the preferred approach is to use a robust multivariate nonlinear regression algorithm to approximate the response surface based on the experimental points available [20]. This approximate response surface, along with the error bounds provided by some algorithms, can be used to predict the most valuable regions of interest by sampling the surface between experimental points. The use of appropriate regression methods may also make it possible to detect likely discontinuities or highly nonlinear responses via hill-climbing, if the regression method is robust in regions of rapid variation of the figure of merit (FOM).

6.3 EXAMPLE APPLICATION: HETEROGENEOUS CATALYST DISCOVERY SCREENING

The domain of catalysis and materials discovery involves phenomena that may be approximated as locally continuous, yet are discontinuous on a global scale. Indirect modeling of catalyst properties through correlation of experimental variables with a performance measure can lead to continuous behavior over a sufficiently small range of parameters. Over a larger parameter range, discontinuities appear due to the interaction of thermodynamic and kinetic factors in a catalytic process, such as phase transitions, creation and destruction of surface defects, and sintering. The catalyst formulation is often modified by pretreatment and reaction conditions, and the active catalyst is formed in situ. Additionally, a kinetic model of catalytic performance on a single catalyst can be very complex, involving the combination of many competing reaction mechanisms. It is the interplay of catalyst modifications and the process conditions that result in identification of a catalyst breakthrough.

Because optimal performance across several related catalytic materials stems from the interaction of thermodynamic and kinetic events, achieving such performance requires investigation of a variety of interrelated variables. These variables include but are not limited to:

- Pretreatment
- Time on stream
- Composition, concentration, and rate of feed
- Operating temperature
- Pressure of operation.

In addition, a complex set of catalyst-specific variables must be considered:

- Phase
- Composition
- Surface and crystal morphology
- Active sites
- Diffusion and mass-transfer properties.

Although the chemical or physical impact of variables such as pretreatment conditions can be detected in catalyst performance, performance is only an indirect measure of the underlying changes that have taken place. This is exemplified in the conversion of hydrocarbons over zeolite Y. In order to make a stable, active acid catalyst from zeolite Y, the material must undergo:

- Ammonium exchange
- Heat treatment in the presence of water vapor (>400°C)
- Ammonium exchange
- Heat treatment > 400°C.

These steps create Bronsted acidity by removal of sodium ions and creation of protons through the thermal decomposition of ammonium cations. Additionally, aluminum is removed from the zeolite framework, which results in changes in the strength of the resulting acidity, a contraction of the crystalline unit cell, and changes in the adsorption properties of the catalyst. The result is abrupt, or discontinuous, changes in catalyst performance as a function of variations in treatment parameters.

Once a lead formulation and corresponding process parameters have been identified, optimization experiments can be carried out in the neighborhood of the lead. Optimization experiments vary the relevant parameters over much smaller ranges and necessarily avoid discontinuities identified during discovery. Under these conditions, assumptions may hold about local continuity in material responses over the regions of interest. Combinatorial optimization experiments can be conducted not only to optimize material formulations and process conditions but also to optimize manufacturing recipes and conditions [21].

6.3.1   Operational Use of the Optimal Coverage Approach

The optimal coverage DOE system performs experiment design within an experimental region of interest defined by the scientist. A suite of design tools supports the most labor-intensive aspects of this process, allowing the scientist to concentrate as much as possible on the logic of the experiment. The implementation has been focused on tools to incorporate the types of constraints needed in the domain of heterogeneous catalysts and materials, but the approach is completely general. The elements of the optimal coverage DOE system are outlined in Figure 6.6. The DOE system has been implemented using a Java application server to provide a Web-based user interface, integration with the optimal coverage software, and integration with the database.

The scientist begins by choosing the set of chemical species to vary within the experiment, and the stock reagents that will be used to deliver those species, as shown in Figure 6.7. Other experiment particulars, such as pipetting equipment and sample identification schemes, are also selected. All available options are provided via graphical menus, which are automatically populated from a database that reflects the laboratory's current inventory of chemicals and equipment.

The scientist is then allowed to graphically enter and edit experimental constraints, which take the form of constraint equations. The available constraint variables are inferred from the previously chosen chemical species. Several forms of constraint are supported, including fixed-value constraints for mass and mass percentage, min/max constraints on mass or mass ratios, and mass ratio equations, as shown in Figure 6.8.

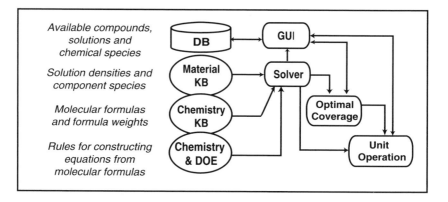

Figure 6.6   Elements of the optimal coverage DOE system. DB: database; KB: knowledge base; GUI: graphical user interface.

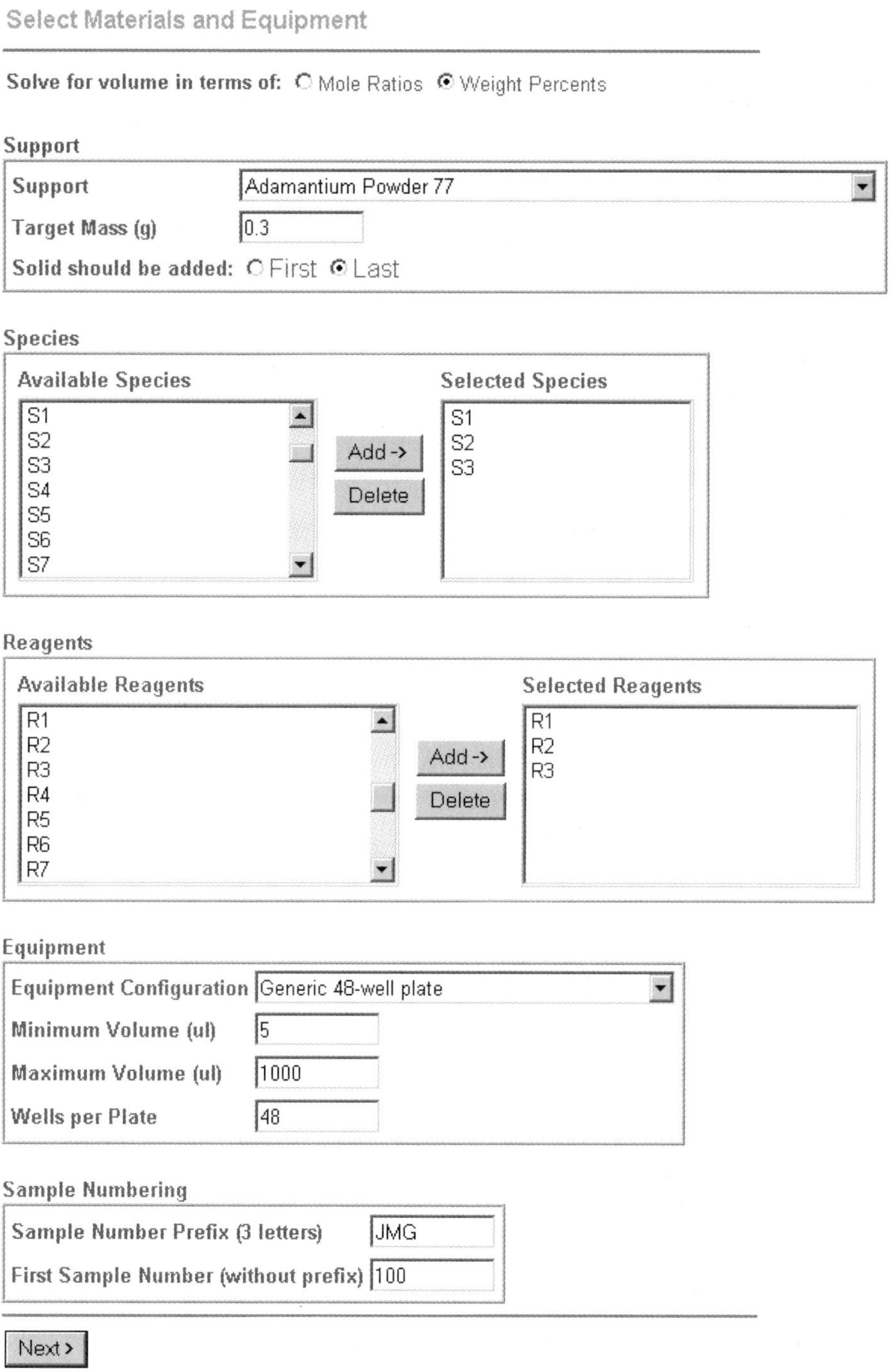

Select Materials and Equipment

Solve for volume in terms of: ○ Mole Ratios ⦿ Weight Percents

Support

Support	Adamantium Powder 77 ▼
Target Mass (g)	0.3
Solid should be added:	○ First ⦿ Last

Species

Available Species

S1
S2
S3
S4
S5
S6
S7

Add ->
Delete

Selected Species

S1
S2
S3

Reagents

Available Reagents

R1
R2
R3
R4
R5
R6
R7

Add ->
Delete

Selected Reagents

R1
R2
R3

Equipment

Equipment Configuration	Generic 48-well plate ▼
Minimum Volume (ul)	5
Maximum Volume (ul)	1000
Wells per Plate	48

Sample Numbering

Sample Number Prefix (3 letters)	JMG
First Sample Number (without prefix)	100

Next >

Figure 6.7   Graphical user interface to DOE system: experimental setup screen.

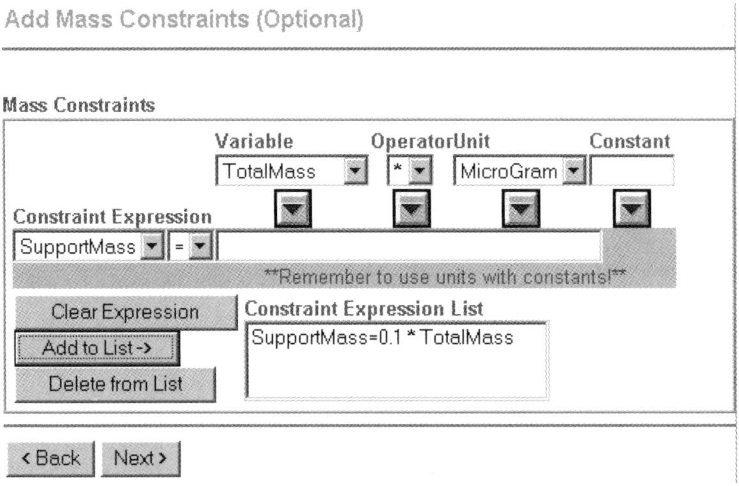

Figure 6.8 Graphical user interface to DOE system: constraint screen.

The system of constraint equations defined by the scientist need not be complete, in the sense of being immediately solvable. The DOE suite includes an expert system-based solving engine that is capable of generating missing equations from a knowledge base of chemical properties, such as molecular weights and stock solution densities. For instance, a solution density allows the solution's mass to be mathematically related to its volume: $M = V*D$. These "connecting" equations fill in any missing mathematical relations between the constraint equations provided by the scientist, and allow the DOE system to solve for reagent delivery volumes in terms of the desired species mass or mass ratios. All equations are maintained internally with correctly matching units via the application of another knowledge base that understands unit conversions. This allows the scientist to work in whichever units are convenient, and even refer to various equivalent units (e.g., grams, kilograms, or pounds) without the labor of manually checking unit conversions.

Determining the set of desired masses or ratios for the species of interest is the core of the experiment-design process. The DOE system also allows the scientist to directly enter any particular experimental points that are desired, perhaps as references or for validation. These two methods can also be used together: the scientist can specify certain experimental points of particular interest, and then allow the DOE system to distribute the remaining points optimally. In this case, the optimal coverage algorithms will take the fixed points into account when determining the optimal placement of the remaining points. The number of points to be placed depends on the number of fixed points, the number of replicate points, and laboratory constraints, such as the number of wells in a plate. Figure 6.9 shows how this information is entered.

Once the experimental points have been defined, the DOE system applies the

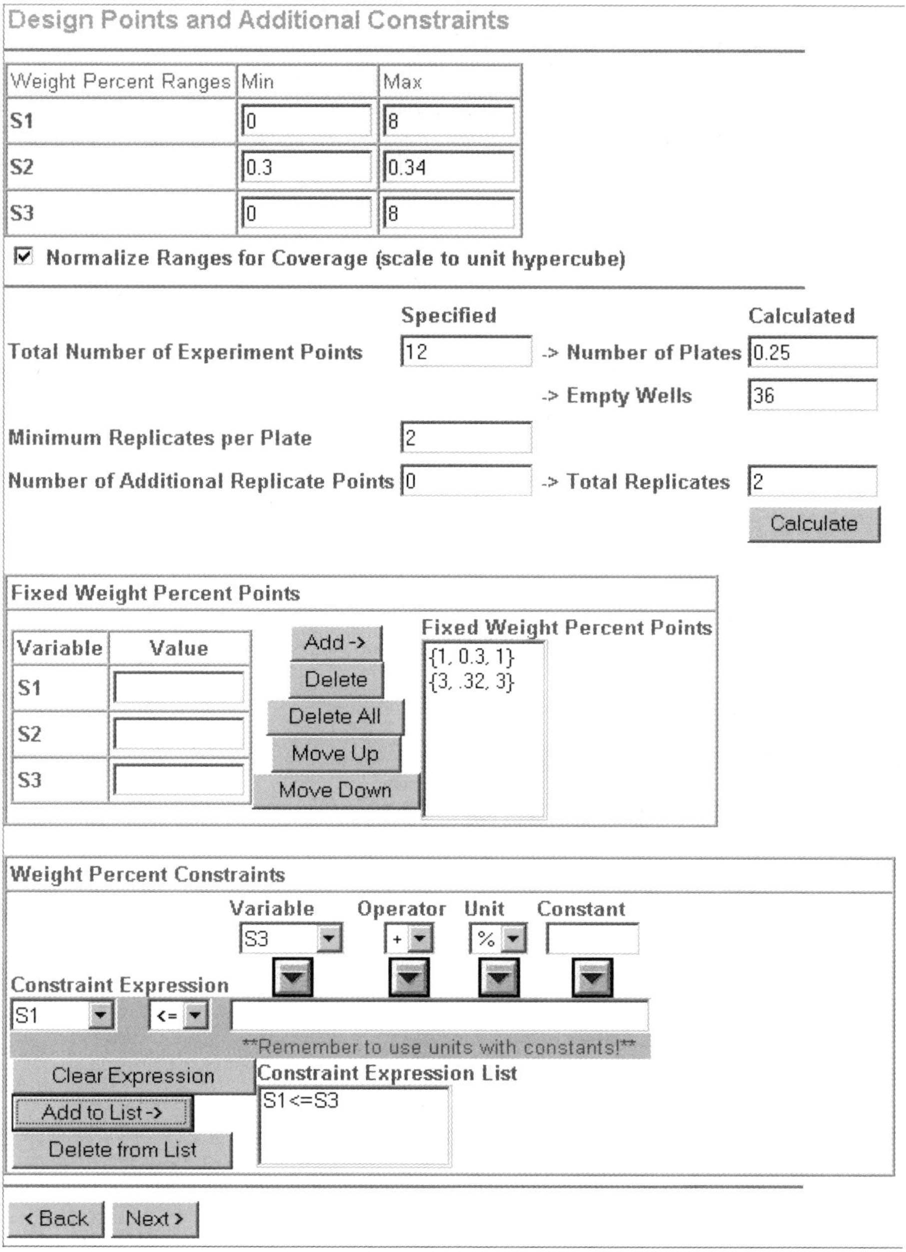

Figure 6.9 Graphical user interface to DOE system: setting fixed points.

equations automatically solved in previous steps to transform these points into the corresponding stock solution delivery volumes. These delivery volumes effectively *operationalize* the experiment: they specify exactly how much of the appropriate solutions are needed to obtain the desired experimental masses and/or mass ratios, under the constraints given by the scientist. Figure 6.10 shows the resulting experiment design with fixed points, optimally placed points, and replicate points. For this

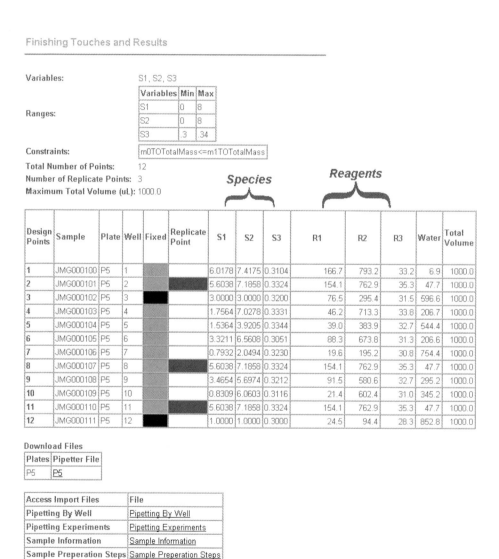

Figure 6.10   Graphical user interface to DOE system: final design.

example, the experiment space is the weight percent of three species, but what is needed in the laboratory is a control file for the liquid delivery system specifying the volumes of the three reagents and water for each well in a randomized plate.

6.3.2 Example Experiment

Sulfated zirconia catalysts have been the subject of intense investigation over the last 10–15 years [22,23]. This system responds well to the addition of various catalyst modifiers, including Pt, Fe, Mn, and many other transition metals. We decided to examine a subset of this chemistry with the modifiers platinum and nickel [24]. The three compositional variables chosen to study and their ranges are Pt: 0.05–1 wt %; SO_4: 2–10 wt %; and Ni: 1–10 wt %. In this particular case, it was not considered necessary to limit the experimental space further. The optimal coverage routine was used to generate the experimental design with 48 distinct points. No replicates or reference points were used in this case study, but these capabilities are available. It is generally desirable to run replicate samples in each experiment. A common practice in our lab is to prepare 10–20% of the total number of samples as replicates of a reference composition. This reference composition can be repeated across multiple experiments to provide a check on catalyst preparation and testing reproducibility. A plot of the design points in three-dimensional space is presented in Figure 6.11.

The design was converted to specific instructions for each piece of automated catalyst synthesis equipment. The instruction files generated in this manner were downloaded to each instrument and the experimental protocol was carried out in a modular fashion. A description of the general protocol follows.

The catalyst samples were made starting with zirconium hydroxide that had been prepared by precipitating zirconyl nitrate with ammonium hydroxide at 65°C. The zirconium hydroxide was dried at 120°C, ground to 40–60 mesh, and then weighed into a 48-well microtiter plate (target 400–450 mg/well). Solutions of ammonium sulfate and nickel nitrate were prepared and transferred to a microtiter plate with an automated pipetter. The zirconium hydroxide was transferred to the solution plate.

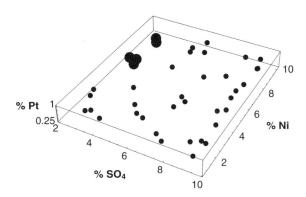

Figure 6.11 Experiment design for 48 sulfated zirconia samples. Large dots indicate highest FOM.

The samples were agitated briefly and then dried with 80–100°C air. The samples were then calcined in a muffle oven at 650°C for 2 hours in air. Forty-eight solutions of chloroplatinic acid were prepared in a new microtiter plate. The solids were added, then the samples agitated and dried as before. The samples were finally calcined at 500°C in air for 2 hours.

Catalyst activity was checked in a 48-reactor parallel test unit. Sulfated zirconia catalysts are generally tested for either butane or pentane isomerization. We chose to test the catalysts using pentane isomerization at low pressure. The catalyst samples were pretreated at 450°C in air and 200°C in hydrogen to remove water and reduce the metal components. The feed composition was ~10 mol % pentane in hydrogen. Gas chromatographic data were collected at temperatures of 120, 150, and 180°C and weight hourly space velocities of 0.8, 1.6, and 2.4 h^{-1} for a combined total of nine conditions. An experimental FOM was computed for all data points.

6.3.3 Experiment Design and Results

The design for the experiment just described is shown in Figure 6.11. Here the three principal composition variables (Pt, Ni, and SO_4) are shown on the three axes on the same scale. The 48 experiment points are distributed in all three dimensions.

The FOM is, in general, a nonlinear function of the composition variables Pt, SO_4, and Ni. To facilitate the visualization and analysis of the reactor data, a nonlinear multivariate regression is formed to assign approximate FOM values throughout the entire range of the composition variables. Figure 6.12 presents the

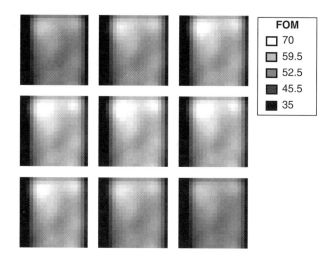

FOM
☐ 70
▨ 59.5
▨ 52.5
▧ 45.5
■ 35

Figure 6.12 Results of multivariate nonlinear regression performed on the FOM from the experiment in Figure 6.11. The *x*- and *y*-axis in each thumbnail represent SO_4^- and Ni concentration, respectively. The Pt level increases from the upper left to the lower right thumbnail by rows.

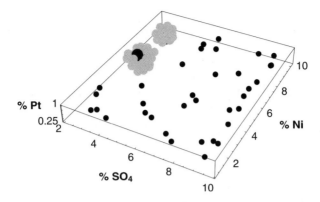

Figure 6.13 The second experiment iteration (grey dots) designed around the maxima identified in the experiment of Figure 6.11.

result of the regression. For this visualization, the regression results have been sampled at nine separate Pt values to create slices of the chemical space parallel to the sulfate/Ni plane. For each level of platinum, there is one density plot of the FOM as a function of Ni (vertical-axis) and SO_4 (horizontal-axis) values. High values of the FOM are bright-colored regions, as indicated in the legend.

From this visualization, it is easily observed that there are particular ranges of Pt, Ni, and SO_4 at which the FOM has its largest values. This region of high performance can be identified analytically and sampled with the optimal coverage algorithm to provide a recommended set of next experiments. However, a simpler approach suffices in this case (as described earlier), which is to isolate a small set of the highest values, and then use optimal coverage to sample a neighborhood of those points. In Figure 6.11 the set of points identified by applying this test to the experiment data are enlarged. In complex situations or in situations where the sample-to-sample FOM variation is small in relative terms, a statistical test can be applied to ensure a strong likelihood of correctly identifying five distinct high values. In this example, it is sufficient to simply select the five highest values. The next round of experiment points is shown in Figure 6.13, including the selected points from the previous experiment. As stated earlier, several iterations typically allow the experimenter to hone in on the combinations of composition variables that provide superior performance.

6.4 CONCLUSIONS

The optimal coverage approach to experiment design offers many advantages for combinatorial materials discovery, as illustrated in the synthetic and experimental examples in this chapter. When coupled with appropriate multivariate nonlinear regression methods, the approach maximizes the probability of discovery in discontinuous and complex spaces, explicitly supporting an iterative experimental approach. The method allows the use of known chemical information to constrain the experiment space and to specify points of particular interest, while maximizing the overall information obtained from an experiment of any given size. The number of experiments can be specified by the researcher, reflecting practical allocations of resources and time. Replicates and fixed points can be incorporated into a design of any size and mapped onto available equipment configurations. The existing optimal coverage DOE tools can be extended through augmentation of the underlying rules and knowledge bases, without altering the design algorithms or primary interface.

ACKNOWLEDGMENT

This work was performed under the support of the U.S. Department of Commerce, National Institute of Standards and Technology, Advanced Technology Program, Cooperative Agreement Number 70NANB9H3035.

REFERENCES

1. Dagani, R. *Chemical and Engineering News,* 1999, 77: 51–60.
2. Dagani, R. *Chemical and Engineering News,* 2000, 78: 66–68.
3. Meredith, J. C.; Karim, A.; Amis, E. J. *Macromolecules,* 2000, 33: 5760–5762.
4. Brocchini, S.; James, K.; Tangpasuthadol, V.; Kohn, J. *J. Biomed. Mater. Res.,* 1998, 42: 66–75.
5. Bem, D.; Bratu, C.; Broach, R.; Lewis, G.; McGonegal, C.; Miller, M.; Moscoso, J.; Murray, R.; Raich A.; Wu, D.; Akporiaye, D.; Karlsson, A.; Plassen, M.; Wendelbo, R. Presented at *COMBI2000—Combinatorial Approaches for New Materials Discovery,* The Knowledge Foundation, San Diego, CA, January 23, 2000.
6. Bricker, M.; Vanden Bussche, K.; McGonegal, C.; Karlsson, A.; Akporiaye, D.; Dahl, I.; Plassen, M. Presented at *COMBI2000—Combinatorial Approaches for New Materials Discovery,* The Knowledge Foundation, San Diego, CA, January 23, 2000.
7. Cawse, J. N. Presented at *COMBI2000–Combinatorial Approaches for New Materials Discovery,* The Knowledge Foundation, San Diego, CA, January 23, 2000.
8. Valler, M. J.; Green, D. *Drug Discovery Today,* 2000, 5: 286–293.
9. Fisher, R. A. *The Design of Experiments,* Oliver and Boyd, Edinburgh, Scotland, 1935.
10. Drain, D. *Handbook of Experimental Methods for Process Improvement,* Chapman & Hall, New York, 1997.
11. Wolf, D.; Buyevskaya, O. V.; Baerns, M. *Appl. Catal. A: General,* 2000, 200, 63–77.
12. Falcioni, M.; Deem, M. W. *Phys. Rev. E,* 2000, 61: 5948–5952.
13. Chen, L.; Deem, M. W. *J. Chem. Inf. Comput. Sci.,* 2001, 41: 950–957.
14. Tobias, R. "Space-filling Experimental Designs for Combinatorial Chemistry," Presented at COMBI2000—Experimental Strategy Workshop, The Knowledge Foundation, San Diego, CA, January 23, 2000.

15. Higgs, R. E.; Bemis, K. G.; Watson, I. A.; Wikel, J. H. *J. Chem. Inf. Comput. Sci.,* 1997, 37: 861–870.

16. Cummins, D. J.; Andrews, C. W.; Bentley, J. A.; Cory. M. *J. Chem. Inf. Comput. Sci.,* 1996, 36: 750–763.

17. Agrafiotis, D. K. *J. Chem. Inf. Comput. Sci.,* 2001, 41: 159–167.

18. Agrafiotis, D. K.; Lobanov, V. S. *J. Chem. Inf. Comput. Sci.,* 1999, 39: 51–58.

19. Harmon, L. A.; Schlosser, S. G.; Vayda, A. J. In *Combinatorial Materials Development,* Malhotra, R. Ed.; ACS Symposium Series 814, American Chemical Society, Washington, DC, 2002, 129–145.

20. Montgomery, D.; Myers, R. H. *Response Surface Methodology,* Wiley, New York, 1995.

21. Holmgren, J. S.; Bem, D.; Gillespie, R.; Bricker, M.; Lewis, G.; Murray, R.; Sachtler, A.; Willis, R.; Akporiaye, D.; Karlsson, A.; Plassen, M.; Wendelbo, R. Presented at *COMBI2002—Combinatorial Approaches for New Materials Discovery,* The Knowledge Foundation, San Diego, CA, January 22, 2002.

22. Song, X.; Sayari, A. *Catal. Rev.—Sci. Eng.,* 1996, 38: 329–412.

23. Corma, A. *Curr. Opin. Solid State Mater. Sci.,* 1997, 2: 63–75.

24. Venkatesh, K. R.; Hu, J.; Tierney, J. W.; Wender, I. U.S. Patent No. 6,184,430, 2001.

CHAPTER 7

COMBINATORIAL MATERIALS DEVELOPMENT USING OVERLAPPING GRADIENT ARRAYS: DESIGNS FOR EFFICIENT USE OF EXPERIMENTAL RESOURCES

JAMES N. CAWSE AND RONALD WROCZYNSKI
GE Global Research

7.1 BACKGROUND

Gradient arrays are now common tools in combinatorial chemistry for discovery of new leads to commercial materials. Although the cost per sample has dropped markedly with new high throughput methods, efficient use of experimental resources is still important. A common approach in solid-state materials studies is examination of a ternary (or higher) materials gradient. This can be done using continuous or point techniques. For example, in continuous composition spread, a single film with a ternary composition spread was generated on a 63×66-mm substrate in one step using sputtering techniques and the electronic properties measured at ~4000 points [1]. It found an excellent dielectric, $Zr_{0.15}Sn_{0.3}Ti_{0.55}O_{2-\delta}$.

A more common theme is a ternary gradient studied at regular intervals [2,3]. Gradients such as 0–100% by 10% intervals or 0–1% by 0.1% intervals are convenient; these generate 66-point triangular arrays (Figure 7.1). Single ternary and quaternary systems can be represented by two-dimensional triangular or three-dimensional tetrahedral graphs, respectively, where each vertex represents one of the factors, each edge represents mixtures of two factors, and each face represents mixtures of three factors. Four-factor combinations in quaternaries are considered interior points. Examples follow.

- The Pt/Pd/In system for cyclohexane dehydrogenation catalysis was studied in the 0–1% range at 0.1% intervals [4].
- The Rh/Pd/Pt system for CO oxidation was studied over the 0–100% range using 15 steps [2].

Experimental Design for Combinatorial and High Throughput
Materials Development, Edited by James N. Cawse.
ISBN 0-471-20343-2 © 2003 John Wiley & Sons, Inc.

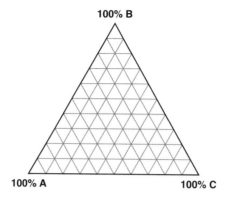

Figure 7.1 A ternary-gradient array from 0 to 100 in 10 steps. (Reprinted, with permission, from Cawse, J.N. *Accounts of Chemical Research,* 2001, 34: 213–221; copyright © 2001. American Chemical Society, Washington, DC.)

The technique can be extended to more complex combinations. A quaternary phase diagram was studied in the Pt/Ru/Os/Ir system for methanol fuel cell catalysis [3].

Experimental designs for these mixture systems have been extensively investigated by Cornell [5], but the underlying purpose of those designs is the fitting of a mathematical model to the experimental space. The relationship between the compositions and the response variable is assumed to be mathematically continuous. The model is assumed to be adequately approximated by a relatively simple (linear, quadratic, or cubic) polynomial. These models can be fit to a design with relatively few points (Figure 7.2). Combinatorial experimentation, by contrast, is a search for a local, unexpected, and strongly nonlinear synergy of the formulation ingredients. This requires a much finer search grid than the simplexes favored by Cornell. The

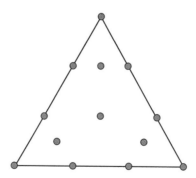

Figure 7.2 Design points for a ternary array to be fit by a cubic model. (Reprinted, with permission, from Cornell, J. A. *Experiments With Mixtures,* 3rd ed.; copyright © 2001. John Wiley & Sons, New York.)

Table 7.1   Samples Required for Ternary, Quaternary, and Pentanary Arrays

Steps	Interval (%)	Ternary	Quaternary	Pentanary
3	50.00	6	10	15
4	33.30	10	20	35
5	25.00	15	35	70
6	20.00	21	56	126
7	16.67	28	84	210
8	14.28	36	120	330
9	12.50	45	165	495
10	11.11	55	220	715
11	10.00	66	286	1,001
13	8.33	91	455	1,820
15	7.14	120	680	3,060
17	6.25	153	969	4,845
19	5.55	190	1,330	7,315
21	5.00	231	1,771	10,626
M	$100/(M-1)$	$M\times(M+1)/2$	$M\times(M+1)\times(M+2)/6$	$M\times(M+1)\times(M+2)$ $\times(M+3)/24$

Note: The number of steps M in the array is related to the interval I by $I = \text{range}/(M-1)$. The range is assumed to be 0–100%.

fineness of the grid is specified by the number of interval steps M; the interval spacing is then equal to the range of compositions divided by $(M-1)$.

Table 7.1 shows the number of sample points required for ternary, quaternary, and pentanary arrays as a function of the intervals examined. Although in some solid-state applications, arrays as large as 25,000 samples per plate are possible [6], most combinatorial systems are more modest, often with 96-sample arrays and a throughput of 1–3 arrays/day. Examination of the chemical space at fine detail will strain the resources of most systems.

In our experience, there is an important advantage to fitting as many comparable systems as possible on a single array. We have found that the between-plate experimental error is often two to ten times larger than the within-plate error. This requires a much higher degree of statistical rigor when comparisons must be extended across multiple plates [7]. Chapter 8 discusses some statistical methods for multiple-plate comparisons. The following approaches are helpful in maximizing the useful comparisons that can be made in a single plate. Sections 7.2 and 7.3 discuss methods for reducing the design space, while the balance of the chapter covers methods to reduce the number of experiments in the design.

7.2   STRATEGIES FOR EFFICIENTLY SAMPLING GRADIENT SYSTEMS

In some chemical circumstances, there are efficiencies to be gained by judicious pruning or combining of arrays. The grouping or classification of materials according to chemical type or reactivity group is a naturally enforced efficiency. For example, in a redox system, materials are grouped into a reductant class and an oxidant class, and in solid-state systems, materials are often organized as donors and acceptors. In polymer systems, in particular condensation polymers, the monomers are similarly divided according to reactant type. For linear polyamides, at least three monomer classes can exist: diacids, diamines, and amino acids. The following discussion will use the case of a partially branched system formed by addition polymerization, in which three classes of monomers are organized as mono-, di-, and trifunctional. These systems are common in paints and coatings.

In this system, we wished to determine the best combination of monomers to optimize the required physical properties, but it was clear that the optimum property could occur at different mono:di:tri ratios with different compounds. Therefore a combined combinatorial strategy was necessary:

- Use diversity strategies to select subsets of each class of monomer, so that the chemical space could be sampled effectively. Strategies for selecting a rational subset of a number of compounds using diversity techniques are being studied extensively and have been reported elsewhere [8,9].
- Use efficient gradient techniques to explore the mono:di:tri space for each combination of compounds.

The number of compounds and the resulting potential number of samples is shown in Table 7.2. If, by using these strategies, we reduce the number of compounds to 5–10 of each type and the number of samples/gradient to 20–30, the total number of samples is then 2500–30,000. This is accessible using high throughput screening technologies. This chapter focuses on methods of minimizing the number of samples used in the required gradients.

7.3 PRUNING STRATEGIES

In a system of this type, it is often chemically unreasonable to require the gradients to reach 100% in all constituents. A far more likely case is one in which

Table 7.2 Number of Compounds and Combinations for an ABC Polymer System

Monomer Type	Compounds
A: Mono-functional monomers (reactive diluents)	20–50
B: Difunctional monomers (oligomers)	40–60
C: Trifunctional monomers (cross-linkers)	20–60
Number of possible combinations	16,000–180,000
Number of samples required for 10% gradient ($M = 11$)	1,000,000–12,000,000

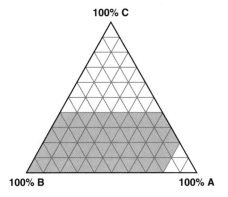

Figure 7.3 A pruned gradient array for $A = 0$–80 and $C = 0$–40.

some of the constituents have limits imposed on the system by prior domain knowledge. In our study, only difunctional monomers are likely to form a useful material over the full 0–100% range. Monofunctional monomers will only form a linear polymer by themselves in addition-type polymerization; they can be limited to the 0–80% range. Trifunctional monomers will form an intractable, highly cross-linked polymer at high levels; they can be limited to the 0–40% range. The resulting chemical space is shown in Figure 7.3; it requires only 27 formulations rather than 66 for the full 10% interval gradient. More examples for various mono:tri ratios are in Table 7.3.

If all three materials can be limited, the resulting space will often be a triangular subset of the full gradient (Figure 7.4). The number of required samples can be determined from Table 7.1, depending on the number of steps M for the triangle.

7.4 EDGE-SHARING STRATEGIES

When there are a number of different chemical entities that can play a single role (such as mono, di, or tri in this study), various types of edge-sharing strategies are

Table 7.3 Samples per Array at 10% Intervals

Maximum Trifunctional	Maximum Monofunctional			
	50%	60%	70%	80%
10%	12	14	16	18
20%	18	21	24	27
30%	24	28	32	35
40%	30	35	39	42

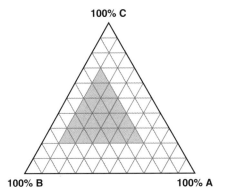

Figure 7.4   A pruned gradient array for $A = 10–70$, $B = 20–80$, and $C = 20–80$.

possible. In the following discussion we will use the notation $A^iB^jC^k$ for materials of types A, B, and C, with i, j, and k indicating representatives of the respective types. In the simplest case, a single example of each of two of the three constituents (such as a the mono- and difunctional monomers) can be studied with a pair of third constituents (such as trifunctional monomers). This can be called an $A^1B^1C^1C^2$ system, and can be shown as two triangles joined at one edge (Figure 7.5); the system can be the whole gradient or pruned, as earlier (Figure 7.6). The improvement for full gradients is given in Table 7.4; for pruned gradients it is given in Table 7.5. Many gradients can share a single A^1B^1 edge, to give $A^1B^1C^1C^2C^3$ systems, and so on. As the number of gradients sharing a single edge increase, the percentage reduction in number of samples increases.

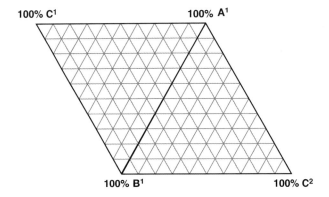

Figure 7.5   A pair of triangular arrays sharing a single edge.

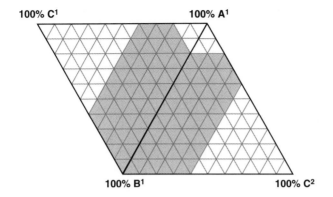

100% C¹ ... 100% A¹
100% B¹ ... 100% C²

Figure 7.6 A pair of pruned triangular arrays sharing a single edge.

7.5 MULTIPLE EDGE-SHARING STRATEGIES

In cases where we wish to inspect a triangular gradient for multiple choices of components, it becomes possible to share more than one edge (Table 7.6). An $A^1B^1B^2C^1C^2$ system could have the $A^1B^1C^1$ triangle sharing edges with an $A^1B^1C^2$ triangle and an $A^1B^2C^1$ triangle. Furthermore, the $A^1B^1C^2$ and $A^1B^2C^1$ triangles each can share an edge with a fourth triangle, $A^1B^2C^2$. In the most extensive version, an $A^1A^2B^1B^2C^1C^2$ system could be sampled by a combination, such as shown in Figure 7.7. At this level of complexity, however, it is far more effective to escape

Table 7.4 Reduction in Samples With Ternary-Gradient Sharing One Edge

M	Unshared	Shared	Improvement (%)
5	30	24	20
6	42	35	17
11	132	120	9
21	462	440	5

Table 7.5 Reduction in Samples for Pruned (40%/80%/100%) Ternary Gradients Sharing a Single Edge

M	$N = 1$	$N = 2$	Improvement (%)	$N = 3$	Improvement (%)	$N = 4$	Improvement (%)
6	10	15	25	20	33	25	38
11	27	45	17	63	22	81	25
21	85	153	10	221	13	289	15

Note: $N = 1$: single gradient; $N = 2$: two gradients sharing an edge, etc. Improvement = reduction in number of required samples

Table 7.6 Ternary Gradients Available With Multiple Edge-Sharing Strategies

Type	Components	Gradients Sampled
Single edge	$A^1B^1C^1C^2$	$A^1B^1C^1$ $A^1B^1C^2$
Double edge	$A^1B^1B^2C^1C^2$	$A^1B^1C^1$ $A^1B^1C^2$ $A^1B^2C^1$ $A^1B^2C^2$
Octahedron	$A^1A^2B^1B^2C^1C^2$	$A^1B^1C^1$ $A^1B^1C^2$ $A^1B^2C^1$, $A^1B^2C^2$ $A^2B^1C^1$ $A^1B^2C^2$
		$A^2B^1C^2$ $A^2B^2C^1$ $A^2B^2C^2$

into the third dimension and combine the ternary gradients into the surface of an octahedron (Figure 7.8). This is the most sample-efficient shape for ternary gradients; its improvement over eight isolated gradients is shown in Table 7.7.

7.6 HIGHER-DIMENSIONAL EDGE-SHARING STRATEGIES

This strategy can be extended to higher-dimensional gradient systems. A quaternary $A^1B^1C^1D^1$ system can be a voracious user of samples (Table 7.1). Quaternary systems can be represented by tetrahedra (Figure 7.9). The triangular faces of pairs of tetrahedra can be shared in a fashion analogous to the sharing of edges of triangles. This will produce similar savings in numbers of samples, for example, in an $A^1B^1C^1D^1D^2$ system (Figure 7.10). The limiting case would be the

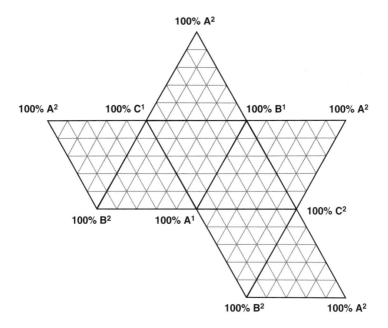

Figure 7.7 Triangular arrays sharing multiple edges.

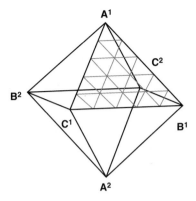

Figure 7.8 Eight triangular arrays folded into an octahedron.

Table 7.7 Samples Required for Triangular and Octahedral Systems

M	Interval (%)	Triangle	8 Triangles	Octahedron	Improvement (%)
3	50.0	6	48	18	63
4	33.3	10	80	38	53
5	25.0	15	120	66	45
6	20.0	21	168	102	39
9	12.5	45	360	258	28
10	11.1	55	440	326	26
11	10.0	66	528	402	24
14	7.7	105	840	678	19
21	5.0	231	1,848	1,602	13

$A^1A^2B^1B^2C^1C^2D^1D^2$ system in which faces can be shared to form a *16-cell*[1] (Figure 7.11) which is the four-dimensional analog of an octahedron [10]. The resultant savings in samples is shown in Table 7.8.

In principle this strategy can be extended to higher dimensions without limit. For any dimension N there is an N-dimensional analog of a tetrahedron (with $N + 1$ cells) and an $N + 1$-dimensional octahedron (with 2^N cells). The efficiency of this face-sharing methodology continues to increase in higher dimensions. This is caused by a little-understood property of high-dimensional solids: they have increasingly large fractions of their volume near their faces. For example, a five-component system (which forms a four-dimensional analog of a tetrahedron), divided into 20% gradient steps, only has a single interior point out of 126.

---

[1]In these higher dimensional figures, a "cell" is analogous to a "face" of a three-dimensional figure. Thus a three-dimensional octahedron has eight two-dimensional triangular faces, while a four-dimensional hyperoctahedron has 16 three-dimensional tetrahedral cells.

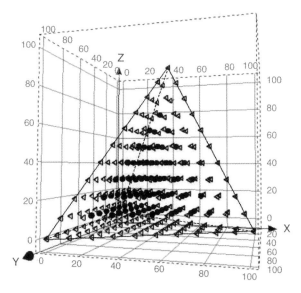

Figure 7.9 A quaternary array as a tetrahedron. $M = 9$. Hollow triangles: ternary systems on the faces, binary systems on the edges, and unary systems on the vertices. Solid spheres: quaternary systems in the interior.

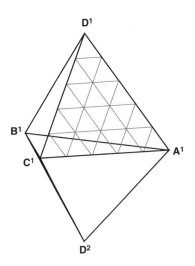

Figure 7.10 Two quaternary arrays joined at their faces.

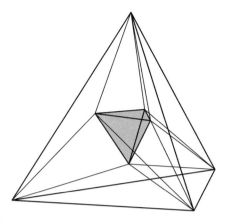

Figure 7.11 The 16-cell, or hyperoctahedron. (Adapted, with permission, from Hilbert, D.; Cohn-Vossen, S. *Geometry and the Imagination.* Chelsea, New York, 1952; copyright © 1952. American Mathematical Society, Providence, RI.)

7.7 EDGE-SHARING STRATEGIES IN NONSYMMETRICAL SYSTEMS

In many practical situations, the systems are more complex and these highly symmetrical solutions will not suffice. There are at least two general situations, and extensions and intermediate cases are possible.

In Case 1, there are more than two examples of a type component, and often there are an unequal number of examples of various components that are of interest. To illustrate, consider a coatings study where monofunctional (type A), difunctional (type B), and trifunctional (type C) reactants are combined in mixtures. Appropriate overlap of edges and faces of the gradients can result in very substantial savings in

Table 7.8 Samples Required for Four-Dimensional Gradients Using Tetrahedral and 16-Cell (hyperoctahedral) Systems

M	Interval (%)	Tetrahedron	16 Tetrahedra	16-cell	Improvement (%)
3	50.0	10	160	32	80
4	33.3	20	320	88	73
5	25.0	35	560	192	66
6	20.0	56	896	360	60
9	12.5	165	2,640	1,408	47
10	11.1	220	3,520	1,992	43
11	10.0	286	4,576	2,720	41
14	7.7	560	8,960	5,928	34
21	5.0	1,771	28,336	21,440	24

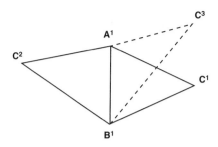

Figure 7.12   An $A^1B^1C^1C^2C^3$ system.

the number of runs required to completely study these gradients. The complexity of these systems, however, makes it more difficult to visualize their geometry. Figures 7.12–7.14 show the progression of complexity from $A^1B^1C^1C^2C^3$ to $A^1B^1B^2C^1C^2C^3$ to $A^1A^2A^3B^1B^2B^3C^1C^2C^3$ systems. Hand calculation of the number of samples that will be needed (and saved) by the used of overlapping edges or faces also becomes difficult.

In Case 2, the components do not belong to defined types and can be interchanged freely. Thus a ternary system can be defined from any three of the four components $ABCD$: ABC, ABD, ACD, and BCD. In simple cases they can be visualized (Figure 7.9, using only the points shown as hollow triangles), but they quickly enter the realm of higher dimensions. Figure 7.15 shows a 96-well plate containing 83 catalyst compositions made up by a 5-metal, 20% gradient overlap of using the scheme shown in Figure 7.16. Based on prior chemical knowledge, only 7 of the 10 ternaries possible in a 5-metal system were sampled. Without the savings in runs from overlapping gradients, the separate triangles would have re-

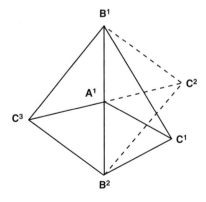

Figure 7.13   An $A^1B^1B^2C^1C^2C^3$ system.

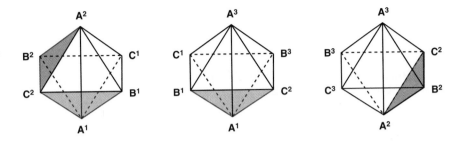

Figure 7.14 An $A^1A^2A^3B^1B^2B^3C^1C^2C^3$ system. Shading indicates two pairs of the faces that are shared between the figures.

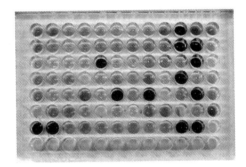

Figure 7.15 An array of catalysts made up from an $ABCDD'$ system. (Photo courtesy of GE Global Research, Niskayuna, NY.)

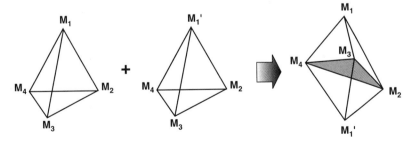

Figure 7.16 Building up an $ABCDD'$ system, with $52 + (52 - 21) = 83$ experiments.

quired 147 samples and would not fit in a single plate. With overlaps, only 83 samples were used, which left space on a 96-well plate for internal standards and some replicates.

7.8   CALCULATION OF SAMPLE POINTS FOR VARIOUS EDGE-SHARING STRATEGIES

7.8.1   Case 1

The calculations for Case 1 cover gradient arrays with multiple components of each type and overlap of edges, faces, and higher-dimensional units. They apply to both unsymmetrical and symmetrical systems. Equations for ternary, quaternary, and pentanary systems are given. The number of intervals M must be the same for all components. Table 7.9 gives definitions of the symbols and some of the preliminary calculations. Tables 7.10 and 7.11 give the savings in runs for representative ternary and quaternary systems, respectively.

Ternary:

$$\text{Number of samples} = V + \prod_{i=1}^{3} n_i \times I_3 + \left[\sum_{i=1}^{3} \frac{1}{n_i} \prod_{i=1}^{3} n_i \right] \times I_2.$$

Quaternary:

$$\text{Number of samples} = V + \prod_{i=1}^{4} n_i \times I_4 + \left[\sum_{i=1}^{4} \frac{1}{n_i} \prod_{i=1}^{4} n_i \right] \times I_3 + \left[\sum_{i=1}^{4} \sum_{j=i+1}^{4} \frac{1}{n_i n_j} \prod_{i=1}^{4} n_i \right] \times I_2.$$

Table 7.9   Terms and Equations for Case 1

Terms and Equations	Definitions
T	Number of types of reagents.
n_T	Number of factors or reagents of type T
M	Number of intervals for each reagent
$Q_T = n_1 \times n_2 \times n_3 \times \cdots \times n_T$	Number of separate T-dimensional gradients
$V = n_1 + n_2 + n_3 + \cdots + n_T$	Number of overlapping vertices
$E_T = \dfrac{(M + T - 2)!}{(M-1)! \times (T-1)!}$	Number of samples for a single gradient
$I_T = \dfrac{(M-2)!}{(M - T - 1)! \times (T-1)!}$	Number of samples needed for the T-way combinations of factors of a single space with T types and M intervals
$Q_T \times E_T$	Number of samples without overlap

Table 7.10 Number of Sample Points for Ternary Systems With and Without Overlap[a]

n_1	n_2	n_3	M	Interval (%)	Without Overlap	With Overlap	Improvement (%)
2	2	2	3	50	48	18	62.5
3	3	3	3	50	162	36	77.8
4	4	4	3	50	384	60	84.4
5	5	5	3	50	750	90	88.0
3	2	2	3	50	72	23	68.1
4	2	2	3	50	96	28	70.8
5	2	2	3	50	120	33	72.5
10	2	2	3	50	240	58	75.8
10	5	5	3	50	1,500	145	90.3
2	2	2	6	20	168	102	39.3
3	3	3	6	20	567	279	50.8
4	4	4	6	20	1,344	588	56.3
5	5	5	6	20	2,625	1,065	59.4
3	2	2	6	20	252	143	43.3
4	2	2	6	20	336	184	45.2
5	2	2	6	20	420	225	46.4
10	2	2	6	20	840	430	48.8
10	5	5	6	20	5,250	2,020	61.5
2	2	2	11	10	528	402	23.9
3	3	3	11	10	1,782	1,224	31.3
4	4	4	11	10	4,224	2,748	34.9
5	5	5	11	10	8,250	5,190	37.1
3	2	2	11	10	792	583	26.4
4	2	2	11	10	1,056	764	27.7
5	2	2	11	10	1,320	945	28.4
10	2	2	11	10	2,640	1,850	29.9
10	5	5	11	10	16,500	10,145	38.5

[a]n_1, n_2, and n_3 as defined in Table 7.9.

Pentanary:

$$\text{Number of samples} = V + \prod_{i=1}^{5} n_i \times I_5 + \left[\sum_{i=1}^{5} \frac{1}{n_i} \prod_{i=1}^{5} n_i\right] \times I_4 + \left[\sum_{i=1}^{5} \sum_{j=i+1}^{5} \frac{1}{n_i n_j} \prod_{i=1}^{5} n_i\right] \times I_3 +$$

$$\left[\sum_{i=1}^{5} \sum_{j=i+1}^{5} \sum_{k=j+1}^{5} \frac{1}{n_i n_j n_k} \prod_{i=1}^{5} n_i\right] \times I_2.$$

These equations not only count the number of runs, but actually describe the various combinations and permutations of components needed for the overlapping gradients.

As an illustration, consider the quaternary gradients in an $ABC^1C^2D^1D^2$ system with intervals of 25% ($M = 5$). Each term in the generalized quaternary formula represents the product of the *n-way* combinations of component types times the num-

Table 7.11   Number of Sample Points for Quaternary Systems With and Without Overlap[a]

n_1	n_2	n_3	n_4	M	Interval (%)	Without Overlap	With Overlap	Improvement (%)
2	2	2	2	3	50	160	32	80.0
3	3	3	3	3	50	810	66	91.9
4	4	4	4	3	50	2,560	112	95.6
5	5	5	5	3	50	6,250	170	97.3
10	10	10	10	3	50	100,000	640	99.4
2	2	2	2	4	33	320	88	72.5
3	3	3	3	4	33	1,620	228	85.9
4	4	4	4	4	33	5,120	464	90.9
5	5	5	5	4	33	12,500	820	93.4
2	2	2	2	6	20	896	360	59.8
3	3	3	3	6	20	4,536	1,200	73.5
4	4	4	4	6	20	14,336	2,960	79.4
5	5	5	5	6	20	35,000	6,120	82.5
2	2	2	2	11	10	4,576	2,720	40.6
3	3	3	3	11	10	23,166	11,190	51.7
4	4	4	4	11	10	73,216	31,600	56.8
5	5	5	5	11	10	178,750	71,870	59.8

[a] n_1, n_2, n_3, n_4, as defined in Table 7.9.

ber of ratios needed for the interval chosen for the n-way combination. For example, in the term

$$\left[\sum_{i=1}^{4} \frac{1}{n_i} \prod_{i=1}^{4} n_i \right] \times I_3,$$

where I_3 corresponds to the number of ratios needed for a given interval for a three-way combination of factors, and the rest of the term describes the number of combinations of those factors. Similarly, the I_2-containing term describes the two-ways, and the I_4-containing term the four-ways.

If we substitute A for 1, B for 2, etc., and utilize the last bracketed expression we can derive all of the two-component samples needed for the $ABC^1C^2D^1D^2$ quaternaries. Similarly we can utilize the second bracketed expression for all of the three-component samples and the expression associated with I_4 can be used for all the four-component samples. For example,

$$\left[\sum_{i=1}^{4} \sum_{j=i+1}^{4} \frac{1}{n_i n_j} \prod_{i=1}^{4} n_i \right] = \frac{n_A \times n_B \times n_C \times n_D}{n_A \times n_B} + \frac{n_A \times n_B \times n_C \times n_D}{n_A \times n_C} + \frac{n_A \times n_B \times n_C \times n_D}{n_A \times n_D}$$

$$+ \frac{n_A \times n_B \times n_C \times n_D}{n_A \times n_C} + \frac{n_A \times n_B \times n_C \times n_D}{n_B \times n_C} + \frac{n_A \times n_B \times n_C \times n_D}{n_B \times n_D}$$

$$= (n_A \times n_D) + (n_A \times n_C) + (n_A \times n_B) + (n_B \times n_D) + (n_B \times n_C) + (n_C \times n_D).$$

These formulas describe the two-component combinations needed. Note that for each component C and D there are two choices so that there are actually two possible combinations:

$A - D \rightarrow A^1D^1$	and	A^1D^2	$(n_A \times n_D = 2)$
$A - C \rightarrow A^1C^1$	and	A^1C^2	$(n_A \times n_C = 2)$
$A - B \rightarrow A^1B^1$			$(n_A \times n_B = 1)$
$B - D \rightarrow B^1D^1$	and	B^1D^2	$(n_B \times n_D = 2)$
$B - C \rightarrow B^1C^1$	and	B^1C^2	$(n_B \times n_C = 2)$
$C - D \rightarrow C^1D^1, C^1D^2, C^2D^1$	and	C^2D^2	$(n_C \times n_D = 4)$.

For each binary combination there are three ratio levels ($I_2 = 3$): 75/25, 50/50, and 25/75. This results in 39 binary samples in the experiment.

Similarly, the second bracketed expression yields the following three-component combinations:

$$\left[\sum_{i=1}^{4} \frac{1}{n_i} \prod_{i=1}^{4} n_i \right] = (n_B \times n_C \times n_D) + (n_A \times n_C \times n_D) + (n_A \times n_B \times n_D) + (n_A \times n_B \times n_C).$$

$B\text{-}C\text{-}D \rightarrow B^1C^1D^1, B^1C^1D^2, B^1C^2D^1, B^1C^2D^2$	$(n_B \times n_C \times n_D = 4)$
$A\text{-}C\text{-}D \rightarrow A^1C^1D^1, A^1C^1D^2, A^1C^2D^1, A^1C^2D^2$	$(n_A \times n_C \times n_D = 4)$
$A\text{-}B\text{-}D \rightarrow A^1B^1D^1, A^1B^1D^2$	$(n_A \times n_B \times n_D = 2)$
$A\text{-}B\text{-}C \rightarrow A^1B^1C^1, A^1B^1C^2$	$(n_A \times n_B \times n_C = 2)$.

The three component samples also have three ratio levels ($I_3 = 3$): 50/25/25, 25/50/25, and 25/25/50, resulting in 36 three-component samples in the experiment.

These combinations of components and levels will completely fill the runs needed to construct the four desired quaternaries. The general algorithms should be amenable to development of software to generate the actual runs needed when using overlapping of any general array.

7.8.2 Case 2

The second set covers any dimension of gradient array with multiple equivalent components. Table 7.12 gives definitions of the symbols used and the equations. Table 7.13 gives the savings in runs for representative ternary and quaternary systems. In both cases, the savings can be very substantial, particularly for the quaternary and larger systems.

These types of designs can be combined in various ways. There are too many possibilities, however, for general equations for the number of runs to be written. The equations given in this chapter can be adapted for specific situations.

Table 7.12 Terms and Equations for Case 2

Terms and Equations	Definitions
D	Dimension of gradients being generated
R	Number of equivalent components or reagents
M	Number of intervals from 0 to 100% for each reagent
$E_D = \dfrac{(M + D - 2)!}{(M - 1)! \times (D - 1)!}$	Number of runs needed for a single gradient
$I_D = C(R,D)$	Number of separate gradients
$I_D \times E_D$	Number of runs needed without overlap
$\displaystyle\sum_{n=0}^{D-1} \dfrac{C(R, D - n) \times (M - 2)!}{(M - D + n - 1)! \times (D - n - 1)!}$	Number of runs needed with overlap

Note: $C(x,y)$ = the number of combinations of x items taken y at a time.

Table 7.13 Ternary and Quaternary Systems With Multiple Equivalent Components

D	R	M	Interval (%)	Without Overlap	With Overlap	Improvement (%)
3	4	4	33	40	20	50
3	4	6	20	84	52	38
3	4	11	10	264	202	23
3	5	4	33	100	35	65
3	5	6	20	210	105	50
3	5	11	10	660	455	31
3	6	4	33	200	56	72
3	6	6	20	420	186	56
3	6	11	10	1,320	861	35
4	5	4	33	100	35	65
4	5	6	20	280	125	55
4	5	11	10	1,430	875	39
4	6	4	33	300	56	81
4	6	6	20	840	246	71
4	6	11	10	4,290	2,121	51

Note: D and R as defined in Table 7.12.

7.9 CONCLUSION

Although modern methods of high throughput screening have enormously increased the productivity of the research laboratory, the task of exploring for new materials remains daunting. As Hsieh-Wilson and coworkers put it: "Given approximately 60 elements in the periodic table that can be used to make compositions consisting of three, four, five, or even six elements, the universe of possible new compounds with interesting physical and chemical properties remains largely uncharted" [11]. These methods for modifying and combining gradient arrays will markedly improve the productivity of combinatorial and high throughput screening programs in chemical research and development [12].

ACKNOWLEDGMENT

It is a pleasure to acknowledge the contributions of the Avery Research Center, particularly Billy Yang. Some of the work in this chapter was performed under the support of the National Institute of Science and Technology's Advanced Technology Program, contract number 70NANB9H3038.

REFERENCES

1. Dover, R. B. V.; Schneemeyer, L. F.; Fleming, R. M. *Nature (London),* 1998, 392: 162–164.

2. Cong, P.; Doolen, R. D.; Fan, Q.; Giaquinta, D. M.; Guan, S.; McFarland, E. W.; Poojary, D. M.; Self, K.; Turner, H. W.; Weinberg, H. *Angew. Chem., Int. Ed.,* 1999, 38: 484–488.

3. Reddington, E.; Sapienza, A.; Gurau, B.; Viswanathan, R.; Sarangapani, S.; Smotkin, E. S.; Mallouk, T. E. *Science,* 1998, 280: 1735–1737.

4. Senkan, S.; Krantz, K.; Oztrurk, S.; Zengin, V.; Onal, I. *Angew. Chem., Int. Ed.,* 1999, 38: 2794–2799.

5. Cornell, J. A. *Experiments With Mixtures : Designs, Models, and the Analysis of Mixture Data,* 3rd ed., Wiley, New York, 2001.

6. Danielson, E.; Colden, J. H.; McFarland, E. W.; Reaves, C. M.; Weinberg, H.; Wu, X. D. *Nature (London),* 1997, 389: 944–948.

7. Milliken, G. A.; Johnson, D. E. *Analysis of Messy Data,* Van Nostrand Reinhold: New York, 1984.

8. Clark, R. D.; Cramer, R. D. *CHEMTECH,* 1997, 1997: 24–31.

9. Hassan, M.; Bielawski, J. P.; Hempel, J. C.; Walden, M. *Mol. Diversity,* 1996, 2: 64–74.

10. Hilbert, D.; Cohn-Vossen, S. *Geometry and the Imagination,* Chelsea, New York, 1952.

11. Hsieh-Wilson, L. C.; Xiang, X.-D.; Schultz, P. G. *Acc. Chem. Res.,* 1996, 29: 164–170.

12. Cawse, J. N.; Wroczynski, R.; Yang, B. U.S. Patent Application 09/696071, 2000.

CHAPTER 8

SPLIT-PLOT DESIGNS

MARTHA M. GARDNER AND JAMES N. CAWSE
GE Global Research

8.1 BACKGROUND

8.1.1 Elements of a Split-Plot Design

Split-plot designs, which have been long used in the agricultural arena, are becoming more popular in industrial applications. Consider the following problem. A new engineering thermoplastic is being developed that requires the use of an extruder. The engineers designing the new process need to determine the best screw design as well as the best rotational speed [revolutions per minute (rpm)] for the new material. A standard experimental design was developed that contained two different screw designs, three levels of rpm, and two complete replicates of the original design. Upon examining the completely randomized design, the engineers were dismayed. Considering the length of time it takes to change the screw design (about 1.5–2 h), complete randomization was not feasible. A simpler approach, which would allow one whole replicate of the design to be run in one day, would be to use one screw design and randomly run all three rpm's, then go to the next screw design and randomly run all three rpm's. The entire process could be repeated each of the next two days for the second and third replicates. Essentially the engineers have created a randomization restriction on the screw design yielding three days of effort versus the estimated ten days with the completely randomized design. In this setup, the screw design is called the *main-plot* (or *whole-plot*) *factor,* while the rpm is called the *subplot* (or *split-plot*) *factor.* The use of the word "plot" is reflective of the agricultural use of these designs where different types of crops would be planted across a large plot of land with the plot "split" into smaller areas for treatment with different types of fertilizer, insecticide, etc.

8.1.2 Split-Plot Elements in Combinatorial/HTS Experiments

In the combinatorial/HTS process, the situation is more complex than the extrusion example just given, but still contains classic split-plot elements. There are generally

Experimental Design for Combinatorial and High Throughput
Materials Development, Edited by James N. Cawse.
ISBN 0-471-20343-2 © 2003 John Wiley & Sons, Inc.

two types of variables being considered: formulation variables, which describe the relative percentages of components in the mixture of ingredients, and processing variables, which describe the processing conditions to which different formulations are subjected. The goal is to compare the properties of the samples in the array in order to discover leads—materials whose properties are indicative of commercial potential. As in the preceding example, complete randomization is neither feasible nor efficient, especially when one considers the following setup, which is typical of many HTS processes (Figure 8.1):

- Selection of the actual chemicals to be used in the experiment;
- Introduction of these chemicals into the formulation system, typically by weighing and dissolution to form stock solutions;

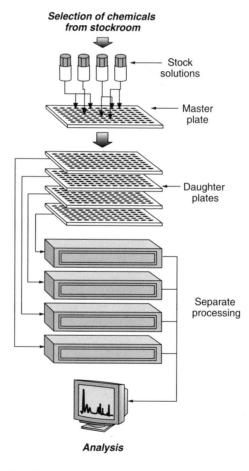

Figure 8.1 Flowchart of a common type of combinatorial experiment.

- Combination of aliquots of those stock solutions into formulations or mixtures in a geometrical array (master plate), typically by the use of a pipetting robot;
- Production of identical copies of the master plate (daughter plates);
- Processing each array to a final composition;
- Analysis of the properties of the resulting final composition.

In general, a specific combination of processing conditions must be applied across a daughter plate, and the efficiency results from using many different formulations across the daughter plate. Consequently, processing conditions are treated as main-plot factors and the formulation combinations as the subplot factors. This type of approach has been used by Cawse [1] and Reis et al. [2].

8.2 ADVANTAGES OF SPLIT-PLOT EXPERIMENTS

The split-plot setup has many advantages, although the statistical analysis of such designs is more complex than the completely randomized design (as will be shown in Section 8.3). The most obvious advantage is in the efficiency of experimentation. The faster experiments can be completed, the faster new leads can be obtained and commercialization can occur. From a combinatorial/HTS viewpoint, there is an additional advantage of detecting higher order interactions than what are typically sought in a traditional experimental design. It is possible that the area of interest can only be found through certain combinations of numerous formulation components and several processing conditions. Overall, this can result in the need to estimate three-way or even four-way or higher-order interactions among all these variables. The following discussion shows how it is possible to structure experiments so that these interaction effects are estimable.

8.2.1 Observable Effects

A recent paper by Bisgaard [3] gives a foundation for this type of approach in the context of split-plot designs. Bisgaard introduced the concept of two arrays set up as $2^{k-p} \times 2^{q-r}$ split-plot experiments with confounding within, but not between, the two arrays. The experimental structures of Bisgaard's examples are generally the Taguchi inner/outer array form, but are run in a split-plot fashion. This is very similar to the combinatorial/HTS approach since the formulation space can be thought of as an inner array and the processing conditions as an outer array, yet the overall experiment is run in a split-plot fashion. In Bisgaard's paper, the effects of interest tend to be the two-way interactions between the inner array and outer array factors. For example, let A, B, and C represent inner array factors and P, Q, and R represent outer array factors. Then in a $2^{3-1} \times 2^{3-1}$ split-plot experiment (Figure 8.2), the confounding patterns are shown in Table 8.1, where l_n is a calculated effect. Note that the inner/outer crossterms such as AP are not additive with any other two-way or lower-order terms.

In combinatorial/HTS experimentation, the two-way interactions shown in Bisgaard's work are of less interest; the high-value leads are expected to be the 3-factor

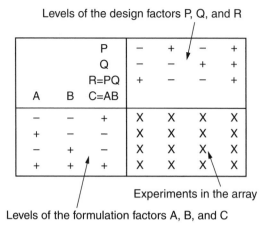

Levels of the design factors P, Q, and R

Experiments in the array

Levels of the formulation factors A, B, and C

Figure 8.2 Structure of a $2^{3-1} \times 2^{3-1}$ split plot experiment. (Adapted, with permission, from Bisgaard, S. *Journal of Quality Technology,* 2000, 32: 39–56; copyright © 2000. American Society for Quality, Milwaukee, WI.)

or higher interactions [4]. The Bisgaard approach can be easily extended to detect these interactions. It has been found that when arrays of higher resolution than IV (main effects confounded with three-way interactions and two-way interactions confounded with other two-way interactions) are used for the formulation factors, it becomes possible to observe these desired three-factor interactions, which include at least two formulation variables along with a processing variable. The large numbers of samples inherent in CHTS experimentation allow easy access to these higher-resolution designs. Thus, in the $2^{5-1} \times 2^{3-1}$ split-plot experiment shown in Figure 8.3 the three-way interactions between two formulation factors and one process factor are not confounded with main effects, 2-factor interactions, or other 3-factor interactions. This is a general situation that can be extended to designs with larger numbers of process and formulation factors. Representative situations where the resolution of the formulation factor design is V (main effects confounded with four-way interactions and two-way interactions confounded with three-way interactions) and the process design is III (main effects confounded with two-way interactions) are given in Table 8.2.

Table 8.1 Confounding of the $2^{3-1} \times 2^{3-1}$ Split-Plot Experiment

$l_1 \rightarrow A + BC$	$l_6 \rightarrow R + PQ$	$l_{11} \rightarrow BQ$
$l_2 \rightarrow B + AC$	$l_7 \rightarrow AP$	$l_{12} \rightarrow BR$
$l_3 \rightarrow C + AB$	$l_8 \rightarrow AQ$	$l_{13} \rightarrow CP$
$l_4 \rightarrow P + QR$	$l_9 \rightarrow AR$	$l_{14} \rightarrow CQ$
$l_5 \rightarrow Q + PR$	$l_{10} \rightarrow BP$	$l_{15} \rightarrow CR$

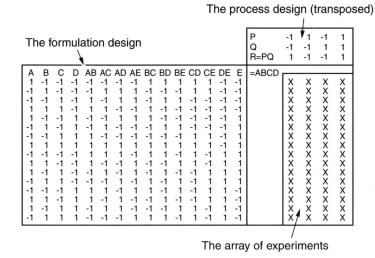

Figure 8.3 Structure of a $2^{5-1} \times 2^{3-1}$ split-plot experiment.

8.2.2 Setting up the Design

The process of setting up such an experiment, using a $2^{5-1} \times 2^{3-1}$ system as an example, is as follows:

1. An experimental space that comprises a set of formulation factors and processing factors is identified. High and low levels of the factors are identified. The high and low levels are then coded as +1 and −1, respectively.

2. Objectives of the experiment are identified. Both degree and types of interactions to be investigated are determined. Identified interactions may be formulation-only interactions, process-only interactions, or formulation * process interactions.

3. The resolutions of the formulation design and the process design, which will allow the interactions identified in step 2 (particularly the formulation *

Table 8.2 Formulation Designs of Resolution V and Process Designs of Resolution III

Formulation Designs		Process Designs	
Number of Samples in Formulation Design	Number of Formulation Factors	Number of Arrays in Process Design	Number of Process Factors
16	5	4	3
32	6	8	5
64	8	16	10

process interactions) to be observed without confounding by lower-order interactions are ascertained. Table 8.3 shows the highest observable formulation/process interactions, where $F * P$ is an interaction of a single formulation factor with a single process factor, and so on.

4. Once the appropriate resolution has been selected for the process design and the formulation design, experimental arrays of the correct resolution are generated in coded form. This is a routine process using software such as Minitab® [5] or Design-Expert® [6].

5. The formulation design and process design are crossed with each other, producing a matrix of the type shown in Figure 8.3. The process factor design is transposed from a standard array (in which each factor and its levels form a column) to the array shown in Figure 8.3 (in which they form a row).

6. The experiment is performed to produce response data.

7. Results of the experiment are evaluated by the following process:

   (a) The measured results of the experiments performed in step 6 are arranged in a matrix such that each result is in the position corresponding to the X that identified its experimental conditions in Figure 8.3. This is the results matrix **Y**.

   (b) A column of ones, **I,** is added to the coded process design (Figure 8.4a) and the coded formulation design (Figure 8.4b). The arrays are then complete Hadamard matrices [7]. These new arrays, of dimension $p \times p$ and $f \times f$, are called the **P** array and the **F** array, respectively.

   (c) A matrix of effects **X** can be then be found by the matrix equation (see Appendix 8.A for derivation):

$$\mathbf{X} = (pf)^{-1}\mathbf{F^{T}YP}. \tag{8.1}$$

The observable effects from this experiment are shown in Figure 8.5. They consist of the main effects, two-way interactions, and (formulation × formulation × process) three-way interactions. From a statistical point of view, it is important to note that, as Bisgaard stated, "the sub-plot contrasts (effects) always will be estimated with greater precision than the whole plot contrasts" [3]. Thus the three-way

Table 8.3   Formulation/Process Interactions Observable Without Confounding from Lower-Order Interactions

Resolution of the Formulation Design	Resolution of the Process Design		
	III	V	VII
III	$F * P$	$F * P * P$	$F * P * P * P$
V	$F * F * P$	$F * F * P * P$	$F * F * P * P * P$
VII	$F * F * F * P$	$F * F * F * P * P$	$F * F * F * P * P * P$

a.

I	P	Q	R= PQ
1	-1	-1	1
1	1	-1	-1
1	-1	1	-1
1	1	1	-1

b.

I	A	B	C	D	AB	AC	AD	AE	BC	BD	BE	CD	CE	DE	E
1	1	-1	-1	-1	-1	-1	-1	-1	1	1	1	1	1	-1	-1
1	-1	1	-1	-1	-1	1	1	1	-1	-1	-1	1	1	-1	-1
1	-1	-1	1	-1	1	-1	1	1	-1	1	1	-1	-1	-1	-1
1	1	1	1	-1	1	1	-1	-1	1	-1	-1	-1	-1	-1	-1
1	1	-1	-1	1	-1	-1	1	1	1	-1	-1	-1	-1	-1	1
1	-1	1	-1	1	-1	1	-1	-1	-1	1	1	-1	-1	-1	1
1	-1	-1	1	1	1	-1	-1	-1	-1	-1	-1	1	1	-1	1
1	1	1	1	1	1	1	1	1	1	1	1	1	1	-1	1
1	-1	-1	-1	-1	1	1	1	-1	1	1	-1	1	-1	1	1
1	1	1	-1	-1	1	-1	-1	1	-1	-1	1	1	-1	1	1
1	-1	1	1	-1	-1	1	-1	1	-1	1	-1	-1	1	1	1
1	1	-1	1	-1	-1	-1	1	-1	1	-1	1	-1	1	1	1
1	-1	1	1	-1	-1	-1	1	-1	1	-1	1	-1	1	1	-1
1	-1	-1	-1	1	1	1	-1	1	1	-1	1	-1	1	1	-1
1	1	1	-1	1	1	-1	1	-1	-1	1	1	-1	1	1	-1
1	1	-1	1	1	-1	1	1	-1	-1	-1	1	1	-1	1	-1
1	-1	1	1	1	-1	-1	-1	1	1	1	-1	1	-1	1	-1

Figure 8.4 (a) A 4×4 Hadamard matrix P, (b) a 16×16 Hadamard matrix F.

interactions will be estimated with the maximum precision possible from the experiment.

The calculated values in the effects matrix $\mathbf{X}$ are then examined for statistical significance (using $\alpha = 0.05$). If all of the effects observed in the experiment are caused simply by random processes (the null hypothesis), the effects will fit to a normal distribution and form a relatively straight line in a normal probability plot. Any effects that fall off the line by more than three to four standard deviations can be interpreted to have been caused by nonrandom processes (Figure 8.6) [8]. These are likely to be the leads of interest. Section 8.3 details the appropriate error assignments in either testing the effects for significance or constructing normal probability plots of effects.

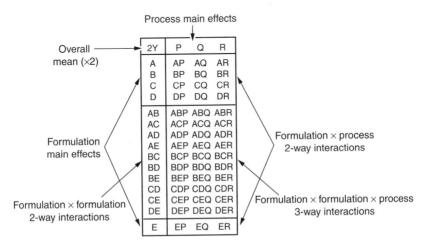

Figure 8.5 Observable effects and interactions in a $2^{5-1} \times 2^{3-1}$ split-plot experiment.

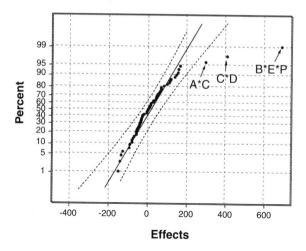

Figure 8.6 Use of a normal probability plot to detect significant effects. The dotted line shows a 2 standard deviation interval around the straight line.

8.3 STATISTICAL ANALYSIS OF SPLIT-PLOT EXPERIMENTS

As stated earlier in Section 8.2, the analysis of split-plot experiment requires special care. Although some basics will be covered here, many articles and texts are available for further details and examples. Milliken and Johnson [9] and Montgomery [10] are excellent resources.

8.3.1 Assignment of Experimental Error

In all the following cases, the focus is on the construct of the appropriate F statistics for testing the various effects. The reader should refer to various references given for actual formulas for calculation of sums of squares. In instances where appropriate, SAS® code using PROC MIXED is given.

Basic Split-Plot Analysis Consider the extrusion example from Section 8.1. Each of the three days can be treated as a block, the screw design is the main-plot factor (two levels), and the rpm is the subplot factor (three levels). This setup supports the following model:

$$Y_{ijk} = \mu + B_i + M_j + BM_{ij} + S_k + BS_{ik} + MS_{jk} + BMS_{ijk}, \tag{8.2}$$

where

$$\mu = \text{overall mean}$$

$$B_i = \text{effect of } i\text{th block}; \; i = 1, 2, 3$$

$$M_j = \text{effect of } j\text{th main-plot treatment}; j = 1, 2$$

$$S_k = \text{effect of } k\text{th subplot treatment}; k = 1, 2, 3$$

$$BM_{ij} = \text{block by main-plot interaction}$$

$$BS_{ik} = \text{block by subplot interaction}$$

$$BMS_{ijk} = \text{block by main-plot by subplot interaction.}$$

An appropriate analysis of variance (ANOVA) table for the extrusion problem, assuming that day (block) is a random effect and that screw design and rpm are fixed effects, is shown in Table 8.4.

Basic Split-Plot Analysis With Pooled Subplot Errors   Many authors, including Cornell [11] and Steel and Torrie [12] propose that the two error terms in the subplot partition be combined to form one pooled estimate of error for testing all subplot effects. The main reasoning is that if the number of blocks (and/or number of subplot treatments) is small, pooling yields a more powerful test of statistical significance for the subplot treatment effects [11]. However, Montgomery [10] advises against pooling if there is an expected interaction between the block and the other factors in the experiment.

The revised model, assuming that the block by subplot and block by subplot by main-plot interactions are pooled with the overall error term, is as follows:

$$Y_{ijk} = \mu + B_i + M_j + BM_{ij} + S_k + MS_{jk} + \varepsilon_{ijk}, \tag{8.3}$$

where

$$\mu = \text{overall mean}$$

$$B_i = \text{effect of } i\text{th block}$$

$$M_j = \text{effect of } j\text{th main-plot treatment}$$

$$S_k = \text{effect of } k\text{th subplot treatment}$$

$$BM_{ij} = \text{block by main-plot interaction—used as main-plot error}$$

$$\varepsilon_{ijk} = \text{pooled error—used as subplot error.}$$

It should also be noted that one of the important underlying assumptions is that the main-plot and subplot errors are uncorrelated.

The corresponding ANOVA table (shown in Table 8.5) would be as follows, still assuming that day (block) is a random effect and that screw design and rpm are fixed effects.

For this example, output could be produced using the following SAS® code [13]:

```
Proc MIXED data = split;
```

Table 8.4 Basic Split-Plot ANOVA

Source of Variation	Degrees of Freedom	Mean Square	F Statistic
Block (B)	$i - 1 = 2$	SS_B/df_B	
Screw (M)	$j - 1 = 1$	SS_M/df_M	MS_M/MS_{BM}
Block * screw (BM— main plot error)	$(i - 1)(j - 1) = 2$	SS_{BM}/df_{BM}	
rpm (S)	$k - 1 = 2$	SS_S/df_S	MS_S/MS_{BS}
Block * rpm (BS)	$(i - 1)(k - 1) = 4$	SS_{BS}/df_{BS}	
Screw * rpm (MS)	$(j - 1)(k - 1) = 2$	SS_{MS}/df_{MS}	MS_{MS}/MS_{BMS}
Block * screw * RPM (BMS—subplot error)	$(i - 1)(j - 1)(k - 1) = 4$	SS_{BMS}/df_{BMS}	
Total	$N - 1 = 17$	N/A	

```
Class Screw Block;
Model y = Screw|RPM;
Random Block Screw*Block;
Run;
```

Extension to Combinatorial and HTS Experimentation The preceding procedure can be extended to a slightly more complex, and combinatorial/HTS-related example. Assume there are *five* formulation factors and *three* process variables to investigate. Also assume that the design around the formulation variables is a Resolution V 2^{5-1} Fractional Factorial Design with 16 runs, while the design for the processing variables is a full factorial with 8 runs. The full crossed design then has 16 * 8 = 128 runs. Both types of designs can be developed using software such as Minitab® [5] or Design-Expert® [6]. As discussed in the beginning of this chapter, the process variables are the main-plot factors and the formulation variables are the subplot factors. Also assume the entire experiment will be replicated (by arrays) for a total of 256 runs. Model (2) is the appropriate model, and the resulting ANOVA (shown in Table 8.6) would appear as follows (where $F * P$

Table 8.5 Basic Split-Plot ANOVA With Pooled Subplot Errors

Source of Variation	Degrees of Freedom	Mean Square	F Statistic
Block (B)	$i - 1 = 2$	SS_B/df_B	
Screw (M)	$j - 1 = 1$	SS_M/df_M	MS_M/MS_{BM}
Block * Screw (BM— main-plot error)	$(i - 1)(j - 1) = 2$	SS_{BM}/df_{BM}	
rpm (S)	$k - 1 = 2$	SS_S/df_S	MS_S/MS_E
Screw * rpm (MS)	$(j - 1)(k - 1) = 2$	SS_{MS}/df_{MS}	MS_{MS}/MS_E
Subplot error	$j(i - 1)(k - 1) = 8$	SS_E/df_E	
Total	$N - 1 = 17$		

Table 8.6 $2^{5-1} \times 2^3$ Split-Plot ANOVA

Source of Variation	Degrees of Freedom	Mean Square	F Statistic
Block (B)	1	SS_B/df_B	
Process main effects	3	SS_P/df_P	MS_P/MS_{BP}
Process * process interactions	3	SS_{PP}/df_{PP}	MS_{PP}/MS_{BP}
Process * process * process interaction	1	SS_{PPP}/df_{PPP}	MS_{PPP}/MS_{BP}
Block * process (BP–main-plot error)	7	SS_{BP}/df_{BP}	
Formulation main effects	5	SS_F/df_F	MS_F/MS_E
$F * F$ interactions	10	SS_{FF}/df_{FF}	MS_{FF}/MS_E
$F * P$ interactions	15	SS_{FP}/df_{FP}	MS_{FP}/MS_E
$F * F * P$ interactions	30	SS_{FFP}/df_{FFP}	MS_{FFP}/MS_E
$F * P * P$ interactions	15	SS_{FPP}/df_{FPP}	MS_{FPP}/MS_E
$F * F * P * P$ interactions	30	SS_{FFPP}/df_{FFPP}	MS_{FFPP}/MS_E
$F * P * P * P$ interactions	5	SS_{FPPP}/df_{FPPP}	MS_{FPPP}/MS_E
$F * F * P * P * P$ interaction	10	SS_{FFPPP}/df_{FFPPP}	MS_{FFPPP}/MS_E
Subplot error (E)	120	SS_E/df_E	
Total	255	N/A	

is an interaction of a single formulation factor with a single process factor, and so on).

Note that the process effects include the main effects of the main-plot factors, two-way interactions of main-plot factors, and the three-way interactions of the main-plot factors. The formulation effects include the main effects, two-way interactions, and interactions between all these effects with all the effects listed earlier for the processing factors.

In some cases, it may be more appropriate to use a mixture design around the formulation factors, rather than one of the standard nonmixture designs, such as Factorials or Central Composites. This tends to be the case when it is expected that the relative proportion of ingredients to each other, rather than their absolute amounts, will significantly affect the response. In many of the authors' combinatorial and HTS studies, the formulation factors represent either qualitative factors, such as catalyst type, or the quantitative percentages of the factors are so small, that the absolute amounts can be used appropriately. The reader is directed to Cornell [14] for an in-depth look at mixture designs and to Design-Expert software [6] for software development of these designs. The reader is also directed to Cornell [11] for an example of mixture designs with nonmixture process variable designs in a split-plot setup. In his paper, Cornell covers the structure of these designs, as well as calculation of sums of squares and F-statistic construction for the different effects. More recently, Kowalski et al. [15] have introduced new models and estimation methods for these types of experiments.

Now assume that instead of using a full factorial for the processing variables, a Resolution III fractional factorial was used instead. Assuming the use of the Resolution V 2^{5-1} Fractional Factorial Design for the formulation factors and two total replicates of the entire experiment, the total number of runs will be cut from 256 to 128. Even though the number of runs is cut in half, the three-way interaction effects between two formulation components and one processing variable are still estimable. The resulting ANOVA table (shown in Table 8.7) is very similar to the preceding one, but with fewer degrees of freedom available to estimate effects for the process variables.

Handling of Nonreplicated Split-Plot Experiments Assume that it was not possible to repeat the preceding experiment, so there are no blocks in this experiment. Notice that this makes the estimate of the main-plot error no longer estimable. This is usually an undesirable situation, and therefore, at least three blocks (or replicates) are usually recommended for the split-plot setup. However, in the event that replication is not feasible, normal probability plots are recommended to assess the importance of effects. The key is that rather than have one normal probability plot for all effects together, there should be one normal probability plot for the main-plot effects and a separate normal probability plot for the subplot effects. See Bisgaard [3] for additional discussion around this point.

8.3.2 Other Types of Split-Plot Designs

The notion of a split-plot can be easily extended to other types of experimental setups. In this next section, the concepts of split-split-plots and repeated-measures designs are introduced.

Split-Split-Plot Consider the basic scenario given in Section 8.1, with screw design and rpm as the two factors. Screw design is very difficult to change, while rpm is easy, so there is one level of randomization restriction. Now suppose that an additional factor, barrel temperature, is included in this study. Completely randomizing all barrel temperature and rpm combinations within a screw design, however, is not a feasible approach. It is much more feasible to set one temperature and perform all

Table 8.7 $2^{5-1} \times 2^{3-1}$ Split-Plot ANOVA

Source of Variation	Degrees of Freedom	Mean Square	F Statistic
Block (*B*)	1	SS_B/df_B	
Process main effects (*M*)	3	SS_M/df_M	MS_M/MS_{BM}
Block * process (*BM*—main-plot error)	3	SS_{BM}/df_{BM}	
Formulation main effects	5	SS_F/df_F	MS_F/MS_E
F * *F* interactions	10	SS_{FF}/df_{FF}	MS_{FF}/MS_E
F * *P* interactions	15	SS_{FP}/df_{FP}	MS_{FP}/MS_E
F * *F* * *P* interactions	30	SS_{FFP}/df_{FFP}	MS_{FFP}/MS_E
Subplot error (*E*)	60	SS_E/df_E	
Total	127	N/A	

rpm's, then change the temperature and run all rpm's. In this instance a second level of randomization restriction has been introduced, and the result is called a split-split-plot structure. Assuming that the entire experiment is repeated three times, one possible ANOVA table structure is shown in Table 8.8.

It should be noted that as done in the earlier split-plot examples, certain error terms can be pooled within specific levels of the structure. In the preceding example, the block * temp (BS) and block * screw * temp (BMS) effects could be combined into one term for the subplot error. Also, the block * rpm (BP), block * screw * rpm (BMP), block * temp * rpm (BSP), and block * screw * temp * rpm ($BMSP$) could be combined into one term for the sub-subplot error. This yields the following model:

$$Y_{ijkl} = \theta + B_i + M_j + BM_{ij} + S_k + MS_{jk} + BMS_{ijk} + P_l + MP_{jl} + SP_{jl} + MSP_{jkl} + \varepsilon_{ijkl}$$
(8.4)

The corresponding SAS® code would be:

```
Proc MIXED data = spltsplt;
 Class Block Screw Temp RPM;
 Model y = Screw|Temp|RPM;
 Random Block Screw*Block Screw*Temp*Block;
Run;
```

Table 8.8 Split-Split-Plot ANOVA

Source of Variation	Degrees of Freedom	Mean Square	F Statistic
Block (B)	$a-1$	SS_B/df_B	
Screw (M)	$b-1$	SS_M/df_M	MS_B/MS_{BM}
Block * screw (BM— main-plot error)	$(a-1)(b-1)$	SS_{BM}/df_{BM}	
Temp (S)	$(c-1)$	SS_S/df_S	MS_S/MS_{BS}
Block * temp (BS)	$(a-1)(c-1)$	SS_{BS}/df_{BS}	
Screw design * temp (MS)	$(b-1)(c-1)$	SS_{MS}/df_{MS}	MS_{MS}/MS_{BMS}
Block * screw * temp (BMS— subplot error)	$(a-1)(b-1)(c-1)$	SS_{BMS}/df_{BMS}	
rpm (P)	$d-1$	SS_P/df_P	MS_P/MS_{BP}
Block * rpm (BP)	$(a-1)(d-1)$	SS_{BP}/df_{BP}	
Screw * rpm (MP)	$(b-1)(d-1)$	SS_{MP}/df_{MP}	MS_{MP}/MS_{BMP}
Block * screw * rpm (BMP)	$(a-1)(b-1)(d-1)$	SS_{BMP}/df_{BMP}	
Temp * rpm (SP)	$(c-1)(d-1)$	SS_{SP}/df_{SP}	MS_{SP}/MS_{BSP}
Block * temp * rpm (BSP)	$(a-1)(c-1)(d-1)$	SS_{BSP}/df_{BSP}	
Screw * temp * rpm (MSP)	$(b-1)(c-1)(d-1)$	SS_{MSP}/df_{MSP}	MS_{MSP}/MS_{BMSP}
Block * screw * temp * rpm ($BMSP$—sub-subplot error)	$(a-1)(b-1)(c-1)(d-1)$	SS_{BMSP}/df_{BMSP}	
Total	$N-1$	N/A	

In a combinatorial and HTS process, an example of the application of a split-split-plot design would be as follows. Suppose one has five formulation factors, three process factors, as well as a substrate factor for a coating development project. When setting up the actual flow of experimentation, the following occurs. At one combination of process factors, all combinations of formulation factors are run on one substrate and then all combinations of formulation factors are run on the other substrate. Then the processing factors are changed to the second set of conditions and the whole formulation and substrate combination flow is repeated. This procedure continues until all desired combinations of processing conditions have been covered. In this case, the processing conditions are the main-plot (or whole-plot) factors, the substrate is the subplot (or split-plot) factor, and the formulation factors are the sub-subplot (or split-split- plot) factors.

Repeated Measures   Another manifestation of the split-plot design is the repeated-measures design. Very often, there is interest in the performance of materials over time. There are many types of responses that fall into this group. One example would be the effectiveness of a reaction over time. In this case, time is the whole-plot factor in a split-plot experiment. The main difference that can arise has to do with the error structures. In the standard split-plot situation, it is assumed that the errors are uncorrelated; however, in the repeated-measures view, there is no such assurance that this is indeed the case. It is easy to imagine that a previous measurement on a same sample would affect the next measurement, thus inducing a correlation structure among the errors. Fortunately, it is easy enough to try different error structures and use comparison criteria to determine the best model. Of course, any time there is physical, rather than empirical, knowledge of how measurements might be correlated over time, that knowledge should be taken into consideration.

Consider the following case of investigating the effect of temperature on a certain reaction. The temperature will be applied for a specified amount of time and the reaction will be monitored over a one-week time period at daily intervals. Assume there are three temperatures of interest and that the entire study is repeated three times.

The statistical model for this would be

$$Y_{ijk} = \mu + \alpha_i + d_{ij} + \tau_k + \alpha/\tau_{ik} + \varepsilon_{ijk}, \tag{8.5}$$

where

$i = 1, \ldots, 3$ temperatures

$j = 1, \ldots, 3$ samples

α_i, τ_k, and $\alpha\tau_{kj}$ are fixed parameters such that the mean for the ith temperature at time k is $= \mu_{ik} = \mu + \alpha_i = \tau_k + \alpha\tau_{ik}$

$d_{ij} =$ random effect associated with the jth sample at temperature i

$\varepsilon_{ijk} =$ random error associated with the jth sample at temp i at time k

If the d_{ij} are multivariate normal $(0, \sigma_S^2 I)$ and the ε_{ijk} are multivariate normal $(0, \sigma_T^2 I)$, then the resulting variance/covariance structure is called compound symmetric [16].

One example of corresponding SAS® code would be

```
Proc MIXED data=repeat method=ml;
 Class sample temp time;
 Model y = temp time temp*time;
 Repeated/Type=CS subject = sample(temp);
Run;
```

The type statement tells what kind of underlying covariance structure to use. CS stands for compound symmetry and is the simplest structure. There are other types of covariance structures that can be used, such as AR(1), unstructured, etc. The Aikake information criterion (AIC) that is output can be used to decide which is the best fit. The reader is referred to Littell et al. [16] for more details.

SUMMARY

Since combinatorial experiments, in which there is interest in process factors as well as formulation factors, are inherently split-plot designs, it makes good sense to use the inherent properties of these designs to the best advantage. Arranging the formulation and process designs as shown in this chapter allows the estimation of commercially interesting high-order interactions with relative ease.

ACKNOWLEDGEMENTS

The authors thank Dr. Richard Taylor and Dr. William Tucker for their thorough reviews of the manuscript, which greatly improved its presentation, as well as Dr. Erin Blankenship for numerous discussions around the use of PROC MIXED in SAS® for split-plot designs. Some of the work in this chapter was performed under the support of the National Institute of Science and Technology's Advanced Technology Program, contract number 70NANB9H3038.

APPENDIX 8.A

Derivation of Equation (8.1)

The matrix equation for a standard linear least-square model, $\mathbf{Y} = \mathbf{X}\boldsymbol{\beta} + \boldsymbol{\varepsilon}$ [17] can be extended to the split-plot situation, as shown in Figure 8.2, by renaming the $\mathbf{X}$ matrix $\mathbf{F}$ and postmultiplying by the transpose of process factors matrix $\mathbf{P}$:

$$Y = F\beta P^T + \varepsilon. \tag{8.A.1}$$

Then if **b** is the least-square estimator of the effects matrix β, we can generate the least-square equation by premultiplying both sides of the equation by F^T and post-multiplying by **P**:

$$F^T F b P^T P = F^T Y P. \tag{8.A.2}$$

However, **F** and **P** are complete Hadamard matrices [7], and for any Hadamard matrix **H** of dimension $h \times h$,

$$(H^T H) = (H H^T) = hI, \tag{8.A.3}$$

where **I** is the identity matrix [18].

Therefore equation (8.A.4) simplifies to

$$(f)b(p) = F^T Y P, \tag{8.A.4}$$

where f and p are the dimensions of **F** and **P**, so

$$b = (pf)^{-1} F^T Y P. \tag{8.A.5}$$

REFERENCES

1. Cawse, J. N. U.S. Patent Application 09/681222, 2001

2. Reis, C.; de Andrade, J. C.; Bruns, R. E.; Moran, R. C. C. P. *Anal. Chim. Acta,* 1998, 369: 269–279.

3. Bisgaard, S. *J. Qual. Technol.,* 2000, 32: 39–56.

4. Cawse, J. N. *Acc. Chem. Res.,* 2001, 34: 213–221.

5. Minitab Inc., 12th ed., State College, PA, 1999.

6. Stat-Ease, 6th ed., Stat-Ease, Inc: Minneapolis, MN, 2000.

7. Diamond, W. J. *Practical Experiment Designs for Engineers and Scientists,* Lifelong Learning Publications: Belmont, CA, 1981.

8. Montgomery, D. C. In *Design and Analysis of Experiments,* Wiley, New York, 1984, 357ff.

9. Milliken, G. A.; Johnson, D. E. *Analysis of Messy Data,* Van Nostrand Reinhold, New York, 1984.

10. Montgomery, D. C. *Design and Analysis of Experiment,* 3rd ed. Wiley, New York, 1991.

11. Cornell, J. A. *J. Qual. Technol.,* 1988, 20: 2–23.

12. Steel, R. G. D.; Torrie, J. H. *Principles and Procedures of Statistics: A Biometrical Approach,* McGraw-Hill, New York, 1980.

13. Latour, D.; Latour, K.; Wolfinger, R. D. "Getting Started with PROC MIXED," SAS Institute Inc., 1994.

14. Cornell, J. A. *Experiments With Mixtures: Designs, Models, and the Analysis of Mixture Data,* 2nd ed. Wiley, New York, 1990.

15. Kowalski, S. M.; Cornell, J. A.; and Vining, G. G. *Technometrics,* 2002 44: 72–79.

16. Littell, R. C.; Milliken, G. A.; Stroup, W. W.; Wolfinger, R. D. "SAS System for Mixed Models," SAS Institute Inc., Cary, NC, 1996.

17. Draper, N.; Harry, S. *Applied Regression Analysis,* 2nd ed., Wiley-Interscience, New York, 1981.

18. Craigen, R. In *The Handbook of Combinatorial Designs,* Coburn, C. J., Dinitz, J. H., Eds., CRC Press, New York, 1996, 371–377.

CHAPTER 9

AN EVOLUTIONARY STRATEGY FOR THE DESIGN AND EVALUATION OF HIGH THROUGHPUT EXPERIMENTS

DORIT WOLF
Degussa

MANFRED BAERNS
ACA-Berlin

9.1 INTRODUCTION

Genetic algorithms (GA) are a global optimization method for functions with no a priori known parametric form. All the information a genetic algorithm needs is values of the function (fitness or objective function) to be optimized for particular input data. These input data sets ("individuals") are generated by simulating biologically motivated evolutionary principles such as selection, crossover, and mutation to find the optimal solutions for a given task. The differences of GA, with respect to conventional optimization strategies are:

- Parameter values for optimization are encoded by a gene-analogous structure that allows evolutionary operations;
- Different solutions of the optimization problem are considered simultaneously (population of solutions) allowing a parallel search through the multidimensional solution space;
- No derivatives are required for optimum search;
- Stochastic elements are used. This, however, does not lead to a random search, but rather to an intelligent search through the solution space. Due to a higher probability of selection and a replication rate of those individuals related to a higher fitness, the search is rapidly concentrated on regions that appear to be successful with respect to the aim of the optimization.

These features lead to the very broad application of the GA strategy even to very complex optimization problems [1].

Experimental Design for Combinatorial and High Throughput
Materials Development, Edited by James N. Cawse.
ISBN 0-471-20343-2 © 2003 John Wiley & Sons, Inc.

The application of genetic algorithms is recommended for systems where only poor empirical and theoretical information on the structure–performance relationships of materials is available. Then, the large number of composition and preparation variables that have to be taken into account leads to a combinatorial explosion of the parameter space. This situation is a serious problem in the development of solid catalytic materials, since a complex interplay between reaction conditions (pressure, concentration of reactants, temperature, reaction time of reactants, etc.), solid-phase properties (bulk and surface structure, electronic state of active sites, density of active sites, porosity, etc.), and catalytic properties (activity and selectivity) must be taken into account when designing and evaluating high throughput (HT) experiments. For such situations, a rigorous combinatorial approach considering all possible combinations of parameters, as well as methods of factorial design of experiments, results in a tremendous experimental effort. Even with high throughput techniques, the required test capacities are difficult to achieve.

Activities in the field of development and application of genetic algorithms encompass a large variety of concepts and approaches. This results from the deficiency in theoretical recommendations on how to construct a proper genetic algorithm and from the variety of problem classes such as engineering design optimization, modeling of nonlinear dynamical systems, optimization of scheduling activities, etc., to which genetic algorithms are applied [1–5]. Thus, the application point of view often dominates and defines GA principles.

The problem of finding the proper setup for a genetic algorithm and its adaptation during the convergence process will appear as a central thread of the following paragraphs, where the emphasis is put on the design of a genetic algorithm for the development of solid catalysts, its testing, and the phenomenological understanding of the algorithm's performance.

For illustration, an example was chosen that reflects the real-life problems of screening for heterogeneous catalysts (the partial oxidation of ethane to acetic acid over mixed-oxide catalysts) and, hence, indicates the challenges related to an efficient search strategy. For readers who are interested in the fundamentals of genetic algorithms and evolutionary strategies, references [1, 2, 6–11] are recommended.

9.2 SETUP OF A GENETIC ALGORITHM TAKING CATALYST FUNCTIONALITY INTO ACCOUNT

The terminology of GAs is inspired by terms of biologic evolution. The most important terms and their explanation are summarized in Table 9.1.

As with the terminology, the basic steps of GAs relate to principles of biological evolution:

1. Starting from an incidental population of individuals or chromosomes, respectively, the fitness of each of the individuals is evaluated.
2. New chromosomes or children are created by mating chromosomes of the current population by applying mutation and crossover. The probability of

Table 9.1 Important Terms Related to Genetic Algorithms

Term	Meaning
Population	Amount of structures (alternative solutions)
Individual	Structure representing the elements of a solution
Fitness or performance	Quality of the solution with respect to the objective of optimization
Generation	Iteration step
Parents	Those individuals that were chosen for reproduction
Descendants, children	Solutions obtained by parents' replication
Chromosome	See individual, usually represented by a string variable
Gene	Digit (binary or other) belonging to a string variable or chromosome, respectively
Allele	Value of a digit
Genotype	Solution encoded by a string variable
Phenotype	Decoded solution
Metagenesis	Process of transition from the current to the next generation

parent individuals to be selected for mating depends on their fitness values. Individuals with high fitness or performance are usually selected more frequently.

3. Members of the population are deleted to make room for the new chromosomes, which are inserted into the population, forming a new generation.

4. The fitness values of the new chromosomes are evaluated.

5. The process cycle starts again with step 2 until optimization is reached.

9.2.1 Encoding of Parameters in Catalyst Development and Optimization

In order to simulate biological evolution based on GAs for the purpose of catalyst discovery, a gene-analogous encoding structure must be established first. Originally, GAs were developed for engineering purposes to solve complex numerical systems with highly correlated parameters [9]. These algorithms were based mostly on binary encoding of the numerical parameters. With such an encoding structure genetic operators like crossover and mutation can proceed as illustrated in Figure 9.1. Crossover therefore includes the exchange of string segments (genes) of two parent individuals, while mutation occurs if a gene that was chosen randomly is altered. In both cases the modification of the genotype (binary string) leads to a change in the phenotype of the individuals (numerical value) as well.

This principle, however, leads to a restriction that is not very helpful in describing the qualitative and quantitative properties of solid catalysts. Encoding catalyst properties must take into account reasonable joint functional units representing typical catalyst component types, such as supports, different active components, and dopants, which, again, can be categorized with respect to their typical functions in a catalytic reaction cycle (redox site or acid/base site, adsorption sites for particular

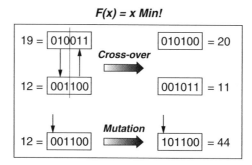

Figure 9.1 Scheme of the basic GA operators based on binary code.

reactant molecules, etc.). These functional units of a catalyst then can be considered as genes of a chromosome. That is, if the binary encoding restriction is abandoned, one is free to set up a modular–hierarchical encoding structure that takes the different functional units of catalytic material into account, with each of these categories in turn consisting of a pool of potential elements or compounds.

A first-encoding structure for solid catalysts was suggest by Wolf et al. [12], which was later extended to a hierarchical structure, as illustrated in Figure 9.2.

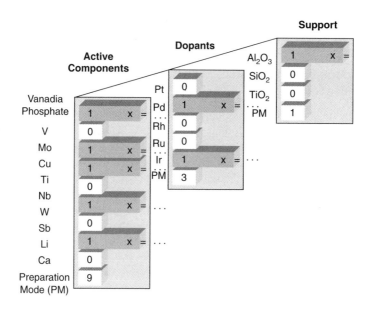

Figure 9.2 Principle of encoding a catalyst composition taking into account different functional types of components, with x describing component amounts (by user´s definition: mass or molar fractions or concentrations).

9.2.2 Genetic Operators

For the encoding structure just presented, genetic operators such as mutation and crossover can be applied separately for each component type, as illustrated in Figure 9.3. For crossover, different modes were suggested in the literature [7, 13, 14], such as on-point, uniform, or multipoint or shuffle crossover related to different numbers of crossover points and modes of exchanging the crossover segments. The question of which setup is the most efficient one is, in principle, related to the present application. A large number of crossover and mutation points leads to a strongly exploratory feature of the evolutionary strategy. In an early state of search this is the desired feature. However, as soon as promising combinations of catalyst genes are found, a large number of such breakpoints rather disturbs and slows down the optimization process. Thus, one must either find a compromise leading to a constant number of breakpoints within the whole search process or a smart way of adapting the number of crossover points to the convergence step. This becomes a setup parameter of the GA process.

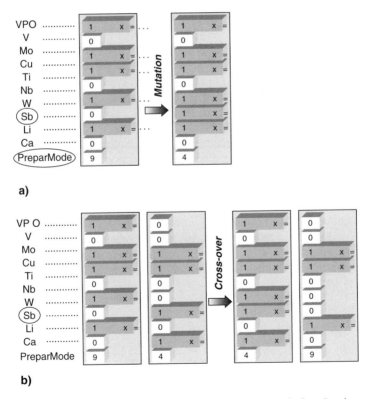

a)

b)

Figure 9.3 Principle scheme of mutation and crossover for a certain functional type of catalyst component. (a) Mutation at the Sb gene location and at preparation-mode location, and (b) crossover at the Sb gene location.

A quantitative mutation operation can be introduced as well. Here, one can contemplate implementation of a whole variety of numerical search strategies, such as a simplex method, bisection, or stochastic methods (random search), and even straightforward gradient methods. In the prototype version, a random search method [12], which can be described by the following formulas, was suggested:

$$\sigma_i = (-1)^t \frac{x_i^n}{d} \tag{9.1a}$$

$$x_i^{n+1} = x_i^n + \sigma_i \tag{9.1b}$$

where σ is the step of a quantitative change; d is the parameter determining the step size; t is the parameter determining the direction of a quantitative change; and x_i^n, x_i^{n+1} are, respectively, the old and new quantitative values with which component i appears in the population.

Quantitative mutation is based on vectors of real numbers where x_i represents the values for concentrations, mass, or molar fraction of the particular components in the catalytic material. The control vector determines the extent and direction of the change in concentration. Quantitative mutation occurs by varying the parameters, x_i, in a certain range according to Eq. (9.1). This range of concentration depends on the type of component (main component, dopant, or carrier) and must be specified by the user. A random integer number (0; 1), t, controls the direction of the concentration change.

Like the number of crossover points, the frequencies W_{cross}, W_{quant}, W_{mut} of the different genetic operators (crossover, quantitative, and qualitative mutation) are setup parameters that should change with the progress of convergence. A suggestion on how to incorporate this feature into the algorithm was made in [12]. This suggestion resulted from the idea that the evolutionary algorithm must include a wide range of material properties, even at an advanced state of optimization. Diversity of materials in the population makes it highly likely that catalytic performance will improve and prevents early local convergence. Since the difference between the mean and best value of catalytic performance in the population (P_{mean}, P_{best}) is a criterion for diversity, a relationship between the probabilities W_{cross}, W_{quant}, W_{mut} and the performance values P_{best} and P_{mean} can be used in a self-adapting sense. We have chosen a relationship between the operators which can be expressed by equations (9.2a)–(9.2c):

$$W_{quant. \, mut} = \frac{B \cdot P_{mean}}{P_{best}} \tag{9.2a}$$

$$W_{qual. \, mut} = A \cdot \frac{P_{best} - B \cdot P_{mean}}{P_{best}} \tag{9.2b}$$

$$W_{cross} = C \cdot \frac{P_{bext} - B \cdot P_{mean}}{P_{best}} \tag{9.2c}$$

where A and B are control parameters in the 0 to 1 range that determine the influence of each of the operators (crossover, quantitative or qualitative mutation) in the optimization process.

9.2.3 Initialization of Test Populations and a Scheme of Metagenesis

After the encoding structure and the setup of the genetic operators have been established, a proper mode of initializing a test population and selecting individuals from this test population in order to modify them by the genetic operators must be defined. If this question is focused on HT screening based on parallel testing of material populations with a constant test capacity, the transition from one material generation to the next during the search and optimization process should, in principle, deal with a constant population size. This usually corresponds to the parallel test capacity. The first generation of materials to be tested in parallel will be generated randomly within the boundaries of the parameter space defined by the user. When creating the subsequent generations of materials, the performances of the preceding generation must be determined and evaluated. Parent individuals will be selected from the test population based on performance information, in order to use the information about their properties (their genetic code) for the genesis of children that represent the next generation. The number of parent reproductions must again correspond to the test capacity. This setup ensures efficient usage of the test capacity, since in each generation only modified materials with new properties are considered for further testing. The mode of transition from a parent to a child generation is illustrated in the following scheme (Figure 9.4).

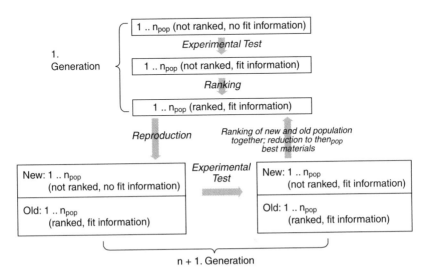

Figure 9.4 Metagenesis scheme for testing of solid materials.

To ascertain that information on the properties of individuals with high performance is, in principle, maintained in the youngest generation, the transition from one generation to the next goes via a transient state where the individuals of both the parent and the child generation are ranked with respect to the catalyst performance. This, of course, supposes that the tests of all individuals have been finished. After ranking, only the best 50% of both generations, corresponding again to the test capacity, are used for further reproduction.

9.2.4   Mode of Selection

The question on how the individuals are selected for the actual reproduction process remains unanswered. A large variety of selection modes have been suggested in the literature [8, 13–16]. The main concepts are illustrated in Figure 9.5.

Roulette selection is a very common variant of selection, wherein the probability for an individual to be selected is proportional to the ratio of its own performance value to the sum of performances of all individuals in one generation. This relationship can be illustrated by the projection of performances onto a roulette wheel. However, there are cases where roulette selection might not be appropriate. In particular, if the performance values differ by only a few percentages for all individuals, the best and the worst materials would have nearly the same probability of selection. If, however, these small differences must be considered as significant for performance discrimination, the respective differences in selection probability should be ascertained as well. This requirement can be fulfilled by the so-called linear ranking selection where the probability of selection is proportional to the rank number and not the performance value of an individual.

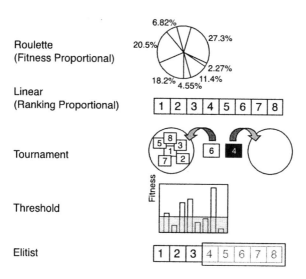

Figure 9.5   Illustration of different selection modes.

Tournament selection is also a selection mode that differentiates the selection probability by the ranking position rather than by the actual performance value. Compared to linear ranking, tournament selection offers a higher probability that worst individuals will be selected, and hence can be considered to be a mode with lower selection pressure. The selection modes that increase the selection pressure, and hence speed up the rate of convergence compared to the three former strategies, are threshold and elitist. These modes restrict the individuals allowed to be selected to the best ones with respect to their performance value (threshold) or rank in the population (elitist). Individuals with high performance are favored for selection to a much greater extent.

Again, the development of a portfolio of selection strategies was motivated by the different application cases. The proper selection strategy depends on the actual problem to be solved, and consequently must be chosen by the user, who should use the preinformation on the scaling of his search problem and on the characteristics of the actual objective function (catalyst activity, selectivity, yield, rate of deactivation, etc.), which also must be carefully chosen. In the prototype GA version for catalyst optimization suggested in [12], the selection mode comprises the following definitions:

- Fitness-proportional selection for quantitative mutation,
- Random selection among the whole population for qualitative mutation,
- Crossover between one catalytic material selected in a fitness-proportional manner and the other catalytic material chosen randomly.

We never can be sure that we have chosen the optimal set of criteria. However, based on model functions representing typical cases of catalyst development, we can learn about the limits of the applied strategy in finding the optimum and about ways to overcome them. This is illustrated in the following section.

9.3 CASE STUDY FOR PHENOMENOLOGICAL UNDERSTANDING OF THE PERFORMANCE OF THE ALGORITHM

In order to obtain a phenomenological understanding of the GA performance for catalytic material optimization, the influence of the setup parameters of a prototype GA on the rate and certainty of convergence are analyzed.

The following performance study was focused on the questions of how a GA is able to learn the functionality of catalyst components in a self-adaptive autonomic way and how this process can be influenced by the setup parameters of the GA. An instructive criterion for performance evaluation can be derived from a plot of the best catalyst performance and the mean catalyst performance vs. number of catalyst generations (Figure 9.6). The distance between both performance values indicates the state of convergence, which is complete as soon as both values are the same. The faster this happens, the faster the GA works. In general, the difference between the mean performance and the best performance is an indicator of the potential for

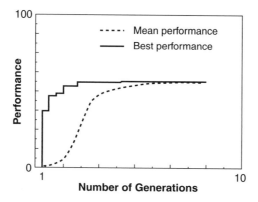

Figure 9.6 Principal scheme of a convergence plot.

innovation that is still hidden in the population. Thus, by this plot, the user receives a continuous feedback as to when the search can be finished.

In order to evaluate the GA performance, one needs either a real experimental database for which the global optimum is known and that can be interpolated by a numerical function (multivariable functions, neural networks, etc.) or a hypothetical test function that reflects the known relationships and problems of catalyst material development. The first approach is probably closer to reality and would reveal the shortcomings and restrictions of GA application much better than hypothetical examples. However, few extensive databases are obtained under the same boundary conditions, or are complete with respect to the relevant search parameters in the open literature. This approach probably will be possible in the near future, since experimental HT techniques will provide these data. Until then, one must make do with hypothetical test functions as used in [12], which should be demanding with respect to the search effort. The test example should result in a large search space of complex topology. For example, several local optima and discontinuities should exist, reflecting the formation of different catalytically active solid phases, whose stability depends on the concentration of compounds and preparation conditions. The example should deal with the search for a catalyst type with complex composition, which includes different functionality of components, such as catalyst supports, active component, and promoters.

For illustration, the search for catalysts that are active and selective in acetic acid production via partial oxidation of ethane was simulated by using a test function that reflects the following relationships known from literature [17]:

- Different mixed oxide systems (V-Mo-Nb, V-Mo-W, VPO) are found to produce acetic acid by partial oxidation of ethane;
- Dopant components (either Re or Pd) significantly improve the selectivity to acetic acid;
- The optimal concentration of the dopant component Pd in the catalyst with

the best performance is very low (10^{-5} mol %). This low amount ensures that the dopant is incorporated in its nonmetallic state into the mixed oxides matrix, which acts as a host structure. Too high a concentration of the dopant metal leads to the formation of segregated metal clusters that act as combustion catalysts, and hence, diminish the selectivity;

- The global optimum of the search space corresponds to the catalyst composition $Mo_1V_{0.25}Nb_{0.12}Pd_{0.0005}O_x$ with an 8.7% yield of acetic acid. Two local optima exist for the mixed oxide systems, V-Mo-Nb and V-Mo-W-Re, leading to a 6.4% yield of acetic acid for standard reaction conditions;

- In order to achieve the proper host structure of the mixed-oxide system, V-Mo-Nb, a special mode of preparation (choice of precursor and calcination procedure) is required.

These boundary conditions were expressed by the following arbitrary expression:

No. mixed oxide systems

$$P_k \sum_{j=1}^{4} \sum_{i=1} A_{i,j}(x_{1,j}^{a_{i,j}} \cdot x_{2,j}^{a_{i,j}} \cdot x_{3,j}^{a_{i,j}} \cdot x_{4,j}^{a_{i,j}}) = Y(\text{AcOH})$$

$$x_{\text{Pd}} > \text{limit} \Rightarrow S(\text{AcOH}) = 0$$
$$P_k = f(\text{Preparation mode}),$$

where $A_{i,j}$, $a_{i,j}$ are adjustable parameters; $x_{1,j} \ldots, x_{4,j}$ are molar fractions of components in a certain mixed oxide system (here, the number of components per catalyst is limited to a maximum number of four).

Since the example includes catalyst components of different functionality, the encoding structure as well as the setup of the first catalyst generation must be carefully chosen. This can be illustrated by the following aspect of our example: Usually, the amount of the main components is in the 1–100% range, and hence covers only one order of magnitude, while the dopant concentrations are not very well known and cover a wide range. Accordingly, during initialization of the catalyst population it is recommended that the main components be linearly distributed, while an exponential distribution covers the search space for dopant components much better. This indicates that the quantitative distribution mode must be adapted to the functional type of the various components, i.e., preknowledge of the catalyst functionality must be considered when starting the GA.

For the search space of the present example (a total of 10 main components, 2 dopant components, 15 modes of preparation), the influence of the population size on the rate and certainty of convergence was studied. The result is shown in Figure 9.7.

The test simulation shows that there is an unstable range up to a population size of about 200. Above the size limit of 200, no further change in the best fitness and number of generations required up to convergence can be observed. This is a much higher critical population size compared to our former study [12], where the stable

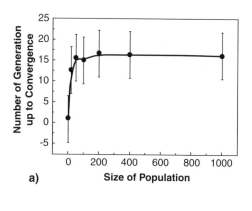

a)

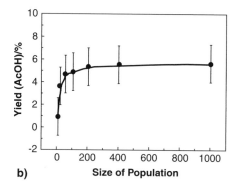

b)

Figure 9.7 (a) Number of generations, and (b) best catalyst performance, depending on the size of the population (catalysts tested in parallel).

range was obtained above 50 catalysts within one generation. This discrepancy results from the different search-space size, which is much larger in the present example due to the larger numbers of potential components and preparation modes compared to those considered in [12]. Hence, it must be concluded that the critical population size is system dependent and has to be preestimated for each new setup of optimization parameters.

The number of crossover points can be studied as being firmly based on test simulations (Figure 9.8). The variation in the number of generations required for convergence increases probably because of the algorithm's capability for exploring the parameter space. Thus, for the present example, the final performance value also rises with the increasing number of crossover points.

Unfortunately, a preliminary conclusion of this analysis was that the simulation used did not find the global optimum (yield of acetic acid = 8.7%) for the present setup of control parameters. When analyzing the reasons for this disappointing result, one must recall that the dopant component, Pd, is removed from the population

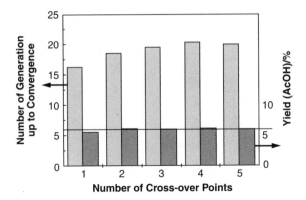

Figure 9.8 Influence of the number of crossover points on rate and certainty of convergence.

at a very early stage in the optimization process, where the optimal host structure (mixed oxide system V-Mo-Nb) was still not found. This is the only combination of elements that allows Pd to be incorporated into the nonmetallic state. In this situation, the GA learns first that Pd is a poison that leads to combustion rather than being a positive promoter. Consequently, catalysts including palladium are related to low fitness values and a low rank position in the first generations. Hence, they are rapidly excluded from reproduction. An approach that might lead to a more successful convergence to the global optimum would be to use different operator frequencies for each component type at different states of convergence. This is especially true for dopant components, because a high frequency of mutation at an advanced state of convergence when the host structure has been found might help to complete the optimal host–dopant combination. Actually, this approach would imitate a human stepwise strategy in catalyst screening. Indeed, if such an approach is simulated for the test example, the global optimum can be found, as indicated in Figure 9.9. For the "dopant mode," a jump of the best fit can be observed after a preliminary convergence to a local optimum. The final level of the best and mean yields of acetic acid at 8.6% corresponds to the optimum catalyst composition, which includes traces of palladium as a dopant component.

The preceding case studies indicated the following aspects:

- An optimal setup of control parameters must be found for each new problem in catalyst development. Using test functions to simulate certain aspects of the special problems of catalyst development is one strategy for estimating optimal control parameters, such as size of population, mode of selection, and relative frequency of genetic operators.

- A much more fundamental understanding of the performance of GAs is required to improve the efficiency of the search process.

- The evolutionary/GA strategy appears to be suitable for an initial explorative

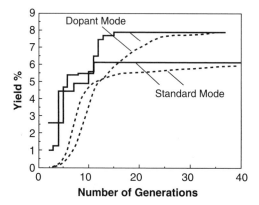

Figure 9.9 Comparison of convergence plots for the standard mode of the GA (equality of all component types with respect to the setup of the frequencies of GA operators), and the dopant mode (increasing probability of mutation of the dopant components with the advanced state of convergence).

search through a large-parameter space and for finding optimal multiple combinations of compounds. However, finding catalysts with complex functionality first requires the conception of the potential role of catalyst components within the multicomponent system.

- The implementation of this knowledge in the search strategy requires a suitable encoding strategy that reflects the functionality of catalyst components and that should present a compatible interface with the database systems that store the information obtained during the search process.

REFERENCES

1. Davis, L., Ed. *Handbook of Genetic Algorithms,* Van Nostrand Reinhold, New York, 1991.

2. Nissen; V. *Evolutionäre Algorithmen;* Deutscher Universitätsverlag, Bamberg, Germany, 1994.

3. Weuster-Botz, D.; Pramatarova, V.; Spassov, G.; Wandrey, C. *J. Chem. Technol. Biotechnol.,* 1995, 64: 386–392.

4. Weuster-Botz, D.; Wadrey, C. *Process Biochem.,*1995, 30: 563–571.

5. Weuster-Botz, D. *J. Biosci. Bioeng.,* 2000, 90: 473–483.

6. Holland, J. H. *Adaption in Natural and Artificial Systems,* The University of Michigan Press, Ann Arbor, 1975.

7. De Jong, K. *An Analysis of the Behaviour of a Class of Genetic Adaptive Systems,* Ph.D. Dissertation, University of Michigan, Ann Arbor, 1975.

8. Goldberg, D. E. *Genetic Algorithms in Search, Optimization and Machine Learning,* Addison-Wesley, Reading, MA, 1975.

9. Schwefel, H.-P. *Numerical Optimisation of Computer Models,* Wiley, Chicester, UK, 1981.

10. Bäck, T.; Schwefel, H.-P. *Evol. Comput.,* 1975, 1: 1–23.

11. Fogel, D. B. *Evolving Artificial Intelligence,* Ph.D. Dissertation, University of California, San Diego, 1992.

12. Wolf, D.; Buyevskaya, O.; Baerns M. *Appl. Catal.,* 2000, 200: 63.

13. Eshelman, L. J.; Caruana, R. A.; Schaffer, J. D. *Proc. 3rd Int. Conf. on Genetic Algorithms,* Schaffer, J. D., Ed., Morgan Kaufmann, San Mateo, CA, 1989, 10–19.

14. Ackley, D. H. *A Connectionist Machine for Genetic Hill Climbing,* Kluwer, Boston, 1987.

15. Grefenstette, J. J.; Baker, J. E. *Proc. 3rd Int. Conf. on Genetic Algorithms,* Schaffer, J. D., Ed., Morgan Kaufmann, San Mateo, CA, 1989, 20–27.

16. de la Maza, M.; Tidor, B. *Proc. 5th Int. Conf. on Genetic Algorithms,* Forrest, S., Ed., Morgan Kaufmann, San Mateo, CA, 1993, 124–131.

17. Linke, D.; Wolf, D.; Baerns, M.; Timpe, O; Schlögl, R.; Dingerdissen, U.; Zeyss; S. *J. Catal.,* 2002, 205: 16–31.

CHAPTER 10

ARTIFICIAL NEURAL NETWORKS IN CATALYST DEVELOPMENT

MARTIN HOLEŇA AND MANFRED BAERNS
Institute for Applied Chemistry Berlin-Adlershof

10.1 INTRODUCTION

Basic steps in the optimization of catalyst composition for a specific reaction have been described in the preceding chapter, which comprised

- Selection of primary elements having potential catalytic properties needed based primarily on prior art and fundamental knowledge;
- Synthesis of a first generation of catalytic materials by systematic or randomized qualitative and quantitative combination of the selected elements, whereby randomization is preferred when the number of elements exceeds 3 to 4;
- Catalytic testing of the first generation of materials;
- Designing subsequent generations by applying an evolutionary approach based on the performance of the preceding generations.

The likelihood of finding the best or close to best composition, i.e., the maximum performance for the catalytic reaction, is rather high. Nevertheless, we cannot ignore the fact that the global optimum in catalyst composition has not been discovered.

An alternative approach is the use of neural networks that approximately describe the relation between input data, i.e., composition of the catalytic material and/or its physical and/or chemical properties as well as catalytic testing conditions, and the output data that are most importantly the catalytic performance, but that also may be the deactivation behavior of the catalytic material. Such networks, the bases and application of which are explained below, may be used if a true relationship exists, at least in principle, by which input (factors) and output (results) data can be correlated.

In a broader sense, both evolutionary strategies and neural network approaches are of a heuristic nature: knowledge is acquired in a logical but nondeterministic

Experimental Design for Combinatorial and High Throughput
Materials Development, Edited by James N. Cawse.
ISBN 0-471-20343-2 © 2003 John Wiley & Sons, Inc.

manner and relationships between input and output variables are of a semiquanitative nature. They may be supported by theoretical concepts that cannot, however, currently be used for designing a real complex catalyst for use in practice. Therefore, the application of heuristics is still and will be crucial in catalyst design. Neural networks also can be used to support the evolutionary process if sufficient data are already available for designing them.

In this chapter the applicability of neural networks for establishing relationships between input data (composition of catalytic materials) and output data (catalytic performance of the materials) is dealt with. The procedure is illustrated by the search for an optimum catalyst composition in the oxidative-dehydrogenation of propane; the catalyst compounds consisted mainly of transition metal oxides. Numerous input and output data were obtained by the previously mentioned evolutionary approach and were used for setting up a suitable neural network.

The area of artificial neural networks is highly interdisciplinary, combining knowledge from computer science, mathematics, and biology. That area, which is beginning to also have its place in catalysis, emerged mainly due to the following influences:

- Mathematical modeling of neurons and neural systems;
- Connectionism, which views large networks of simple elements changing their state through mutual interactions as a crucial condition for the existence of any intelligence;
- Discrepancy between the sequential and algorithmic ways of information processing in traditional computers, and the parallel and associative ways in which information is processed in biological systems, as well as between the digital representation of information used in the former and the analog representation typical for the latter;
- Universal parallel computers and general methods of parallel computation.

Merging all these sources led to the development of artificial neural networks (ANNs) as *distributed computing systems attempting to implement* a greater or smaller part of *the functionality characterizing biological neural networks.*

Basic concepts pertaining to artificial neural networks are presented in Section 10.2. Then it is explained in Section 10.3 how a network approximating an unknown dependency of catalyst performance on its composition and reaction conditions can be constructed, while it is shown in Section 10.4 how knowledge about that dependency can be extracted from the constructed network.

10.2 BASIC CONCEPTS

10.2.1 Neurons and Connections

Undoubtedly the most basic concepts pertaining to artificial neural networks are those of a *neuron,* whose biologically inspired meaning is some elementary signal processing unit, and of a *connection* between neurons that enables the transmission of signals between them. Mathematically, a set of neurons is simply a nonempty finite set V, and each connection between two different neurons u, v in V is an oriented edge between u and v, i.e., either the connection (u,v) from u to v, or the connection (v,u) in the opposite direction. Let C denote the set of all connections between neurons in the ANN considered. Thus in mathematical terms, C is a binary relation on V and (V,C) is an oriented graph. The relation C allows us to define, for each neuron v from V, the set $i(v)$, *input set* of v, as the set of all neurons u from which a connection (u,v) exists in C, and the set $o(v)$, *output set* of v, as the set of all neurons u to which a connection (v,u) exists in C.

Actually, the set of connections cannot be an arbitrary binary relation C on V. Besides the requirement that any connected neurons must be mutually different, which mathematically speaking means the *antireflexivity* of C, the graph (V,C) is usually required to fulfill the following condition, called *condition of nonredundancy*: For each neuron v from V, at least one of the sets $i(v)$ and $o(v)$ is nonempty.

In addition to signal transmission between different neurons, signal transmission between neurons and the environment can also take place. To represent that transmission, it is possible to use an additional element ¤ not belonging to V and an additional set of edges E between V and $\{¤\}$ such that each edge $(¤,v)$ from E represents the fact that a neuron v is able to receive signals from the environment, and each edge $(u,¤)$ represents the fact that a neuron u is able to send signals to the environment. The resulting triplet (V,C,E) is usually called the *architecture* of the ANN.

Also the relation E cannot be fully arbitrary. Usually it is restricted by the following two complementary requirements:

1. Each neuron that sends signals to other neurons must first have received some signal, from one or more other neurons or from the environment;
2. Each neuron that has received signals from other neurons must subsequently send some signal, to one or more other neurons or to the environment.

The architecture (V,C,E) induces two partitions in the set V, one of them corresponding to different possibilities of receiving signals by a neuron:

$I_1 = \{v : v$ receives signals from other neurons, but not from the environment$\}$,

$I_2 = \{v : v$ receives signals from the environment, but not from other neurons$\}$,

$I_3 = \{v : v$ receives signals both from other neurons and from the environment$\}$,

the other corresponding to different possibilities of sending signals:

$O_1 = \{v : v$ sends signals to other neurons, but not to the environment$\}$,

$O_2 = \{v : v$ sends signals to the environment, but not to other neurons$\}$,

$O_3 = \{v : v$ sends signals both to other neurons and to the environment$\}$.

Parts I_2 and O_2 of those partitions can be simply expressed using the input set $i(v)$ and output set $o(v)$ of a neuron v:

$$I_2 = \{v : i(v) \text{ is empty}\} \qquad O_2 = \{v : o(v) \text{ is empty}\}.$$

Moreover, in terms of those two partitions, the condition of nonredundancy can be reformulated as a requirement that I_2 and O_2 are disjoint.

In addition to the requirements on C and E just mentioned, another requirement usually restricts the architecture of an artificial neural network, namely, that I_3 and O_3 are empty. While the former requirements are inspired by real, biological neural networks, the last one has been introduced for purely technical reasons, to make a formal description of the ANN easier. On the other hand, that condition does not actually impose any real restriction on the ANN, in the sense that any architecture in which it does not hold can be extended to an architecture in which it holds, while between any two neurons from the non-extended architecture, a connection exists in the extended architecture if and only if it already exists in the nonextended one.

In the case of a neural network architecture in which I_3 and O_3 are empty, the set I_2 is usually denoted simply as I, and its elements are called *input neurons* of the network. Similarly, the set O_2 is usually denoted as O, and its elements are called *output neurons* of the network. The remaining neurons, those from the set $H = V - I - O$, are called *hidden neurons* (Figure 10.1)

10.2.2   Network Architecture

Although the partition of the set of neurons into input, hidden, and output neurons still allows a wide variety of architectures, the architecture of nearly all kinds of ANNs frequently encountered in practical applications is basically the same, namely a *layered architecture*. That architecture is characterized by the following properties:

- The set of neurons V is partitioned into $L + 1$ layers, $V_0, V_1, \ldots, V_L$, in such a way that $V_0 = I, V_L = O$. Hence, if H is nonempty, then $L > 1$ and the layers $V_1 V_L{-}1$ partition H, they are called *hidden layers*.
- If two neurons u and v are connected, then u belongs to some layer $V_k, k = 1, \ldots, L - 1$, whereas v belongs to V_{k+1}. If in addition all neurons of any layer $V_k, k = 1, \ldots, L - 1$ are connected to all neurons of the following layer, V_{k+1}, then we speak about a *fully connected* layered architecture.

In spite of being actually portioned into $L + 1$ layers, a neural network with such an architecture is conventionally called an *L-layer network* (this is due to the fact that only hidden and output neurons can be assigned so-called somatic mappings, see below). In particular, a *one-layer network* is a layered neural network without hidden neurons, while a *two-layer network* is a neural network in which only connections from input to hidden neurons and from hidden to output neurons are possible.

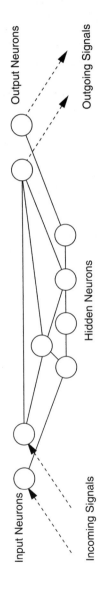

Figure 10.1 A simple generic ANN architecture.

The most prominent example of one-layer networks is the *perceptron,* the earliest representative of ANNs in the sense outlined earlier (proposed by Rumelhart in 1958). Its multilayer extension, the multilayer perceptron (MLP) is the kind of ANN that is nowadays most often encountered in practical applications. The MLP will be described in some detail in the following section. Another very popular kind of multilayer network is the two-layer radial basis function network, which, like the MLP, is used for its ability to approximate very general mappings. Among the one-layer kinds of neural networks, a type called *associative memory* is very useful due to its ability to associate high-dimensional input patterns with low-dimensional output patterns. All these networks are mostly used with a fully connected architecture. A very important example of networks whose architecture is inherently not fully connected is the *self-organizing map* (also called *Kohonen net*), which has a remarkable ability to cluster data, especially in situations when the exact number of clusters is not known in advance. Another example is stacked multilayer perceptrons, which in their output layer linearly combine several fully connected multilayer perceptrons.

Two illustrative multilayer networks with fully connected architectures are depicted in Figure 10.2. The meanings of the labels assigned to the input and output neurons are explained in Section 10.3.

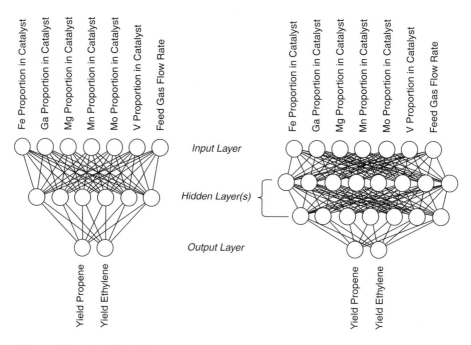

Figure 10.2   An MLP with one layer of hidden neurons (left), and an MLP with two layers of hidden neurons (right).

10.2.3 Neuron Activity

Another very basic concept pertaining to artificial neural networks is the activity of a neuron. This concept serves as a means to describe the propagation of signals from the input neurons to the output neurons. The activity of an input neuron can be thought of as coding, in some appropriate units, the intensity of the signal that the neuron currently receives. Similarly, the activity of an output neuron codes the intensity of the signal the neuron currently sends. Activities of hidden neurons cannot be directly observed, which is why they are called "hidden." In agreement with biological neural networks, the activity of a neuron v is assumed to be time-dependent; thus, it can be in general represented as a real function z_v on some set T representing time. For example, T can be the set of nonnegative reals, the set of nonnegative integers, or some bounded interval of reals or integers. Very often, z_v is required to have a restricted range, typically to the interval $[0,1]$ or $[-1,1]$. The $z(t) = (z_v(t))_{v \in V}$ system of activity values at time t is often called the *state* of the ANN at t.

When studying the propagation of signals from the input neurons to the output neurons, two important aspects must be taken into consideration:

1. Whether the delay of signals due to the final velocity of their transmission must be taken into account. In those kinds of ANNs, in which that delay is considered, it always has a crucial importance (e.g., recurrent neural networks, time-delay neural networks, Hopfield nets, or various kinds of networks based on the adaptive resonance theory). However, such kinds definitely do not belong to those most often encountered in applications. For the purpose of this book, the final signal-transmission velocity can be neglected.

2. Whether the network is in a steady state, in which particular signals received by the input neurons always lead to the same signals sent by the output neurons, or whether it evolves in time, allowing a particular input signal to be answered by different output signals at different times. Although neural networks in steady states are very important (Section 10.4 actually deals with such networks), it is the capability to evolve that is the most distinguishing feature of ANNs, a feature that allows them to adapt to the received signals, as well as to outside feedback. Due to this adaptation aspect, the evolution of neural networks is usually called *learning* or *training*. Although several kinds of ANN learning exist (supervised/unsupervised, discrete/continuous), for the purpose of this chapter, we need to understand only one of them, *supervised discrete learning*. Let us look at that particular kind of learning in some detail.

10.2.4 Learning

In a network without delays, signal transmission is simply a mapping of the received signals to the sent signals, in other words, a mapping of activities of the input neurons to activities of the output neurons. Such a mapping is usually said to be computed by the network. If an ANN has n_I input neurons and n_O output neurons, then each mapping, F, computed by the network maps some previously fixed subset of the n_I-dimensional space into the n_O-dimensional space. However, the network cannot compute an arbitrary mapping of that kind, but only as a mapping, F, that reflects the architecture of the neural network in the sense that:

1. The mapping, F, is gradually composed from a number of simple mappings, each of which is coupled either with a particular hidden or output neuron, and is then called *somatic mapping,* or with a particular connection between neurons, in which case it is called *synaptic mapping* (this terminology is inspired by biological neural networks). Each somatic and each synaptic mapping can be chosen only from some prescribed set of particular simple mappings. Moreover, the same set of simple mappings is typically used for all synaptic mappings composing F, whereas in the case of somatic mappings, differences may eventually exist only between mappings coupled with hidden neurons and those coupled with output neurons. Examples are given in the next section (see also Figures 10.3 and 10.4).

2. The way in which F is composed from somatic and synaptic mappings precisely reflects the structure of the network connections.

If we denote with $\mathscr{F}$ the set of all mappings from n_I-dimensional input space into the n_O-dimensional output space that fulfill restrictions 1 and 2, then learning without delay in networks is nothing other than time evolution of the computed mapping within the set $\mathscr{F}$. In the case of supervised discrete learning, that evolution has several specific features:

- The term *discrete learning* means that the evolution of the computed mapping takes place in a sequence of separate time instants, at which the network computes a sequence of mappings $F^{(0)}, F^{(1)}, F^{(2)}, \ldots$.

- The term *supervised learning* means that an outside feedback is available in the form of a finite number of pairs (x_1, y_1), (x_2, y_2), $\ldots$, (x_p, y_p), called *training pairs, training data, learning data,* and the like, such that for each $j = 1, \ldots, p$, x_j is an n_I-dimensional vector and y_j is an n_O-dimensional vector of activities desired by the output neurons, provided the activities of the input neurons form the vector x_j. Since the activities of the input and output neurons are coding, respectively, the received and sent signals, we can also say that y_j is the vector of signals desired to be sent provided that the vector

Figure 10.3 Synaptic mapping coupled with a connection (u, v) of an MLP.

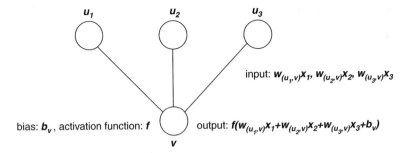

Figure 10.4 Somatic mapping coupled with a hidden neuron v of an MLP.

of signals x_j is received. Recall that the actual activities of the output neurons when the activities of the input neurons are x_j, i.e., the signals really sent when the vector of signals x_j is received, form the vector $F(x_j)$.

- In mathematical terms, the objective of supervised discrete ANN learning is to evolve the sequence $F^{(0)}, F^{(1)}, F^{(2)}, \ldots$ of computed mappings in such a way that for each $j = 1, \ldots, p$, the disagreement between vector y_j (the desired activities) and the vector $F(x_j)$ (the actual activities of the output neurons) decreases. Suppose we can quantify the disagreement between y_j and $F(x_j)$ with the expression $D(y_j, F(x_j))$, where D is a nonnegative function of pairs of n_0-dimensional vectors. Then ideally, the sequence $F^{(0)}, F^{(1)}, F^{(2)}, \ldots$ should evolve according to the condition:

Decrease the sequence of disagreements

$$D(y_j, F^{(0)}(x_j)), \; D(y_j, F^{(1)}(x_j)), \; D(y_j, F^{(2)}(x_j)), \; \ldots \text{ for each } j = 1, \ldots, p.$$

Unfortunately, if the sequence $D(y_j, F^{(0)}(x_j)), \; D(y_j, F^{(1)}(x_j)), \; D(y_j, F^{(2)}(x_j)), \ldots$, for some $j = 1, \ldots, p$, decreases close to its minimum, then necessarily the sequence $D(y_{j'}, F^{(0)}(x_{j'})), \; D(y_{j'}, F^{(1)}(x_{j'})), \; D(y_{j'}, F^{(2)}(x_{j'})), \ldots$ must increase for at least one j' different from j, unless there exists a mapping F in the set $\mathscr{F}$ that exactly fits all the training pairs $(x_1, y_1), (x_2, y_2), \ldots, (x_p, y_p)$. And exactly fitting all available training data is usually connected with irrelevant values $F(x)$ for x different from $x_1, x_2, \ldots, x_p$, i.e., with the inability to capture general dependencies reflected in the training pairs. Examples of that phenomenon, called overtraining, are presented in the next section. That is why instead of the preceding condition, the sequence $F^{(0)}, F^{(1)}, F^{(2)}, \ldots$ is generally evolved according to the following weaker condition: Decrease a sequence of weighted combinations of disagreements

$$c_1 D(y_1, F^{(0)}(x_1)) + c_2 D(y_2, F^{(0)}(x_2)) + \cdots + c_p D((y_p, F^{(0)}((x_p)),$$
$$c_1 D(y_1, F^{(1)}(x_1)) + c_2 D(y_2, F^{(1)}(x_2)) + \cdots + c_p D((y_p, F^{(1)}((x_p)),$$
$$c_1 D(y_1, F^{(2)}(x_1)) + c_2 D(y_2, F^{(2)}(x_2)) + \cdots + c_p D((y_p, F^{(2)}((x_p)), \ldots,$$

where $c_1, c_2, \ldots, c_p$ are positive real constants, indicating a relative importance of the individual training pairs $(x_1, y_1), (x_2, y_2), \ldots, (x_p, y_p)$, respectively, for the evolution of the sequence $F^{(0)}, F^{(1)}, F^{(2)}, \ldots$. In particular, if all training pairs are considered equally important, then that condition takes the form: Decrease the sequence of summed disagreements

$$D(y_1, F^{(0)}(x_1)) + \cdots + D((y_p, F^{(0)}((x_p)), D(y_1, F^{(1)}(x_1)) + \cdots$$
$$+ D((y_p, F^{(1)}((x_p)), D(y_1, F^{(2)}(x_1)) + \cdots + D((y_p, F^{(2)}((x_p)), \ldots.$$

- Learning ends in some step s, i.e., after the evolution through a sequence of computed mappings $F^{(0)}, F^{(1)}, \ldots, F^{(s)}$, either due to fulfilling some prescribed stopping criterion, or through a deliberate break from outside. A stopping criterion is typically some condition on:

  - The step s;

  - The considered weighted combination of disagreements

    $$c_1 D(y_1, F^{(s)}(x_1)) + c_2 D(y_2, F^{(s)}(x_2)) + \cdots + c_p D((y_p, F^{(s)}((x_p));$$

  - The difference of such combinations between the current and the previous step,

    $$c_1[D(y_1, F^{(s-1)}(x_1)) - D(y_1, F^{(s)}(x_1))] + c_2[D(y_2, F^{(s-1)}(x_2)) - D(y_2, F^{(s)}(x_2))] +$$
    $$\cdots + c_p[D((y_p, F^{(s-1)}((x_p)) - D((y_p, F^{(s)}((x_p))];$$

  - The gradient of the weighted combination of disagreements considered, with respect to parameters that parametrize the computed mappings

    $$c_1 \nabla D(y_1, F^{(s)}(x_1)) + \cdots + c_2 \nabla D(y_2, F^{(s)}(x_2)) + \cdots + c_p \nabla D(y_p, F^{(s)}(x_p)),$$

    provided the D is differentiable with respect to parameters;

  - The distance between mappings computed in those steps $dist(F^{(s-1)}, F^{(s)})$, measured with some convenient metrics $dist$;

or a combination of such conditions.

10.2.5 Disagreement Functions

The preceding description of learning indicates the crucial importance of the concept of disagreement between vectors in the output space, in particular between the vector $F(x_j)$, $j = 1, \ldots, p$, of the actual activities of the output neurons and the vector y_j of their desired activities. It is this concept, captured by means of a nonnegative disagreement function D on pairs of vectors in the n_O-dimensional output space, with which we conclude this section. Quite a large number of such functions has already been proposed, for the practical reason that most of them are applicable only in specific situations. Nevertheless, we begin with two functions that are universally applicable. The most frequently encountered disagreement function is based simply on the usual (i.e., Euclidean) distance in the output space, i.e., on the distance induced by the Euclidean norm $\| \; \|$. From a computational point of view, however, more advantageous than the Euclidean norm itself is its square, since it can be quickly computed as the sum of squares of individual components. Therefore, this most common disagreement function is called sum of squared errors (SSE):

$$(\forall \, \hat{y}, y \in \mathfrak{R}^{n_O}) \; \text{SSE}(\hat{y}, y) = \| \hat{y} - y \|^2 = \sum_{i=1}^{n_O} (\hat{y}_i - y_i)^2,$$

where $(y)_i$ and $(\hat{y})_i$ for $i = 1, \ldots, n_O$ denote the ith component of the vectors y and $\hat{y}$, respectively. Recall that if all training pairs $(x_1, y_1), (x_2, y_2), \ldots, (x_p, y_p)$ used for supervised learning are considered equally important, then the condition determining how the sequence $F^{(0)}, F^{(1)}, F^{(2)}, \ldots$ evolves is the decrease in the sequence of summed disagreements. That sequence normalized with the number p of training pairs is, in the case of the disagreement function SSE, called mean sum of squared errors (MSE) of p pairs of vectors in the n_O-dimensional output space:

$$(\forall \, \hat{y}_1, \ldots, \hat{y}_p, y_1, \ldots, y_p \in \mathfrak{R}^{n_O}) \; \text{MSE}((\hat{y}_1, y_1), \ldots, (\hat{y}_p, y_p))$$

$$= \frac{1}{p} \sum_{j=1}^{p} \| \hat{y}_j - y_j \|^2 = \frac{1}{p} \sum_{i=1}^{n_O} \sum_{j=1}^{p} ((\hat{y}_j)_i - (y_j)_i)^2.$$

Needless to say, the usual Euclidean norm is not the only norm in the n_O-dimensional output space. In particular, replacing the squares in the definition of SSE with a general qth power for q of at least 1 leads to the disagreement function sum of qth powers of errors (SPE):

$$(\forall \, \hat{y}, y \in \mathfrak{R}^{n_O}) \; \text{SPE}(\hat{y}, y) = \sum_{i=1}^{n_O} ((\hat{y})_i - (y)_i)^q,$$

which is, like the disagreement function, SSE, the qth power of a norm on the n_O-dimensional output space. If the activities of the output neurons can be interpreted as a probability distribution on some finite set, which is usually the case if the network is used to classify inputs into n_O classes, then in addition to the sum of squares, the information–theoretic disagreement function, relative entropy (RE), is frequently encountered:

$$(\forall \, \hat{y}, y \in \Pi(n_O) \cap (0,1)^{n_O}) \; \text{RE}(\hat{y}, y) = \sum_{i=1}^{n_O} \left[y_i \log \frac{y_i}{\hat{y}_i} + (1 - y_i) \log \frac{(1 - y_i)}{(1 - \hat{y}_i)} \right],$$

where $\Pi(n_O)$ denotes the set of all discrete probability distributions on n_O points.

Moreover, a separate disagreement function is available for the most often occurring situation, i.e., that the desired probability distribution y is degenerated, or equivalently that the network input should be classified into exactly one class—a function called classification figure of merit (CFM):

$$(\forall \hat{y} \in \Pi(n_O))(\forall y \in \Pi(n_O) \cap \{0,1\}^{n_O}) \; \mathrm{CFM}(\hat{y},y) = \frac{\alpha}{n_O - 1} \sum_{\substack{i=1 \\ i \neq i_1}}^{n_O} \frac{1}{\beta + e^{-\gamma(\hat{y})_{i_1} - (\hat{y})_i)^2}},$$

where i_1 denotes the component of y that equals 1, and α, β, γ are positive constants. Finally, in the case of a network with only one output neuron and with the desired activities assuming only two opposite values, say -1 and 1 (for example, if the network is required to make some yes/no decision), rapid learning is possible using the zero-one disagreement function (ZOD):

$$(\forall \hat{y} \in \Re)(\forall y \in \{-1,1\}) \; \mathrm{ZOD}(\hat{y},y) = \max\{0, -y \; \mathrm{sign} \; \hat{y}\}.$$

Readers who would like to learn more about the concepts presented and/or the aspects of the mathematical background of artificial neural networks not covered here should consult some theory-oriented ANN monographs, such as [1–4].

In the following two sections, we demonstrate the usefulness of those concepts on two applications in combinatorial catalysis: approximation of unknown dependencies, and extraction of knowledge contained in experimental data. Readers interested in other applications of artificial neural networks in chemistry are referred to the overview books [5,6], the overview papers [7–9], and relevant parts of the collections [10–12].

10.3 APPROXIMATION OF UNKNOWN FUNCTIONS WITH MULTILAYER PERCEPTRONS

10.3.1 Properties of the Multilayer Perceptron

As already mentioned in the preceding section, MLPs are the ANNs most often used in practical applications. Besides its multilayer architecture with at least one hidden layer, an MLP is distinguished by the following features:

1. Its synaptic mappings are constant multiplications, i.e., the synaptic mapping coupled with any connection (u,v) is the multiplication of the input value in u with some weight $w_{(u,v)}$ assigned to (u,v) (Figure 10.3).
2. A somatic mapping coupled with a hidden neuron (Figure 10.4) is chosen among mappings that can be composed of:
   (a) The summation of the results of the synaptic operations coupled with the incoming connections, and of some constant b, called *bias*;
   (b) A nonlinear measurable function of one variable, applied to the result of that summation and called *activation function*; the same activation function f is usually used in all somatic mappings coupled with hidden neurons.

3. A somatic mapping coupled with an output neuron is either composed in the same way as in the case of hidden neurons, or it is chosen among summations like feature 2(a).

4. Usually, there is a restriction on the nonlinearity of the activation function, f, most often the restriction that f should be bounded both from below and from above, typically through the bounds 0 and 1, or -1 and 1. Functions with such a nonlinearity are called *sigmoidal functions*. Their most common examples among continuous functions are **arctg**, typically in the modifications $2 \cdot$ **arctg$/\pi$** or **arctg$/\pi + 1$**, the *logistic activation function*,

$$f(x) = \frac{1}{1 + e^{-x}},$$

and, among noncontinuous functions, the **step function**

$$f(x) = \begin{cases} 1 & \text{if } x \geq 0 \\ 0 & \text{if } x < 0. \end{cases}$$

In the sequel, the architecture of an MLP with n_I input neurons, one hidden layer of n_H neurons and n_O output neurons will be denoted n_I–n_H–n_O. Taking into account features 1–3, an MLP with the architecture n_I–n_H–n_O whose somatic mappings coupled with the hidden neurons include an activation function, f, whereas somatic mappings coupled with the output neurons do not include any activation function, computes a mapping, F, with the following components F_i, $i = 1, \ldots, n_O$:

$$F_i(x) = w_{2,i}^{\mathrm{T}}\big(f(w_{1,1}^{\mathrm{T}}x + b_{1,1}), \ldots, f(w_{1,n_H}^{\mathrm{T}}x + b_{1,n_H})\big) + b_{2,i},$$

where $w_{1,1}, \ldots, w_{1,n_H}$ are n_I-dimensional weight vectors, $w_{2,i}$ is an n_H-dimensional weight vector, and $b_{1,1}, \ldots, b_{1,n_H}, b_{2,i}$ are scalar biases. Similarly, the architecture of an MLP with n_I input neurons, n_1 neurons in the first hidden layer, n_2 neurons in the second hidden layer, and n_O output neurons will be denoted n_I–n_1–n_2–n_O. An MLP with the architecture n_I–n_1–n_2–n_O, whose somatic mappings coupled with the hidden neurons include an activation function f, while somatic mappings coupled with the output neurons do not include any activation function, computes a mapping F with the components F_i, $i = 1, \ldots, n_O$:

$$F_i(x) = w_{3,i}^{\mathrm{T}}\left(\begin{array}{l} f(w_{2,1}^{\mathrm{T}}(f(w_{1,1}^{\mathrm{T}}x + b_{1,1}), \ldots, f(w_{1,n_1}^{\mathrm{T}}x + b_{1,n_1})) + b_{2,1}, \ldots, \\ f(w_{2,n_2}^{\mathrm{T}}(f(w_{1,1}^{\mathrm{T}}x + b_{1,1}), \ldots, f(w_{1,n_1}^{\mathrm{T}}x + b_{1,n_1})) + b_{2,n_2}) \end{array} \right) + b_{3,i},$$

where $w_{1,1}, \ldots, w_{1,n_1}$ are n_I-dimensional weight vectors, $w_{2,1}, \ldots, w_{2,n_2}$ are n_1-dimensional weight vectors, $w_{3,i}$ is an n_2-dimensional weight vector, and $b_{1,1}, \ldots, b_{1,n_1}, b_{2,1}, \ldots, b_{2,n_2}, b_{3,i}$ are scalar biases.

10.3.2 Approximation Properties of MLPs

The usefulness of MLPs lies in their ability to approximate very general mappings. Numerous results concerning the universal approximation property of MLPs have been published (see, e.g., [13–20]. Here, we formulate that property only for two particular classes of functions, which, however, presumably cover all functions commonly encountered in catalysis.

1. Let X be a bounded compact subset of the space of n_I-dimensional real vectors (i.e., a subset that is bounded and with each sequence of contained points also contains all limit points of that sequence), G be a general continuous mapping of X into the space of n_O-dimensional real vectors, and $d > 0$ be the prescribed distance limit. Then there exists a number n_H and an MLP with one hidden layer of n_H neurons, a continuous activation function, and the property that each component F_i, $i = 1, \ldots, n_O$, of the mapping F computed by that MLP approximates the corresponding component G_i of G within the distance d in the space of continuous functions on X. Hence,

$$\max_{x \in X} |F_i(x) - G_i(x)| < d.$$

This result is practically important in the sense that it is believed to cover all continuous functions commonly encountered in catalysis.

2. Let X be a measurable subset of the space of n_I-dimensional real vectors (X may even be the whole space), μ be a finite nonnegative measure on X (for example, a probability measure), and $q \geq 1$. Further, let G be some general mapping of X into the space of n_O-dimensional real vectors such that for each its component G_i, $i = 1, \ldots, n_O$, $|G_i|^q$ is integrable with respect to μ, and $d > 0$ again be a prescribed distance limit. Then there exists a number n_H and an MLP with one hidden layer of n_H neurons such that each component F_i, $i = 1, \ldots, n_O$, of the mapping F computed by that MLP approximates the corresponding G_i within the distance d in the space $L_q(\mu)$ of functions on X with μ-integrable qth power of the absolute value. Hence,

$$\left(\int_X |F_i(x) - G_i(x)|^q \, d\mu \right)^{1/q} < d.$$

This result is, in turn, practically important in the sense that it is believed to cover all measurable functions commonly encountered in catalysis (both those functions that are continuous and those that are not).

From a mathematical point of view, the latter approximation result is actually a generalization of the former. Indeed, if G is a continuous mapping on a compact subset of a n_O-dimensional space, then each component G_i, $i = 1, \ldots, n_O$, is $|G_i|^q$ integrable for any nonnegative q with respect to any finite nonnegative measure μ. Among the possible combinations of q and μ, there is one we consider so important that we now explicitly reformulate the approximation result for this case. It is the case where $q = 2$ and μ are the discrete empirical probability measure given by the first components of the training pairs $(x_1,y_1),(x_2,y_2), \ldots , (x_p,y_p)$. Since that measure assigns to each n_1-dimensional vector x the probability

$$\mu(x) = \frac{1}{p}\mathbf{card}\{j|1 \leq j \leq p \text{ and } x = x_j\},$$

where **card** denotes the number of elements in a set (in mathematical terms, its cardinality), this setting turns the general approximation result into a simple formulation:

$$\text{MSE}((F_i(x_1),G_i(x_1)), \ldots , (F_i(x_p),G_i(x_p))) = \frac{1}{p}\sum_{j=1}^{p}\text{SSE}(F_i(x_j),G_i(x_j)) < d.$$

It is this formulation of the universal approximation property that motivates the prevailing use of the SSEs as the disagreement function in MLP training.

10.3.3 Use of MLPs in Catalysis Research

In connection with genetic algorithms, we already have mentioned the crucial role played in catalysis by functions that describe the dependency of product yields, catalyst activity, reactant conversions, and product selectivities on catalyst composition, as well as on other catalyst properties and reaction conditions. Therefore, it is not surprising that the MLP-based approximation has been used in catalysis most often for this kind of function. Probably the first who reported such an application of MLPs, to data on oxidative dehydrogenation of ethylbenzene and on oxidation of butane, were Hattori and Kito [21] and Kito et al. [22], who already had an earlier experience with using them for estimating the acid strength of mixed oxides [23,24]. To mention several other examples: Sasaki et al. applied MLPs to data on decomposition of NO into N_2 and O_2 [25], Hou et al. applied them to data on acrylonitrile synthesis via propane [26,27], Sharma et al. applied them to Fisher–Tropsch synthesis data [28], and Huang et al. applied them to data on oxidative coupling of methane [29].

Here we present a similar case study based on our own experimental data on 211 catalysts of different compositions for the oxidative dehydrogenation of propane to propene. The best of these catalysts have been recently written about [30–32], so, in accordance with the purpose of this book, we will concentrate on methodological issues. We used the commercial data mining system Clementine [33] to preprocess and select those data.

All of the catalysts considered consist of elements taken from the pool, Fe, Ga, Mg, Mn, Mo, V, and share certain properties as well as certain reaction conditions in which they were used; see Table 10.1. As to properties and reaction conditions that vary between individual catalysts, they can be divided into two groups:

Table 10.1 Shared Properties and Reaction Conditions of the Considered Catalysts for the Oxidative Dehydrogenation of Propane

Property/Reaction Condition	Shared Value
Material of support	α-Al_2O_3
Diameter of support particles	1 mm
Mass fraction of support	70%
Pool of constituting elements	Fe, Ga, Mg, Mn, Mo, V
Reaction temperature	500°C
Proportions of feed gas components $C_3H_8/O_2/N_2$	3/1/6
Reaction pressure	1 bar

1. Those that can be in the context of catalyst testing are viewed together as independent variables. In the available data, there are seven of them, and all are listed in Table 10.2. We emphasize that viewing them all together as independent variables does not exclude the existence of dependences between them. Indeed, each element proportion in Table 10.2 depends on all the remaining element proportions due to the restriction of summability of all element proportions to 100%. In terms of the encoding of those independent variables that is introduced in Table 10.2, that restriction can be expressed as

$$(x)_1 + \cdots + (x)_6 = 1.$$

2. Catalyst characteristics that are supposed to depend on other characteristics and on reaction conditions. In our experiments, the following properties of that kind have been measured:

   (a) The *selectivities* of propene, ethylene, carbon monoxide, and carbon dioxide,

   from which two additional have been derived:

   (b) The *conversions* of propane and oxygen;

   (c) The *yields* of propene and ethylene.

Table 10.2 Independent Variables Individually Varied Among the Considered Catalysts for the Oxidative Dehydrogenation of Propane

Variable	Minimal Domain[a]	Encoded As
Fe proportion in catalyst [mol %]	0–69	$(x)_1$
Ga proportion in catalyst [mol %]	0–84	$(x)_2$
Mg proportion in catalyst [mol %]	0–100	$(x)_3$
Mn proportion in catalyst [mol %]	0–64	$(x)_4$
Mo proportion in catalyst [mol %]	0–48	$(x)_5$
V proportion in catalyst [mol %]	0–78	$(x)_6$
Feed-gas flow rate [mL/$g_{cat} \cdot$ h]	10–150	$(x)_7$

[a]According to data.

Among the dependent properties, we are most interested in propene and ethylene yields. That is why we have chosen their dependency from the characteristics listed in Table 10.2 to constitute the mapping to be approximated by an MLP. Consequently, the perceptron is required to compute a mapping F from the space of seven-dimensional real vectors to the space of two-dimensional real vectors, such that:

- The domain of F covers the product of minimal domains of all independent variables (see Table 10.2);
- For each $x = ((x)_1, \ldots, (x)_7)$ from the domain of F, whose individual components $(x)_1, \ldots, (x)_7$ encode the individual independent variables according to Table 10.2, the corresponding function value of F is

$$F(x) = (F_1(x), F_2(x)) = ((y)_1, (y)_2),$$

where
- $(y)_1$ encodes the yield of propene;
- $(y)_2$ encodes the yield of ethylene.

10.3.4 Design Decisions

The fact that our MLP has to compute a mapping from a seven-dimensional space to a two-dimensional space implies that it needs to have seven input neurons and two output neurons. The following design decisions remain to be selected:

1. A particular number n_H of hidden neurons;
2. A particular activation function;
3. A particular disagreement function;
4. A particular training method, i.e., a particular method of evolving the sequence $F^{(0)}, F^{(1)}, F^{(2)}, \ldots$ of computed mappings according to the principles described in the preceding section.

As for the disagreement function, the approximation results just recalled suggest that it is advantageous to choose the mean-square error. Also the choice of an activation function is not a serious problem, since no assumptions about the approximation were made in the approximation results just given, i.e., they hold independently of a particular choice. We have chosen the logistic activation function, for the following reasons:

- Its range is the interval [0,1], which is exactly the same as the range of the yields;
- We use a training method that requires the MSE to be at least twice differentiable with respect to weights and biases, so the activation function must be an at least twice differentiable function of one variable. We will return to that point later, when we discuss training methods.

On the surface, the approximation results seem to suggest that it also will be easy to choose an appropriate number of hidden neurons; since they guarantee a sufficiently close approximation from a certain number of hidden neurons upwards, we could simply decide to chose that number large enough. Or even better, we could choose two hidden layers with a sufficiently large number of neurons in each of them. Such a decision, however, would completely neglect the fact that the information on which network learning will be based does not cover the complete dependency of the propene and ethylene yields on the independent variables from Table 10.2 (if we had such information, then we actually would not need any approximation), but is restricted only to the training pairs $(x_1,y_1),(x_2,y_2), \ldots, (x_p,y_p)$. That restriction includes two important dangers:

1. The computed mapping F, resulting from network learning, can precisely fit the training pairs, although they reflect not only the sought dependency, but also noise due to random influences, such as measurement errors.

2. In addition to precisely fitting the training pairs $(x_1,y_1),(x_2,y_2), \ldots, (x_p,y_p)$, the computed mapping F can assume totally irrelevant values $F(x)$ for x different from $x_1, x_2, \ldots, x_p$. Thus instead of approximating the desired dependency, F can actually approximate a mapping that only partially describes what the training pairs used for its learning were, for example,

$$F_i(x) = \begin{cases} (y_j)_i\left(1 - \dfrac{|x - x_j|}{d}\right) & \text{if } |x - x_j| < d \quad \text{for some } j = 1, \ldots, 211, \\ 0 & \text{otherwise, } i = 1,2 \end{cases}$$

where $d > 0$ is sufficiently small. The phenomenon that a neural network learns to approximate only a description of the training pairs instead of generalizing to the unknown dependency that the training pairs reflect is called *overtraining, overlearning,* or *overfitting*. Overtraining is possible because of a property of the measure μ underlying the disagreement function MSE, namely, that μ is nonzero only for those sets in the input space of the neural network that contain any of the values $x_1, x_2, \ldots, x_p$, but is zero for all remaining sets in the input space.

10.3.5   Overtraining

Obviously, to approximate a function like the one in the preceding section, a large number of free parameters is needed, i.e., of weights and biases. Recalling from before that the number of weights and biases is proportional to the number of hidden neurons, we see that to avoid the high risk of overtraining, we have to avoid too large a number of hidden neurons. This is very helpful to realize when searching an appropriate number of hidden neurons: Although we do not know in advance which number of hidden neurons is too high for our data and for the dependency we want to approximate, we can stop increasing the number of hidden neurons in a particular layer if several recent increases resulted in higher and higher overtraining.

Two questions immediately emerge: How do we recognize that overtraining occurs? How do we recognize that it is increasing? Suppose that in addition to the pairs (independent variables from Table 10.2, propene and ethylene yields) used for training, we have other such pairs available as test data. In other words, suppose we have two separate sets of catalysts, one for training the network and one for testing it. It is immaterial whether the test catalysts were already available during network training or whether they were obtained only later through additional catalyst experiments. Provided that the dependency of the yields on the independent variables is the same for both sets, all differences between the MSE value for the training and test data are due to noise and overtraining. If, in addition, we assume that the noise distribution is the same, we can draw the following conclusions from the MSE values obtained for training and test catalysts:

- An MSE value for test data lower than, equal to, or slightly higher than that for the training data can be explained through noise and indicates absence of overtraining;
- An MSE value for test data substantially higher than that for the training data indicates overtraining;
- If the MSE value for the training data during training decreases while the MSE value for the test data simultaneously increases, then overtraining increases;
- In order to compare different MLPs with respect to their ability to approximate the desired dependency of the yields on the independent variables, we have to base our comparison on the MSE value for the test data.

10.3.6 Validation

The sets of training and test catalysts can be most easily obtained by splitting the set of all available catalysts. That method is, however, quite sensitive to the assumptions of the equal dependency of the yields on the independent variables and of equally distributed noise for both sets, even in the case of a random split. If we want to avoid that sensitivity, we can use the following more sophisticated method, called *k-fold cross validation*:

- Divide the set of all available catalysts randomly into *k* parts of approximately equal size.
- Train *k* neural nets with the same architecture, using one specific part of the obtained partition as test catalysts for each of them and all the remaining *k*–1 parts as training catalysts.
- To assess the appropriateness of the considered architecture with respect to its ability to approximate the desired dependency of the yields on the independent variables, average the MSE values for the test data over those *k* trained neural nets.

Observe that in this method the proportion of catalysts used for training is $1-1/k$. Thus the higher k is, the more knowledge about the available catalysts has been used for training the neural network, and is thus inherently incorporated in it. Consequently, the knowledge-richest networks result from a k-fold cross-validation, with k equal to the number of available catalysts, or, in our case study, from a 211-fold cross validation. Since then only one catalyst is left out for testing, this extreme case of a cross-validation is also called leave-1-out validation.

Leave-1-out validation and general cross-validation are excellent methods for comparing different architectures irrespective of overtraining, but they cannot be used in the subsequent phase, when the final MLPs should be trained with the most promising architecture or with each of several promising architectures. At that point, our objective is not merely to disregard overtraining for the purpose of comparison, but to train the network in such a way that overtraining either does not occur or is at least as suppressed as possible. Therefore, we now turn to the last point on our list of design decisions to be selected—the training methods.

10.3.7   Training Methods

Our choice of the MSE disagreement function implies that a training method has to develop the sequence of mappings $F^{(0)}$, $F^{(1)}$, $F^{(2)}$, . . . in such a way that the MSE of $(x_1, y_1), (x_2, y_2), \ldots, (x_p, y_p)$ gets sufficiently small. To make an MSE value as small as possible, the mean squared error has to be minimized with respect to the computed mappings, i.e., with respect to the parameters that determine those mappings, i.e., weights and biases. As we saw earlier, that also means minimizing the sum of squared errors summed over all training pairs. To this end, just about any minimization method can be employed. For a detailed treatment of such methods and of the choice of an appropriate method for a given task, the reader is referred to textbooks on function optimization (e.g., [34–37]). We recall here only the five most basic principles:

1. Existing methods for minimizing functions use three kinds of information: about function values, about gradient, which is opposite to the steepest descent of the function, and about second partial derivatives, which indicate the bending of the function. Kinds two and three can be either analytically calculated or numerically approximated.

2. Information about gradient or second derivatives always directs the method toward a local minimum, or more precisely, toward the local minimum whose attraction area covers the vector of parameters of the starting mapping $F^{(0)}$. Hence, global minima are sought merely by methods that exclusively use function values.

3. Methods using information about gradient are much faster than those using only function values. Even for moderate optimization tasks, the difference between both kinds of methods as to the time needed to get sufficiently close to the desired minimum is typically at least one order of magnitude.

4. Methods relying on second derivatives are even faster than gradient methods, but only when close to the desired minimum, whereas when distant from the minimum, gradient methods are superior to them. Consequently, the fastest minimization methods are those that combine both approaches, relying on second derivatives for proximity to the minimum and on the gradient when far from it.

5. Minimizing a sum of squares is a standard optimization task (called *least-squares optimization*), for which specific variants of various methods have been developed, and with which a large amount of practical experience has been gathered. According to that experience, various variants of the Levenberg–Marquardt method appear to be the most successful in solving this task. One of the earliest combined methods of the kind mentioned [38,39], it switches between the classic Gauss–Newton method based on second derivatives and the steepest descent.

As far as ANNs are concerned, the fact that the search for the most appropriate architecture with the cross-validation or leave-1-out methods typically includes training of thousands of individual networks, i.e., performing thousands of minimizations, makes the speed of a method the most critical choice criterion, thereby excluding methods that use only function values. In the early years of the popularity of MLPs, a steepest-descent variant was most often used. This variant was developed by Rumelhart and co-workers [40] and called back-propagation, due to the stepwise way the components of the gradient are calculated. Modern implementations of MLPs usually also include more sophisticated minimization methods for network training, such as the Levenberg–Marquardt method. The Matlab Neural Network Toolbox, which we have used, implements 14 methods for training MLPs, including 5 variants of back-propagation and 3 variants of the Levenberg–Marquardt method [41]. We started the search for an appropriate MLP architecture with the basic Levenberg–Marquardt method.

10.3.8 Search for the Appropriate MLP

Levenberg–Marquadt Method First, we decided to perform a 10-fold cross-validation for the following architectures:

- All two-layer perceptrons with 1–20 hidden neurons;
- All three-layer perceptrons with 1–20 neurons in the first hidden layer and a number of neurons in the second hidden layer not exceeding that in the first hidden layer (such a heuristic restriction is called *pyramidal restriction*).

Thus, our initial choice included 230 architectures, for each of which 10 individual MLPs had to be trained. Each individual network training was iterated 100 steps long, which made the evolved sequence of computable mappings $F^{(0)}, F^{(1)}, \ldots, F^{(100)}$. The initial mapping, $F^{(0)}$, was obtained through a random initialization of weights and biases.

Results obtained with the basic Levenberg–Marquardt method for two-layer perceptrons are shown in Figure 10.5. To put those results into an appropriate context, we have to complement them with the information that the yields are encoded in %. Then a simple computation shows that the MSE value 1.1 (which is approximately the best value obtained for test catalysts) corresponds to a 0.74% difference between the desired and computed values of both the propene yield and the ethylene yield, or to a 1% difference between the desired and computed values of the propene yield, and a 0.32% difference between the corresponding values of the ethylene yield. Considerable overtraining, which increases with the number of neurons, is apparent in Figure 10.5. This overtraining indicates that our original selection of investigated architectures was sufficiently broad and that we do not need to consider two-layer perceptrons with more hidden neurons. The situation with three-layer perceptrons was similar. We are not going to present complete results here, but we will show overview results for test catalysts later (Figure 10.13).

To provide more insight into the behavior of the basic Levenberg–Marquardt method (as well as of many other training methods, including back-propagation), Figures 10.6 and 10.7 show the time evolution of the MSE values for two individual networks, i.e., their evolution over the 100 training iterations. Although both networks overtrained, the overtraining in Figure 10.6 is approximately constant from the beginning of training, whereas the overtraining in Figure 10.7 occurs only after more than 15 iterations, after which it rapidly increases. Figure 10.7 also immediately suggests how the overtraining in the latter case could be avoided: to stop training the network as soon as overtraining is detected and to use as the final weights and biases those from the last iteration before overtraining starts to increase. This approach, called *early stopping,* is indeed often used to suppress network overtrain-

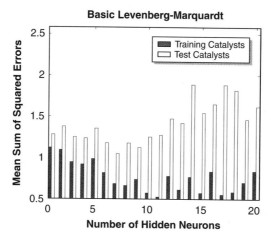

Figure 10.5 Training two-layer perceptrons with the basic Levenberg–Marquardt method. The MSE values depicted have been averaged—separately for training and test catalysts— over the 10 folds of a 10-fold cross validation.

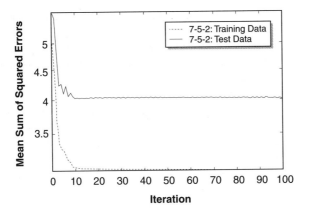

Figure 10.6 Time development of the MSE values during an example MLP training using the basic Levenberg–Marquardt method. The fact that the MSE value for the test data is approximately 25% higher than the MSE value for the training data after the early iterations indicates constant overtraining from the beginning.

ing. However, if test data are used to detect overtraining, then they cannot serve any longer as true test data, because through the stopping condition, they have already been involved in training. Consequently, we need two separate sets of test data for early stopping. Those for overtraining detection are usually called *validation data,* while the test data proper are used for comparing different networks and different architectures. In our implementation, validation catalysts were obtained through randomly splitting off 10% of the training catalysts. An additional stopping condition, based on the validation data, was formulated as follows: Stop if the difference of the MSE value for the validation data between the current and the previous itera-

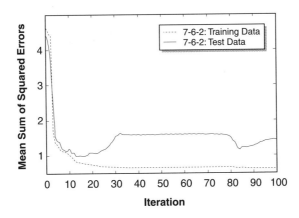

Figure 10.7 Time development of the MSE values during an example MLP training with the basic Levenberg–Marquardt method. Shows an increasing overtraining with a later onset.

tion has been positive for the last 10 iterations. Whenever this condition caused training to stop after producing a sequence of computable mappings $F^{(0)}, F^{(1)}, \ldots,$ $F^{(s)}$, the weights and biases determining $F^{(s-10)}$ were taken as the final weights and biases of the trained network, i.e., $F^{(s-10)}$ became its final computed mapping.

Figure 10.8 depicts an example time development of the MSE values when the early stopping approach is employed in connection with the Levenberg–Marquardt method. The time development ends with the 23rd iteration, because an increase in overtraining was found to start in the 24th iteration That detection was based on the time development of the MSE values for the validation catalysts, and is also in a good agreement with the development of the MSE values for the test catalysts, which indicate a slight increase in overtraining starting at about the 20th iteration. The overall results of applying early stopping to the training of two-layer percep-trons are shown in Figure 10.9. Comparison with Figure 10.5 shows that early stop-ping indeed led to a slightly lower overtraining, and mostly also to slightly lower MSE values for the test catalysts, but the differences are not really substantial. The reason for this is that early stopping is only useful when there is an apparent in-crease in overtraining, as in the situation in Figure 10.7. If the overtraining remains approximately constant, as in Figure 10.6, or if it changes in a more complicated way (e.g., oscillates), then it remains undetected.

Bayesian Regularization   A more sophisticated overtraining suppression method, called *Bayesian regularization,* relies on Bayesian statistics, wherein the weights and biases of an MLP are actually random variables with a particular prob-ability distribution, and that overtraining corresponds to situations where the values of those parameters obtained through MSE minimization are not likely for that dis-tribution [42,43]. The values of the weights and biases corresponding to minimal overtraining then can be obtained by replacing the MSE minimization with an MSE-based statistical parameter estimation. A combination of the Bayesian regu-

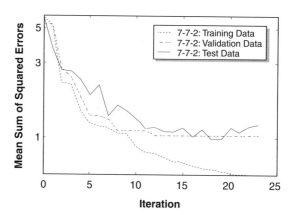

Figure 10.8   Time development of the MSE values during an example MLP training with the Levenberg–Marquardt method when early stopping is used.

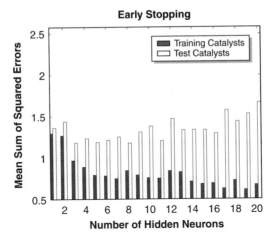

Figure 10.9 Training two-layer perceptrons with the Levenberg–Marquardt method when early stopping is used. MSE values were averaged from a 10-fold cross-validation.

larization and the Levenberg–Marquardt method is also implemented in the Matlab Neural Network Toolbox. This was the last method we employed in our search for an appropriate MLP architecture. As in the case of the basic Levenberg–Marquardt method, we trained each individual network for 100 iterations.

The results for two-layer perceptrons obtained with this method are shown in Figure 10.10. We can see that this method indeed almost completely suppresses

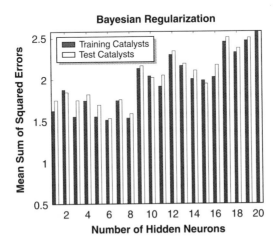

Figure 10.10 Training two-layer perceptrons with a combination of the Bayesian regularization and the Levenberg–Marquardt method. MSE values were averaged from a 10-fold cross-validation.

overtraining. On the other hand, it is not just the MSE values for the training catalysts that are higher than those obtained with both previous methods (this can be expected, since MSE values for the training catalysts substantially lower than the values for the test catalysts in the case of the previous methods were an indication of overtraining), as the MSE values for the test catalysts are also clearly higher. Again, the situation is similar for three-layer networks. As we know, the MSE values for test catalysts are decisive for the comparison of the approximation capabilities of different multilayer perceptrons or different architectures, which means we should not use values obtained by Bayesian regularization for this purpose.

To further illustrate the differences between the three methods used, the results of the two-layer perceptrons are compared separately for the training catalysts (in Figure 10.11) and the test catalysts (in Figure 10.12). Moreover, Figure 10.12 has an additional interpretation: because of the importance of the MSE test-data values in comparing the approximation capabilities of perceptrons with different architectures, this figure serves as an overall comparison of those capabilities between all the two-layer architectures considered. A similar comparison between the three-layer architectures considered is given in Figure 10.13. Finally, Figure 10.14 summarizes 12 two-layer and three-layer architectures, which represent the best 5% of all architectures considered according to the highest average MSE value obtained with any of the methods used. It indicates that two-layer perceptrons are more suitable for our case study than three-layer perceptrons. In addition, those 12 architectures help us decide which of the selected architectures to investigate further using the leave-1-out validation.

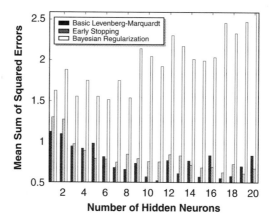

Figure 10.11 Comparison between the basic Levenberg–Marquardt method, the early stopping approach, and the Bayesian regularization of results obtained for the training catalysts when training two-layer perceptrons. MSE values were averaged from a 10-fold cross-validation.

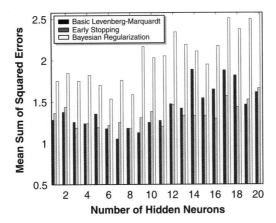

Figure 10.12 Comparison between the basic Levenberg–Marquardt method, the early stopping approach, and the Bayesian regularization of results obtained for the test catalysts when training two-layer perceptrons. MSE values were averaged from a 10-fold cross-validation.

10.3.9 Conclusions

We have drawn the following conclusions from all the results obtained for the 230 initially considered architectures:

1. The most promising approximators seem to be two-layer perceptrons with 3–11 hidden neurons and those from the considered three-layer perceptrons that have 5–15 neurons in the first and 3–8 neurons in the second hidden layer. Therefore, we will compare their approximation capabilities once more, using the MSE values for the test data obtained with the leave-1-out validation.
2. To this end, we will use only the basic Levenberg–Marquardt method and early stopping.
3. On the other hand, we will use Bayesian regularization, which we have found to possess by far the best generalization properties, for the training of the final MLPs with architectures resulting from the comparison at point 1.

The MSE values for the test catalysts obtained with the leave-1-out validation used are shown in Figure 10.15 (two-layer perceptrons), Figure 10.16 (three-layer perceptrons), and Figure 10.17 (best 12 architectures, i.e., best 5% of all initially considered). The best two-layer architecture 7–6–2 (which is at the same time the best among all considered architectures), as well as the best three-layer architecture 7–8–7–2 (which is, however, only the seventh best among all considered architectures) are actually those that were depicted in Figure 10.2. We see that:

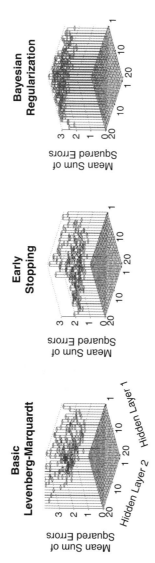

Figure 10.13 Comparison between the basic Levenberg–Marquardt method, the early stopping approach, and the Bayesian regularization of results obtained for the test catalysts when training three-layer perceptrons (1–20 neurons in the first hidden layer and a number of neurons in the second hidden layer not exceeding that in the first hidden layer). MSE values were averaged from a 10-fold cross-validation.

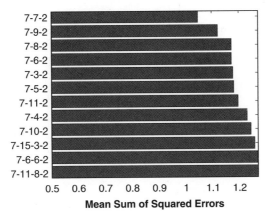

Figure 10.14 Best 12 ($\approx$ 5% of all) architectures, according to results obtained with any considered method for the test catalysts. MSE values were averaged from a 10-fold cross-validation.

- Clearly, the best approximation capabilities for our case study have two-layer perceptrons with 3–6 hidden neurons. According to the leave-1-out validation results, all four of those architectures were among the five best.

- There are only slight differences between the average approximation capabilities of the MLPs and those four architectures. Hence, the final perceptrons can be based on any of them. In addition, it is useful to train several final MLPs for each of those architectures, since the initial computed mapping, $F^{(0)}$, of the evolved sequence $F^{(0)}, F^{(1)}, \ldots, F^{(100)}$ is always obtained through a random initialization of weights and biases, and for different initial map-

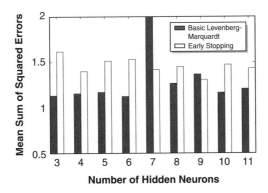

Figure 10.15 Comparison between the basic Levenberg–Marquardt method and the early stopping approach of results obtained for the test catalysts when training two-layer perceptrons. MSE values were averaged from a leave-1-out cross validation.

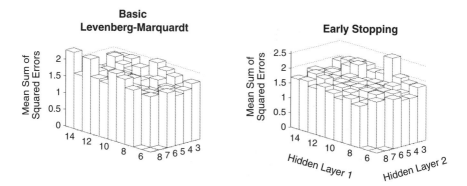

Figure 10.16   Comparison of results obtained for the test catalysts when training three-layer perceptrons. MSE values were averaged from a leave-1-out validation.

pings $F^{(0)}$, the sequence is likely to evolve to different local minima, i.e., to different final computed mappings $F^{(100)}$. The final MLPs can be used individually or in combination, but only if the combination coefficients of the individual combined networks are fixed, e.g., if the combined network computes the mean of the mappings computed by all final MLPs. Combining the final trained networks by means of coefficients that are additional free parameters, as in the case of optimal linear combinations of neural networks [44,45], would introduce an additional source of overtraining, invalidating the results of comparison based on the leave-1-out validation used.

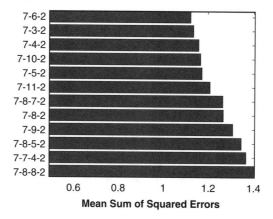

Figure 10.17   Best 12 ($\approx$ 5% of all) architectures, according to results obtained with any considered method for the test catalysts. MSE values were averaged from a leave-1-out validation.

For each of the architectures 7–3–2, 7–4–2, 7–5–2 and 7–6–2, we have trained two individual MLPs on all 211 available catalysts. The use of those final trained MLPs are shown in the last section.

10.4 KNOWLEDGE EXTRACTION FROM DATA WITH ARTIFICIAL NEURAL NETWORKS

The architecture of a trained neural network and the parameters determining the mapping computed by the network inherently represent the knowledge that has been contained in the data used to train the network, knowledge about various relationships between the input and output variables. For example, the architectures, weights, and biases of our final two-layer perceptrons represent relationships between the composition of the 211 catalysts on which they were trained, the feed-gas flow rate in the experiments in which the catalysts were used, and the propene and ethylene yields achieved in those experiments. However, this representation is not very human-comprehensible, since it is very far from the symbolic, modular, and vague way that people themselves represent knowledge. Hence, the larger the amounts of empirical data used to train artificial neural networks, the more important is the application of appropriate knowledge extraction methods. This latter term denotes the ways network architecture and parameters that determine the computed mapping are used to transform the inherent knowledge representation into some human-comprehensible representation, e.g., logical rules or points in the network input space. Those methods are actually only particular representatives of a broad spectrum of methods for knowledge extraction from data, based on various principles and used within the general framework of data mining [46–49].

Before dealing with knowledge extraction from trained neural networks, we point out two important limitations pertaining to knowledge extraction in general, no matter whether it is based on artificial neural networks, on traditional statistical data analysis, or on any other approach:

1. During data acquisition, various kinds of noise are superimposed on relationships and dependences that exist in the real world and are reflected in the data. Although some knowledge extraction methods can filter out some kinds of noise to some extent, no method is immune from all kinds or to any degree of noise.

2. In science, we are primarily interested in causal relationships. However, many relationships existing in the real world are merely phenomenological correlations. Although a deeper causality can often be found behind a phenomenological correlation, such a causality can be hidden at the phenomenological level and, in the best case, discerned only from the extracted phenomenological correlations by using problem-specific scientific methods.

In catalysis, we are primarily interested in two kinds of knowledge:

1. Knowledge about the optimum points of the approximated dependencies, or about the optimum points of the various functions built from those dependences. For example, in our case study, knowledge of the catalyst composition and feed-gas flow rate corresponding to a maximum of the propene yield, to a maximum of the ethylene yield, or to a maximum of the sum of the propene and ethylene yields. Switching from the approximated mapping to its approximation computed by the MLP shifts the search for such optimum points to a standard optimization task. Needless to say, its solution depends on the particular trained network that we use, since as we have seen in the preceding section, different networks are likely to compute different approximations. The results obtained using our eight final trained MLPs are shown in Table 10.3 (for propene yield) and Table 10.4 (for the sum of propene and ethylene yields).

Note that the results in Table 10.3 and Table 10.4 should be interpreted very cautiously. Optimum points are exactly those points in which the approximated dependency and its approximation computed by a neural network are most likely to differ, even when overtraining is low. Therefore, the knowledge about optimum points is usually not reliable enough.

2. Logical rules for the dependencies of the output variables on the input variables, i.e. implications of the general kind:

IF the input variables fulfill an input condition C_I,

THEN the output variables are likely to fulfill an output condition C_O.

During the last 15 years, a number of rule extraction methods have been proposed for trained neural networks, but so far none of them has come into common

Table 10.3 Prediction of the Maximal Propene Yield With the Mappings Computed by the Eight Final MLPs, and With The Average of Those Mappings

MLP	Architecture	$(x)_1$ Fe	$(x)_2$ Ga	$(x)_3$ Mg	$(x)_4$ Mn	$(x)_5$ Mo	$(x)_6$ V	$(x)_7$	$(y)_1$
1	7–3–2	0	70.9	0	23.9	0	5.2	10.0	9.0
2	7–3–2	0	50.0	23.9	7.0	1.8	17.4	10.0	10.1
3	7–4–2	4.2	0	0	95.8	0	0	10.0	9.6
4	7–4–2	0	0	0	0	0	100	69.0	9.7
5	7–5–2	0	100	0	0	0	0	10.0	8.7
6	7–5–2	0	0.77	61.0	12.3	0	26.0	35.0	11.2
7	7–6–2	0	24.6	37.0	9.6	0	28.6	10.0	9.4
8	7–6–2	0	0	5.0	26.2	0	21.6	20.7	10.3
	Mean of 1–8	0	23.4	41.0	8.9	0	26.7	31.4	9.1

Note: $(x)_1$–$(x)_6$ as their proportion in the catalyst; $(x)_7$ = feed-gas flow rate; $(y)_1$ = yield of propene.

Table 10.4 Prediction of the Maximal Sum of Propene and Ethylene Yields With the Mappings Computed by the Eight Final MLPs, and With the Average of Those Mappings

MLP No.	Architecture	$(x)_1$ Fe	$(x)_2$ Ga	$(x)_3$ Mg	$(x)_4$ Mn	$(x)_5$ Mo	$(x)_6$ V	$(x)_7$	$(y)_1 + (y)_2$
1	7–3–2	0	72.7	0	23.5	0	3.7	10.0	10.4
2	7–3–2	0	50.6	24.1	6.7	1.6	16.9	10.0	11.6
3	7–4–2	0	58.2	21.9	0	3.9	16.0	150.0	13.2
4	7–4–2	0	0	0	0	0	100	69.3	10.2
5	7–5–2	0	53.1	33.1	0	0	13.8	150.0	10.4
6	7–5–2	0	1.8	60.6	12.1	0	25.5	34.8	12.3
7	7–6–2	0	27.8	36.5	8.7	0	27.0	10.0	10.3
8	7–6–2	0	59.5	0	0	11.7	28.9	28.2	11.8
Mean of 1–8		0	36.4	32.9	0	10.4	20.4	26.1	10.0

Note: $(x)_1$–$(x)_6$ as their proportion in the catalyst; $(x)_7$ = feed-gas flow rate; $(y)_1 + (y)_2$ = yield of propene + ethylene.

use (cf. the survey papers [50–52]). For our problem, we have used a method that for each output condition of the form

C_O: the value y of the output variables lies in a given rectangular area R

finds an input conditions of the form

C_I: the value x of the input variables lies in a polyhedron P;

hence it creates rules of the form

IF $x \in P$, THEN $y \in R$.

The individual steps of the method are partly explained in [53], partly in [54], so here we outline only its main principles:

- A rectangular area R, which defines the result of an extraction rule, has to be chosen in advance in the output space of a trained MLP. Typically, it is specified by means of an inequality constraint on some of the output variables, e.g.,

$(y)_1 > 9\%$, i.e., propene yield $> 9\%$,

or by means of a combination of such constraints, e.g.,

$(y)_1 > 8\%$ and $(y)_2 > 1.5\%$, i.e., propene yield $> 8\%$ and ethylene yield $> 1.5\%$.

- The sigmoidal activation functions used in the somatic mappings coupled with hidden neurons are approximated with piecewise-linear sigmoidal activation functions. That can be done with arbitrary precision.

- The products of the individual linearity intervals of all the activation functions determine areas in the input space in which the final approximating mapping computed by the MLP is linear.

- In each such area, all points mapped to R form a polyhedron that eventually may be empty or may be concatenated with polyhedra from some of the neighboring areas to a larger polyhedron.

- The union of all the nonempty concatenated polyhedra $P_1, \ldots, P_q$ defines the antecedent (the IF . . . part) of a rule of a combined form

$$\text{IF } x \in P_1 \cup \ldots \cup P_q, \text{ THEN } y \in R,$$

  which is equivalent to a logical disjunction of q rules of the simple form mentioned earlier,

$$\text{IF } x \in P_1, \text{ THEN } y \in R$$

$$\cdots$$

$$\text{IF } x \in P_q, \text{ THEN } y \in R.$$

- To increase the comprehensibility of the extracted rules, visualization using two- or three-dimensional cuts of the set $P_1 \cup \cdots \cup P_q$ can be used. To this end, the values of some of the input variables have to be fixed, so that the number of free input variables is restricted to the dimensionality of the cut. The resulting visualization thus depends not only on the set $P_1 \cup \cdots \cup P_q$ but also on the fixed values. For example, to obtain two-dimensional cuts of the rules extracted in our case study, we have to determine four of the input variables from Table 10.1 (the value of the remaining seventh variable is implied by the constraint $(x)_1 + \cdots + (x)_6 = 1$).

Examples of two-dimensional cuts showing the unions of polyhedra that determine the antecedents of the two combined-form rules extracted in our case study are shown in Figures 10.18 and 10.19. Both figures depict cuts into the same dimensions, but the antecedents of the extracted rules correspond to the increasingly restrictive elements propene yield > 8% and propene yield > 9%, respectively.

Usually, logical rules of the form used earlier,

$$\text{IF } x \in P, \text{ THEN } y \in R,$$

where P is a polyhedron and R is a rectangular area, are the final results of this rule extraction method. However, there is one exception—the case when P is also rec-

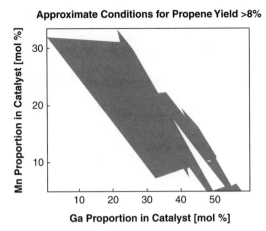

Figure 10.18 A two-dimensional section corresponding to the variables $(x)_2$ and $(x)_4$ of the union of polyhedra from the antecedent of the compact-form rule extracted from trained MLP No. 2 for the consequent $(y)_1 > 8\%$. The values of the input variables $(x)_1$, $(x)_3$, $(x)_5$, $(x)_7$ were fixed to correspond to the maximum propene yield, $(x)_1 = 0$, $(x)_3 = 23.9$, $(x)_5 = 1.8$, $(x)_7 = 10$, and the value of the variable $(x)_6$ is implied by the constraint $(x)_1 + \cdots + (x)_6 = 1$.

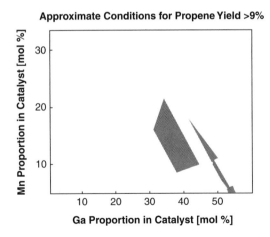

Figure 10.19 A two-dimensional section similar to the one in Figure 10.18, but for the result $(y)_1 > 9\%$.

tangular, or more generally, that P can be approximately replaced with a rectangular area R_I in the input space. Then the preceding rule can be approximately expressed in the conjunctive form:

$$\text{IF } (x)_1 \in I_1 \text{ \& } (x)_2 \in I_2 \text{ \& } \cdots \text{ \& } (x)_{n_I} \in I_{n_I}, \text{ THEN } y \in R,$$

where $I_1, I_2, \ldots, I_{n_I}$ are intervals that represent the projections of R_I into the n_I input dimensions. For example,

$$\text{IF } (x)_1 = 0\% \text{ \& } 15.8\% < (x)_2 < 19.3\% \text{ \& } 37.3\% < (x)_3 < 41.2\% \text{ \& } (x)_4 = 0\%$$

$$\text{\& } 9.7\% < (x)_5 < 11.9\% \text{ \& } (x)_5 = 30 \text{ mL/g}_{\text{cat}} \cdot \text{h, THEN } (y)_1 > 8\%,$$

i.e., IF the Fe proportion in catalyst = 0% and 15.8% < Ga proportion in catalyst < 19.3% and 37.3% < Mg proportion in catalyst < 41.2%, and the Mn proportion in catalyst = 0% and 9.7% < the Mo proportion in catalyst < 11.9% and feed gas flow rate = 30 mL/g$_{\text{cat}}$ · h, THEN propene yields > 8%.

Each such interval can be bounded, unbounded from below or above, or it even can be a complete set of real numbers. However, dimensions for which the corresponding projection of R_I equals the complete real axis are usually not included in the conjunctive-form rules, since they would not provide any new knowledge.

In the preceding outline of the rule extraction method we have employed in our case study, the replaceability of a polyhedron P with a rectangular area R_I is assessed according to the following principles:

1. The resulting dissatisfaction with points that either belong to P but do not belong to R_I or belong to R_I but do not belong to P (i.e., the dissatisfaction with points that belong to the symmetric difference $R_I \Delta P$) has to remain within a prescribed tolerance ε and has to be minimal for R_I among all rectangular areas in the input space.

2. Dissatisfaction with points from $R_I \Delta P$ depends solely on those points and increases with respect to inclusion. Consequently, principle 1 can be expressed as

$$\mu_P(R_I \Delta P) = \min\{\mu_P(R'_I \Delta P): R'_I \text{ is a rectangular area in the input space}\} < \varepsilon$$

where μ_P is some monotone measure on the input space, eventually depending on P.

3. To be eligible for replacement, P has to cover at least one point of the available data.

For principle 2, the most attractive monotone measures, due to their straightforward interpretability, are:

- The joint empirical distribution of the available data about the input variables (thus, in our case study, the joint empirical distribution of the available data about catalyst composition and feed-gas flow rate).
- The conditional empirical distribution of the available data about input variables, conditioned by P.

Unfortunately, in order to check the replacement of polyhedra with rectangular areas using any of these two measures is exponentially complex. More precisely, if n_{FI} denotes the number of free input variables and p denotes the number of all input–output pairs available for training the network, then to check the replaceability of a polyhedron P with rectangular areas using any of those measures μ_P requires that the value $\mu_P(R_I \Delta P)$ be calculated for almost all n_{FI}^p rectangular areas R_I. In our case study, $(n_{FI} = 6, p = 211)$ $n_{FI}^p > 8.8 \cdot 10^{13}$, so using those measures is too expensive. Therefore, we used another measure as μ_P—the product of the marginal conditional empirical distributions of the available data about input variables, conditioned by P. That measure coincides with the conditional empirical distribution of

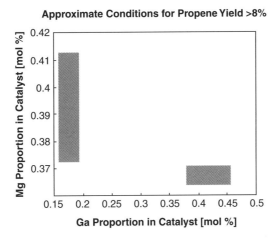

Figure 10.20 A two-dimensional projection of rectangular areas determining the antecedents of the conjunctive-form rules extracted from trained MLP No. 8. The conjunctions in the rules' antecedents involve only the variables $(x)_1, \ldots, (x)_5$ and $(x)_7$, since the value of the variable $(x)_6$ is implied by the constraint $(x)_1 + \cdots + (x)_6 = 1$. The following rules have been extracted:

IF $(x)_1 = 0\%$ **&** $15.8\% < (x)_2 < 19.3\%$ **&** $37.3\% < (x)_3 < 41.2\%$ **&** $(x)_4 = 0\%$

& $9.7\% < (x)_5 < 11.9\%$ **&** $(x)_5 = 30$ mL/$g_{cat} \cdot$ h, THEN $(y)_1 > 8\%$,

IF $(x)_1 = 0\%$ **&** $40.0\% < (x)_2 < 45.7\%$ **&** $36.4\% < (x)_3 < 37.0\%$ **&** $(x)_4 = 0\%$

& $(x)_5 = 0\%$ **&** $(x)_5 = 30$ mL/$g_{cat} \cdot$ h, THEN $(y)_1 > 8\%$.

the available data about input variables if the involved marginal conditional empirical distributions are mutually independent, but it has only a polynomial complexity, since the number of rectangular areas R_I for which the value $\mu_P(R_I\Delta P)$ must be computed is less than $pn_{FI}^2/2$ (in our case study, $pn_{FI}^2/2 = 133,563$).

Conjunctive-form rules are also very convenient from the visualization point of view: because sections of the rectangular areas match the corresponding projections of those areas, the values of no variables need to be fixed. As an example, Figure 10.20 shows two-dimensional projections of rectangular areas determining the antecedents of conjunctive-form rules extracted for the consequent "propene yield > 8%," from one of our final trained MLPs.

REFERENCES

1. Hagan, M. T.; Demuth, H.; Beale, M. *Neural Nework Design,* PWS Publishing, Boston, 1996.

2. Haykin, S. Neural Networks. *A Comprehensive Foundation,* IEEE Press, New York, 1999.

3. Mehrota, K.; Mohan, C. K.; Ranka, S. *Elements of Artificial Neural Networks,* MIT Press, Cambridge, MA, 1997.

4. White, H. *Artificial Neural Networks: Approximation and Learning Theory,* Blackwell Publishers, Cambridge, 1992.

5. Zupan, J.; Gasteiger, J. *Neural Networks for Chemists,* Wiley-VCH, Weinheim, Germany,1993.

6. Zupan, J.; and J. Gasteiger, *Neural Networks in Chemistry and Drug Design: An Introduction,* Wiley-VCH, Weinheim, Germany, 1999.

7. Henson, M. A. *Comp. Chem. Eng.,* 1998, 23: 187–202.

8. Melssen, W. J.; Smits, J. R. M.; Buydens, L. M. C.; Kateman, G. *Chemometrics Intell. Lab. Syst.,* 1994, 23: 267–291.

9. Smits, J. R. M.; Melssen, W. J.; Buydens, L. M. C.; Kateman, G. *Chemometrics Intell. Lab. Syst.,* 1994, 22: 165–189.

10. Clark, J. W., Ed. *Scientific Applications of Neural Nets,* Springer-Verlag, Berlin, 1998.

11. Lisboa, P. G. J., Ed. *Neural Networks: Current Applications,* Chapman & Hall, London, 1992.

12. Murray, A. F., Ed. *Applications of Neural Networks,* Kluwer, Dordrecht, The Netherlands, 1994.

13. Barron, A. R. *Mach. Learn.,* 1994, 14: 115–133.

14. Chui, C. K.; Li, X.; Mhaskar, H. N. *Adv. Comput. Math.,* 1996, 5: 233–343.

15. De Vore, R. A.; Oskolkov, K. I.; Petrushev, P. P. *Ann. Numer. Math.,* 1997, 4: 261–278.

16. Hornik, K. *Neural Networks,* 1991, 4: 251–257.

17. Hornik, K.; Stinchcombe, M.; White, H.; Auer, P. *Neural Comput.,* 1994, 6: 1262–1275.

18. Kůrková, V. *Neural Networks,* 1992, 5: 501–506.

19. Kůrková, V. *Neural Networks,* 1995, 8: 745–750.

20. Leshno, M.; Lin, V. Y.; Pinkus, A.; Shocken, S. *Neural Networks,* 1993, 6: 861–867.

21. Hattori, T.; Kito, S. *Catal. Today,* 1995, 23: 347–355.

22. Kito, S.; Hattori, T., Murakami, Y. *Appl. Catal. A: Gen.,* 1994, 114: L173–L178.

23. Hattori, T. et al. *Stud. Surf. Sci. Catal.,* 1994, 90: 229–232.

24. Kito, S.; Hattori, T., Murakami, Y. *Ind. Eng. Chem. Res.,* 1992, 31: 979–981.

25. Sasaki, M. et al., *Appl. Catal. A: Gen.,* 1995, 132: 261–270.

26. Cundari, T. R.; Deng, J.; Zhao, Y. *Ind. Eng. Chem. Res.,* 2001, 40: 5475–5480.

27. Hou, Z. -Y. et al., *Appl. Catal. A: Gen.,* 1997, 161: 183–190.

28. Sharma, B. K. et al., *Fuel,* 1998, 77: 1763–1768.

29. Huang, K.; Feng-Qiu, C.; Lü, D. W. *Appl. Catal. A: Gen.,* 2001, 219:61–68.

30. Kondratenko, E. V.; Buyevskaya, O. V., Baerns, M. *Top. Catal.,* 2001, 15: 175–180.

31. Langpape, M.; Grubert, G.; Wolf, D.; Baerns, M. *DGMK-Tagungsbericht,* DGMK, Hamburg, 2001, 227–234.

32. Wolf, D.; Buyevskaya, O. V.; Baerns, M. *Appl. Catal. A: Gen.,* 2000, 200: 63–77.

33. SPSS, *Clementine 6.0 Advanced Features Guide,* SPSS Inc., Chicago, 2001.

34. Floudas, C. A. *Nonlinear and Mixed-Integer Optimization. Fundamentals and Applications,* Oxford University Press, Oxford, 1995.

35. Jahn, J. *Introduction to the Theory of Nonlinear Optimization,* Springer-Verlag, Berlin, 1994.

36. Leondes, C. T. *Optimization Techniques. Neural Networks Systmes Techniques and Applications,* Academic Press, Orlando, FL, 1997.

37. Nocedal, J.; Wright S. J., *Numerical Optimization,* Springer-Verlag, New York, 1999.

38. Hagan, M. T.; Menhaj, M. *IEEE Trans. Neural Networks,* 1994, 5: 989–993.

39. Marquardt, D. *SIAM J. App. Math.,* 1963, 11: 431–441.

40. Rumelhart, D. E.; Hinton, G. E.; Williams, R. J. In *Parallel Data Processing,* Vol. 1, Rumelhart, D. E., McClelland, J. L., Eds., MIT Press, Cambridge, MA, 1986, 318–362.

41. Demuth, H.; Beale, M. *Neural Network Toolbox,* Version 4, For Use with MATLAB, The MathWorks Inc., Natick, MA, 2000.

42. Foresee, F. D.; Hagan, M. T. *Proc. 1997 Int. Joint Conf. Neural Networks,* IEEE, Piscataway, NJ, 1997, 1930–1935.

43. MacKay, D. J. C. *Neural Comput.,* 1992, 4: 415–447.

44. Hashem, S. *Neural Networks,* 1997, 10: 599–614.

45. Hashem, S, Schmeiser, B. *IEEE Trans. Neural Networks,* 1995, 6: 792–794.

46. Berthold, M.; Hand D. *Intelligent Data Analysis. An Introduction,* Springer-Verlag, Berlin, 1999.

47. Cios, K.; Pedrycz, W.; Swiniarski, R. *Data Mining Methods For Knowledge Discovery,* Kluwer, Dordrecht, The Netherlands, 1998.

48. Fayad, U. M.; Piatetsky-Shapiro, G.; Smyth, P.; Uthurusamy, R., Eds. *Advances in Knowledge Discovery and Data Mining.* AAAI Press, Menlo Park, CA, 1996.

49. Thuraisingham, B. *Data Mining. Technologies, Techniques, Tools and Trends.* CRC Press, Boca Raton, FL, 1998.

50. Andrews, R.; Diederich, J. Tickle, A. B. *Knowl. Based Syst.,* 1995, 8: 378–389.

51. Mitra, S.; Hayashi, Y. *IEEE Trans. Neural Networks,* 2000, 11: 748–768.

52. Tickle, A. B.; Andrews, R.; Golea, M.; Diederich, J. *IEEE Trans. Neural Networks,* 1998, 9: 1057–1068.

53. Maire, F. *Neural Networks,* 1999, 12: 717–725.

54. Holeňa, M. In *Principles of Data Mining and Knowledge Discovery,* Zighed, D. A., Komorowski, J., Zytkov, J. M., Eds., Springer-Verlag, Berlin, 2000, 440–445.

CHAPTER 11

DESCRIPTOR GENERATION, SELECTION, AND MODEL BUILDING IN QUANTITATIVE STRUCTURE–PROPERTY ANALYSIS

CURT M. BRENEMAN, KRISTIN P. BENNETT, MARK EMBRECHTS, N. SUKUMAR, STEVEN CRAMER, MINGHU SONG, AND JINBO BI

Rensselaer Polytechnic Institute

11.1 INTRODUCTION

Quantitative structure–property relationship (QSPR) models are mathematical relationships that link molecular structures with observable physical properties. In order to develop these models, data mining techniques are used to extract information from "training" sets made up of molecules with known properties. These training data sets can be assembled through combinatorial synthesis and screening, or through the collection of experimental results for a wide variety of molecules for which the property of interest is known. The subject of this chapter relates to the methods used to produce and validate QSPR models that can be used to aid combinatorial library development or as part of a rational design strategy. Besides the collection of experimental data, QSPR model building requires that molecular structures be represented in ways that are interpretable by computer, but retain the essential chemical information pertinent to the observable property under investigation. The art of capturing this essence is contained within the algorithms used to produce molecular "descriptors." Molecular descriptors take many forms, where each kind is designed to encode a part of the whole story of molecular interactions or properties. Several classes of descriptors are discussed in this work, including some of the most modern ones that encode both electronic and spatial information within a single type of descriptor. Almost all current descriptors are computationally derived, but this was not always the case. Any molecular characteristic (either computational or experimental) can potentially be used in this role.

The final component of QSPR modeling is the technique used to actually con-

Experimental Design for Combinatorial and High Throughput Materials Development, Edited by James N. Cawse.
ISBN 0-471-20343-2 © 2003 John Wiley & Sons, Inc.

dense or "mine" the information contained in the molecular descriptors into a model or set of models that can be used for predicting molecular behavior. The kind of computer algorithms capable of doing this are known as "machine learning" methods because they utilize molecular descriptor information from a training data set to generate a model capable of making predictions for a "test" data set consisting of molecules with unknown properties. An important aspect of machine learning is being able to train a model in such a way that it captures the fundamental relationships within the data without causing it to simply "memorize" the training set data. When such memorization takes place, the model is considered to be "overtrained," and is consequently incapable of making good predictions for any case not actually included in the original training data set. For those of us in the business of educating students, this is a familiar scenario that results from "cramming" for examinations.

A number of strategies may be used prevent overtraining, but an important feature for any modeling exercise is the ability to detect when it has taken place. This is the role of cross-validation, or the assessment of whether the model is internally consistent and capable of making predictions on a simulated "test" set for which experimental data are actually available. If a model is able to make reasonably accurate predictions for molecules not used in the original training data, then it is considered to be properly trained. In order to assess internal consistency in an automated manner, a strategy called *bootstrapping* can be used. In its simplest form, this consists of a "leave-1-out" method, in which new models are trained while leaving out each one of the molecules in the training set in turn. This procedure is repeated as many times as there are molecules in the original training set. If the models are relatively unchanged throughout this procedure, it means that no single molecule is biasing the model, and the data are internally consistent. This can be extended to groups of molecules as well. In this case, the original training data are split up into a number of random training and test data sets, where models that are built for each training set are then used to make predictions on the associated test set. The test set results are then accumulated and used to determine if the models are capable of generally representing the physical interactions that are controlling the observed experimental properties. While the state-of-the-art in the field of QSPR falls short of true quantitative accuracy, it is often possible to use QSPR models to both predict and explain the way structural modifications affect molecular properties.

QSAR methods for the analysis of bioactive molecules are a subset of QSPR, and represent the first application of this technology to chemical problems. Early examples of QSPR include the use of relatively simple (two-dimensional topological) molecular descriptors based on the connectivity patterns of their constituent atoms. The resulting data sets would then be analyzed using multiple linear regression techniques (such as least-squares methods) to uncover relationships between descriptor values and the target property under investigation. More sophisticated methods followed, often using descriptors that contained spatial information such as dipole moment, molecular surface area, and functional group properties. More recent developments have provided molecular-property descriptors that encode pertinent properties of molecular electron density distributions. When used alone or together with the more traditional two-dimensional descriptors, these features enable highly predictive partial least-squares (PLS), ANNs, or support vector machine (SVM) regression models of observable physical properties to be created [1].

An important consideration when utilizing large numbers of modern molecular descriptors is finding a way to select a small number of the most important features from a set of several hundred possibilities. Due to the difficulty involved in making the correct choices, the development of new feature selection methods is a current area of research.

Although it might seem problematic at first, it is not unusual to begin the modeling process with more descriptors than experimental cases, a potentially dangerous situation where overdetermined models can be produced. The possibility of producing erroneous models and chance correlations can be minimized by using judicious training techniques for model building that use only part of the experimental data, and the remaining data are treated as if its experimental properties are unknown. The construction of multiple models using different combinations of training and test cases (bootstrapping) also provides a way to determine whether a particular trial set of descriptors can generate predictive models. In some cases, GA techniques are used to select optimal subsets of descriptors during this process.

Once predictive models have been produced and tested, they can be used either in incremental molecular design applications or within virtual high throughput screening (VHTS) processes where molecular structures are automatically proposed and then computationally evaluated. The scope and applicability of the models used in these investigations are dependent on the generality of the descriptors involved and the quality of the machine-learning method used to make the models. Recent investigations have shown that electron-density-based descriptors [such as transferable atom equivalent (TAE) descriptors] [2] extend the predictive range of QSPR models, and that nonlinear regression techniques can further enhance their performance [3].

In the remainder of this chapter, an approach to the QSAR and QSPR process is described in more detail. Section 11.2 provides an overview of molecular optimization and feature selection, while leading-edge methodologies for descriptor generation are discussed in Section 11.3. Two case studies in Sections 11.4 and 11.5 are used to illustrate the modeling process and to show how models can be used for both prediction and explanation of chemical effects. Section 11.6 concludes with a summary and discussion.

11.2 MODERN QSAR AND QSPR AND THEIR ROLE IN MOLECULAR OPTIMIZATION

Central to the goal of computer-assisted molecular design is the ability to build useful mathematical models that quantify the complex relationship between molecular structure and function. The basic elements of this process are unchanged from the intuitive structure/property molecular design concepts that have been used by bench chemists since the field evolved from alchemy. As more sophisticated theories of chemical interactions have emerged, chemistshave created sets of compounds with regular structural modifications in an attempt to probe the sensitivity of experimental properties to such changes. In many ways, this approach is an application of linear free-energy relationship theory, where alterations of molecular structure are considered to represent free-energy increments that can affect the energies of both transition states and equilibria. While systematic laboratory experimentation has enabled chemists to solve many design problems without computational support, a more quantitative and rapid procedure is often desirable. The field of QSAR modeling and the more general field of QSPR analysis were developed to fill this need.

QSAR and QSPR modeling projects are often driven by two opposing requirements: the need to generate models that have a high degree of predictive capability, and the need to extract meaningful chemical information from the resulting models. While these goals may not initially appear at odds, in practice it is usually necessary to optimize one goal at the expense of the other. At the core of this disparity is the need to use numerical descriptors—"features" in machine learning jargon—that represent the structural and chemical properties of candidate molecules. It is generally accepted that if a small group of chemical descriptors is assigned to represent specific chemical interactions, the resulting models will be easier to interpret, but at the expense of model quality [4]. Problems of this sort have been seen from the beginning of QSAR investigations, when certain observable physical properties of molecules (for example, the index of refraction, or the solubility in olive oil) were used as molecular descriptors [5]. Two types of computer-generated descriptors—theoretical linear solvation energy relationship (TLSER) [6] and generalized interaction potential function (GIPF) descriptors [7]—both use small numbers of computed values to represent specific interactions. On the other hand, when an automatic "feature-selection" procedure is used to select a small number of relevant descriptors from a large set of available theoretical features, a predictive model will often result, although it may be more difficult to interpret [8].

Several types of descriptors are used in the following two case studies, and their advantages and drawbacks are discussed. These include traditional two- and three-dimensional descriptors, as well as some modern electron-density-derived descriptors, such as those from TAE and property-encoded surface translator (PEST) calculations [9]. This discussion will omit the use of experimental physical properties (such as refractive index or octanol/water partitioning) as descriptors, and will concentrate on descriptors that can be produced using theoretical methods.

To produce QSPR models, it is necessary to have both good descriptors as well as robust machine learning methods that can capture and exploit the chemical information encoded in the descriptors. Various methods have been utilized for building chemical-property models (see, for example, Table 11.1).

When small numbers of independent descriptors are used to build models, simple linear methods such as linear regression can provide adequate models. In the more general case where large numbers of correlated descriptors are used, extra care must be taken to avoid producing an overdetermined model that lacks predictive power. PLS and SVM regression modeling work well with such large numbers

Table 11.1 QSPR Modeling Methods

Local learning (LL)	Similarity of molecules with parts of the training data
Multiple linear regression (MLR)	Multidimensional least-squares analysis
Principal component analysis (PCA)	Combines descriptors to find relationships that explain linear relationships in the data
Partial least squares (PLS)	Like PCA, but uses the experimental data as well to develop a small set of "latent variables" that explain the data
Artificial neural networks (ANN)	Nonlinear node-based learning system
Support vector machines	Linear or nonlinear classification or regression system

of descriptors, provided appropriate feature-selection procedures are used during model construction.

By removing irrelevant descriptors, models can be greatly improved, but this can also be a source of "soft cheating" that results in a less predictive model. This type of cheating results when the descriptor-selection procedure is allowed to involve any descriptor information from molecules in the unknown part of the data set. A "hard cheat" would result if the feature-selection routine also used the experimental data from the unknowns, but this is much easier to avoid than the "soft cheat."

A theoretical justification for feature selection (or descriptor removal) is provided in the following argument: All molecular descriptors provide information concerning some observable chemical properties, but most descriptors are not general enough to apply to all molecule properties. By eliminating features that are not relevant to a particular property of interest, noise is removed from the descriptors and superior models are produced. Feature selection may either be performed as an inherent part of the regression process, such as in sparse SVM models described later in this work, or as part of a genetic algorithm (GA) driven procedure that uses an objective function to judge which combinations of descriptors give the best models [3] (see the Appendix for more details).

In any quantitative modeling investigation, it is frequently insufficient to create only a regression model of the behavior of interest. This level of investigation can be used for survey purposes, but the validation of modeling results using blind test sets is required to demonstrate their predictive power. Pseudounknowns may also be used in the feature-selection process, where part of the "known" data set is reserved for testing the predictive capabilities of different subsets of descriptors. Final model validation may also be obtained using reserved pseudounknowns.

The two problems used in the case studies involve large-molecule/surface interaction modeling (protein ion-exchange retention-time prediction), and small-molecule/large-molecule interactions (Caco-2 intestinal-wall permeability). The modeling procedures used in these cases are identical to what can be used in any QSPR or materials-by-design investigation. Summaries of the SVM machine learning techniques used are briefly described in the case studies. Additional information is provided in the Appendix.

11.3 DESCRIPTOR GENERATION

Derivations of descriptors that capture key properties of the molecules are the foundation of any successful QSPR method. RECON, wavelet coefficient, PEST, and molecular operating environment (MOE) [10] descriptor generation are discussed in this section.

11.3.1 RECON Descriptor Generation

The TAE/RECON method utilizes a new rapid charge-density reconstruction algorithm that depends upon combining atom charge-density fragments that are precomputed from *ab initio* wave functions. During the development of the TAE method, a library of atomic charge density components (TAEs) was constructed in a form that allows for rapid retrieval of the fragments needed to build a molecule [11,12]. The RECON program reads in a database of molecular structures, and then reconstructs the electron density properties of each one by assigning the closest match from a library of atom types for each atom in the molecule. The algorithm then combines the densities of the atomic fragments to give a large set of new and traditional QSAR descriptors. The CPU and disk resources required for TAE reconstruction are comparable to those utilized by molecular mechanics energy computations.

The TAE/RECON algorithm first reconstructs the molecular electron density distribution, and then represents the surface property distribution by a series of histograms. In Figure 11.1, the local average ionization potential [Politzer ionization potential (PIP)] of the molecule is shown encoded onto its 0.002 e/au³ (electrons per cubic Bohr) electron-density surface. The PIP property is one of ten surface properties used for generating TAE descriptors (see Table 11.2). The distribution of this property is then represented as a histogram such as that shown on the right side

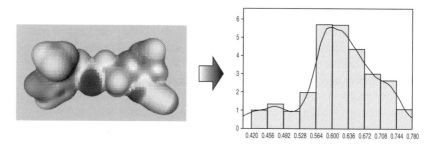

Figure 11.1   TAE PIP surface property.

Table 11.2 TAE Electronic Surface Properties

EP	Electrostatic potential
Del(Rho) $\cdot N$	Electron density gradient normal to 0.002 e/au^3 electron-density isosurface
G	Electronic kinetic energy density $G = (-(\hbar/4m)\int\{\nabla\psi^* \cdot \nabla\psi\}\ d\tau)$
K	Electronic kinetic energy density $K = (-(\hbar/4m)\int\{\psi^* \nabla^2\psi + \psi\nabla^2\psi^*\}\ d\tau)$
Del(K) $\cdot N$	Gradient of K electronic kinetic energy density normal to surface
Del(G) $\cdot N$	Gradient of G electronic kinetic energy density normal to surface
Fuk	Fukui F^+ function scalar value
Lapl	Laplacian of the electron density $= \nabla^2\rho$
BNP	Bare nuclear potential $BNP_{(j)} = \sum\limits_{i=1}^{n} q_i/r_{ij}$
PIP	Local average ionization potential

of the figure. Each bin of the histogram is then used as a descriptor, as well as statistical information such as the minimum, maximum, and average of each surface property. Fixed bin sizes are used to avoid scaling problems when computing descriptors for unknown molecules.

11.3.2 Wavelet Coefficient Descriptors

The surface property distributions utilized in TAE/RECON descriptors can also be characterized by the use of discrete wavelet transforms. In Figure 11.2, the smoothed Politzer ionization potential property density from the earlier figure is represented as a signal vector and subjected to discrete wavelet transformation using the Daubechies D4 wavelet. Wavelet transformation is an efficient numerical procedure performed using a pyramid algorithm that involves a series of iterative multiplications of the Daubechies matrix by permutations of the transforming sig-

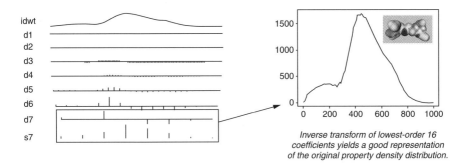

Inverse transform of lowest-order 16 coefficients yields a good representation of the original property density distribution.

Figure 11.2 Wavelet coefficient descriptors. Discrete wavelet transform and inverse transform using only low-order coefficients. An inverse transform of the lowest-order 16 coefficients yields a good representation of the original property-density distribution in Figure 11.1.

nal vector [13–15]. The final result of the transformation is a vector of wavelet coefficients the same size as the input signal, with the different wavelet coefficients representing the shape of the original distribution. Coefficients describing high-frequency variations in the shape of the property distribution can be removed without loss of chemical information.

In Figure 11.2, the top trace is the density distribution of the PIP surface property found on the example molecule. Successively lower traces on the figure (marked d1, d2, d3, ...) illustrate the magnitude of "detail" wavelet coefficients in consecutive iterations of the transformation process. In this example, the original signal contains 1024 points, of which 512 points are represented as detail coefficients in the d1 trace. As seen in the figure, the smoothness of the original signal precludes any contributions from these high-frequency detail coefficients. The trace marked d2 contains 256 coefficients that arose from the second step of the transformation process. Traces d3 and d4 have 128 and 64 coefficients, respectively. In this example, small contributions can be seen in the d4 trace, but it is also clear that most of the information is contained in the d6, d7, and s7 coefficients. If only the last 16 coefficients are used—d7 and s7—the signal can be reconstructed with good fidelity, as shown on the right side of the figure. This reconstruction exercise illustrates that almost all of the surface-property information in the original 1024-point density distribution can be captured using only 16 wavelet coefficients, and suggests that these coefficients should act as effective molecular-property descriptors.

Wavelet coefficient descriptors (WCDs) produced through TAE reconstruction are also additive at the atomic level, and convey more chemical-property information as a result of the data-compression methods used to generate them. In the present incarnation of WCDs, only eight scalar and eight detail coefficients are able to represent each molecular surface property distribution with a high degree of fidelity. These transformed descriptors can be associated with specific locations along the property-value abscissa. Upon inspection of Figure 11.2, it can be seen that the lower set of scalar coefficients (s7) approximates the shape of the property distribution, whereas values from the detail set (d7) can be associated with deviations from the scalar shape. WCDs have been used successfully with both linear (PLS) and nonlinear (ANN, SVM) modeling methods [3,16–18]. Due to the orthogonality of wavelet coefficient vectors, the similarity between molecules can be assessed using a dot product of these vectors. This has been shown in preliminary results to be an important method of molecular classification [19].

11.3.3 Property-Encoded Surface Translator Descriptors

PEST shape/property hybrid descriptor technology, developed in the author's laboratory, allows better representation of intermolecular interactions. Results have shown that the inclusion of PEST descriptors significantly improves QSPR models where intermolecular interactions play an important role in the chemical effects being modeled.

PEST descriptors are generated using TAE molecular surface representations to define property-encoded boundaries for implementing the Zauhar Shape Signature ray-tracing approach to shape/property convolution [20]. The Shape Signature approach seeks to encode the shape of a molecular volume using the distribution of ray lengths obtained by performing a ray-tracing procedure within the molecule from an arbitrary starting position while assuming an infinite reflectivity of the internal surface. The converged ray-length distribution then represents a signature of the original molecular envelope. Zauhar reports that convergence is usually obtained within several thousand ray reflections [20]. If TAE property-encoded surfaces are used instead of Zauhar's original van der Waals spheres, the procedure becomes known as PEST.

The left side of Figure 11.3 illustrates a cutaway molecular envelope in which a ray is initiated and reflected within the molecular volume. The partition plane is part of a binary space-partitioning scheme designed to speed up the ray-tracing procedure. The right side of the figure shows the results of 750 rays, in which a surface property is associated with the endpoint of the rays. In normal use, over 4000 rays are utilized, or the procedure is continued until the shape/property distributions have converged. Such an example is shown in Figure 11.4, in which the benzodiazepine molecule on the left is reconstructed to provide the electrostatic-potential-encoded surface in the middle of the figure. This volume is then subjected to the PEST procedure until the shape/property distribution shown on the right is produced. Since there are ten TAE surface properties to be considered, ten two-dimensional shape/property distributions are generated. An eleventh distribution is also available that illustrates the ray-length/reflection-angle relationship. The latter function is taken directly from Shape Signature methodology. For use as descriptors, the two-dimensional shape/property distributions are mapped onto a 6 × 6 grid to produce 36 PEST descriptors per property. In this project, different combinations of grid resolutions and property-encoding schemes are tested. This would include generating composite property/ray-length distributions, where fundamental pharmacophore information conceptually similar to Cruciani's GRIND descriptors may be available [21].

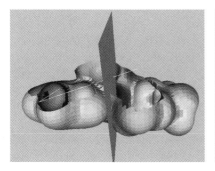

Figure 11.3 Property-encoded ray tracing. Binary space partitioning plane (in gray), and a sparse (750 segment) PEST run.

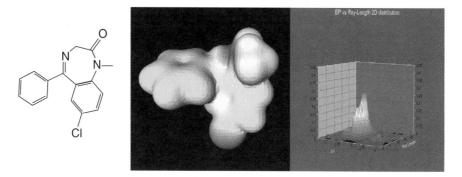

Figure 11.4   Benzodiazapine structure, TAE surface reconstruction, and PEST shape/property signatures.

11.3.4 Molecular Operating Environment Descriptors

The MOE program [10] provides a combination of traditional molecular-property descriptor types that span several classes, including connectivity-based topological two-dimensional and shape-dependent three-dimensional molecular features. The following three classes of descriptors are common to the molecular modeling community:

2D   Topological two-dimensional property descriptors can be produced using only atom-type and connectivity information, and are known to be remarkably effective in a number of applications [22]. The fact that topological descriptors carry no molecular conformation or shape information is both a limitation and a benefit, in that their use reduces the ambiguity that can result from modeling highly flexible molecules. Topological molecular descriptors are defined as numerical properties that can be calculated using a connection table representation of a molecule (e.g., elements, formal charges, and bonds, but not atomic coordinates).

i3D   Internal three-dimensional descriptors use spatial information for each molecule, but are invariant to rotations and translations of the molecule. This class includes descriptors that incorporate quantum mechanical or empirical force-field results, or rely only on the internal coordinates of each molecule. Descriptors such as those obtained from TAE/RECON or PEST also fall into this category, as do potential energy decriptors, surface areas, volumes, dipole-moment, and bulk shape descriptors.

x3D   External three-dimensional descriptors also use three-dimensional coordinate information, but in addition require an absolute frame of reference (e.g., molecules docked into the same receptor). These are less commonly used in QSPR investigations, but figure prominently in comparative molecular-field analysis (CoMFA) ligand/binding site investigations and other "rational" design techniques.

11.4 OVERVIEW OF FEATURE SELECTION

Successful QSPR requires a wise choice of descriptors and robust modeling methodologies. Our strategy is to allow the chemist to produce a very large set of descriptive features, at which point a SVM feature selection strategy is used to identify a small subset of relevant molecular property descriptors. Visualization of the resulting SVM models then allows the chemist to interpret and further refine the feature subset, at which time a nonlinear SVM model may be used to generate a final predictive model. An overview of this process is shown in Figure 11.5. The visualization methods and results are described in the next section.

In this scheme, a partition training set of experimental data is generated by merging the molecular descriptor sets discussed earlier with experimental results. This set is then available for partitioning into many possible random training and validation sets, each of which is called a *fold*. In Figure 11.5, a set of data is split into these two sets, after which the training set and validation sets are used to produce a linear SVM model. The validation set is used to test for predictive capability in order to select the two parameters of the SVM model using a pattern search optimization procedure. This provides a linear-regression modeling result, which is averaged with the results of previous iterations as part of of the bootstrap aggregate or bagging process. Due to the small amount of data, no one model can be considered completely reliable. The use of many iterative examinations of folds allows a robust set of features to be determined.

This feature-selection algorithm can be incorporated into a larger scheme for

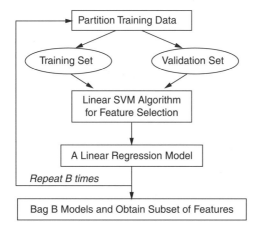

Figure11.5   Feature selection flowchart.

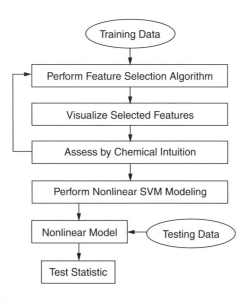

Figure 11.6   Computational chemical-property design.

property prediction, as illustrated in Figure 11.6. In this case, a set of training data is used to perform feature selection, after which the results are used to generate user-friendly graphics that can be used by the chemist to further focus the descriptor set according to chemical intuition. Examples of the use of graphical tools at this stage in the procedure are illustrated in both of the case studies later in the text. After reducing the features to a small, more interpretable set, they can be used to create a nonlinear SVM regression model that can be tested using either pseudounknowns or actual blind data. The predictive power of the model is then finally determined by creating a test statistic that quantifies the error between predicted and actual properties of each molecule.

11.5   CASE STUDY 1: PREDICTION OF PROTEIN RETENTION TIME IN ANION-EXCHANGE CHROMATOGRAPHY USING SUPPORT VECTOR MACHINE REGRESSION AND FEATURE SELECTION

Ion-exchange chromatography is widely used for protein separations due to the large and variable number of differentially ionizable groups present in this class of compounds. Ion-exchange chromatographic separation is a subset of the general field of liquid chromatography, in which different levels of interaction between dissolved materials in a moving phase and the functional groups present on a stationary phase determine the retention time (RT) of a particular protein. For example, in column chromatography the solutes interact with the solid material in the column through a combination of intermolecular interactions, including dipole–dipole attraction, hydrogen bonding, and van der Waals forces as the mobile phase flows down through the column. In ion-exchange chromatography, the attraction for the stationary phase is thought to be mainly due to the electrostatic attraction of positive charges for negative charges, or vice versa. The selectivity of this technique is optimized by varying the composition of the stationary phase and the pH of the mobile phase. Consequently, one of the major challenges in ion-exchange bioseparation is to select the appropriate chromatographic materials for a given biological mixture. It has been suggested that virtual screening of separation materials in a manner that parallels the current QSAR method in drug design would facilitate the selection of proper chromatographic conditions and speed up development processes.

As a result, the level of interest in developing Quantitative Structure-Retention Relationship (QSRR) models based on linear or nonlinear modeling techniques has substantially increased [23–27]. The major aim of these studies is to construct improved QSRR models to predict the retention behavior of solutes in different stationary phases so that the retention process can be better predicted and understood, and to build a valuable chromatographic tool for highlighting the molecular mechanisms of retention. Due to computational bottlenecks in descriptor computation and machine learning algorithms, most current approaches are applicable only for small molecules. Recent research is focused on the adaptation of these techniques to large molecules, such as QSRR models based on a genetic algorithm/partial least-squares approach for macromolecules using the TAE electron-density-derived descriptor technique [2,11,28,29] that utilizes geometric data from protein crystal structures [30]. In this example, we present another modeling method for macromolecules by adapting the SVM [1,31] method to the particular problem of evaluating the retention time of proteins in different ion-exchange conditions. The predictive power of the resulting models is demonstrated by testing them on unseen data that are not used in either descriptor-selection step or model-generation step.

11.5.1   Protein Data and Descriptors

In order to obtain retention times of proteins, Cramer and coworkers carried out linear gradient chromatography using two anion-exchange stationary phases (Q Sepharose and Source 15S) under similar conditions [32]. The crystal structures of 24 structurally diverse proteins were downloaded from the RSCB Protein Data Bank [33], after which the Sybyl v6.5 software [34] was used to preprocess the raw macromolecular structures by eliminating the waters of hydration present in the published protein structures. Table 11.3 provides a list of the names and PDB codes of the 24 proteins. Three proteins that were selected as blind test cases are marked in the list.

A total of 279 MOE and RECON descriptors were computed for these proteins to give a composite set of traditional two- and three-dimensional as well as electron-density-derived TAE descriptors. All computational experiments were carried out on an SGI workstation (Octane/SE). The SVM modeling procedures given earlier were applied to this data set.

Table 11.3 Proteins Used in the Ion-Exchange Chromatography Experiment

PDB Code	Protein Names
1A4L	Adenosine deaminase
1AIV	Conalbumin
1AJC*	Alkaline phosphate
1AO6*	Human serum albumin
1AUO	Carboxylesterase
1AVU	Trypsine inhibitor
1AYY	Glycosylasparaginase
1BEB	Bovine β-lactoglobucin B
1BSO	Bovine β-lactoglobulin A
8CAT	Beef liver catalase
1EG1	Endoglucanase I
1F4H	*E. coli* β-galactosldase
1F6S	α-Lactalbumin
1FWE	Kelbsiella aerogenes urease
3GLY*	Glucoamylase
1IES	Ferritin
4INS	Insulin
1LPN	Lipase
1OVA	Ovalbumin
2PEL	Peanut lectin
3PEP	Pepsin
1QIW	Calmodulin
1UOR	Recombinant human serum albumin
1YGE	Lipoxygenase

Note: Entries marked by * were used in the test set.

11.5.2 Modeling Process

We can illustrate the interactions between a chemist and the SVM modeling process by considering aspects of the human/machine interface. The dimensionality and quantity of modeling data are too large to grasp in tabular form, and simple statistical summaries provide only rudimentary information. Visualization of the boostrap folds allows users to extract information mined from the models and to interact with the modeling process.

First of all, let us show what the prediction looks like using all descriptors available—without using feature selection. In Figure 11.7, each protein example has been plotted using its observed retention time (horizontal axis) versus its corresponding predicted value calculated by a linear SVM regression model. The testing set points are highlighted using large dot symbols. The testing set points were not used in any component of the modeling process. In each case, the middle point of each bar is the mean of predictions for each protein during the bootstrap process, and the length of the bar represents the standard deviation of aggregate predictions for each protein.

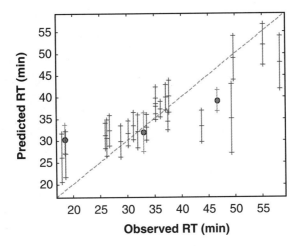

Figure 11.7   The prediction scatter plot by the model using all 279 descriptors.

In the first feature-selection phase, 20 bootstrap iterations of SVM models were constructed based on 20 different partitions of the training data. After feature selection, the aggregate SVM model was found to contain only 15 descriptors with nonzero weights. These 15 descriptors are given in Table 11.4. The scatter plot for the nonlinear SVM based on these 15 descriptors is shown in Figure 11.8. One can see the model is much improved, although one of the validation set predictions has become less accurate.

At this point, the chemist has the choice of accepting these features or further refining the feature-selection procedure using chemical intuition or experience with the descriptors. To assist in this step, there are a number of tools available for visualizing high-dimensional data that allow the chemist to interact with the modeling process. For example, starplots in Figure 11.9 allow us to examine the importance of each chemical-property descriptor throughout each of the 20 bootstrap folds. In this mode, the radius of each starplot (beginning at the 12 o'clock position and proceeding clockwise) represents the weight or importance of each descriptor in each of the 20 bootstrap iterations. In all of the starplot figures, the order of presentation is arranged so that the most significant descriptor with negative weight appears in the upper left-hand corner, and the most significant positive contributor appears on the lower right. The order proceeds from left to right in a columnar fashion. Overall descriptor significance is a function of the sum of ray lengths within each starplot. As shown in Figure 11.9, the SIGIA descriptor is important in all of the folds, indicating that this feature (the surface integral of the G electronic kinetic energy density) represents a chemical interaction that is fundamental to ion-exchange protein binding. Other features with smaller and less symmetrical starplots can also be important, but to a lesser extent. In this example, starplots contained within solid lines contribute to high values of protein retention, while those within dashed lines have

Table 11.4 Definitions of Descriptors Selected in the First Phase of Feature Selection

Descriptor Name (source)	Chemical Information Encoded in These Descriptors
Q.VSA.FPPOS (MOE)	Fraction of positive polar van der Waals surface area
PEOE.VSA.FPPOS (MOE)	The same chemical meaning as Q.VSA.FPPOS while using a different approach to calculate atomic partial charges
VSA.POL(MOE)	Sum of van der Waals surface areas of "polar" atoms
SIEP(TAE)	Surface integral of molecular electrostatic potential
PMIX, PMIZ(MOE)	The x and z components of the principle moment of inertia
PIPAVG(TAE)	Average value on all bins of PIP properties. PIP refers to the local average ionization potential needed to remove an electron from a specific part of a molecule
DIPOLE(MOE)	The first energy derivative with respect to an applied electric field. It is a measure of the asymmetry in the molecular charge distribution. It has three components: DIPOLEX, DIPOLEY, DIPOLEZ along the x,y,z axes as well as a magnitude DIPOLE
SIKIA(TAE)	K electronic kinetic-energy density, which correlates with the presence and strength of Brönsted basic sites *Note:* $K = (-(\hbar/4m)\int\{\psi^*\nabla^2\psi + \psi\nabla^2\psi^*\}\,d\tau)$
PIP1(TAE)	The first histogram bin of PIP property
DEL.K.IA(TAE)	Gradient of the K electronic kinetic energy normal to the molecular surface that describes differences in the polarizability and hydrophobicity of molecular regions. More negative ranges of this function are believed to indicate that certain regions of a molecule are more hydrophobic and also less susceptible to electrophilic attack
SIGIA(TAE)	Derived from the G electronic kinetic-energy density on the molecular surface *Note:* $G = (-(\hbar/4m)\int\{\nabla\psi^* \cdot \nabla\psi\}\,d\tau)$
STD.DIM3 (MOE)	The square root of the third largest eigenvalue of the covariance matrix of the atomic coordinates. A standard dimension is equivalent to the standard deviation along a principal component axis (i3d descriptor).

negative contributions to retention. An unenclosed starplot indicates that sign changes take place within different bootstrap folds. In this case, the DIPOLEX descriptor exhibits this behavior.

By examining the descriptor information in Table 11.4 and the starplots in Figure 11.9, a chemist can further refine the choice of descriptors. By inspection of the starplots, it can be seen that the electrostatic descriptor Q.VSA.FPPOS is complementary with PEOE.VSA.FPPOS. The definitions in Table 11.4 support this observation, since these two descriptors both represent the same kind of electrostatic molecular surface properties [35].

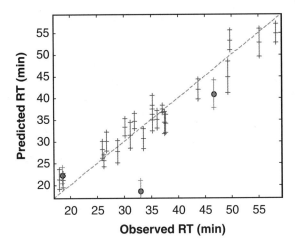

Figure 11.8   The prediction scatter plot by the model using 15 selected descriptors.

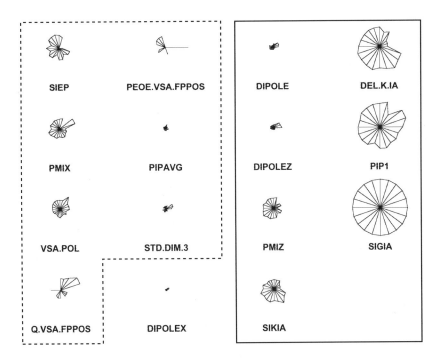

Figure 11.9   The starplots for 15 descriptors selected by feature-selection algorithm. Descriptors within dashed lines decrease retention time, descriptors enclosed by solid lines increase retention time, while the remaining descriptors alternate signs within the bootstrap folds.

Two additional descriptors were dropped: DIPOLEX and STD.DIM3. The DIPOLEX descriptor (Table 11.4) has very small weights and these weights flip signs during the 20 bootstraps, indicating that this feature does not significantly affect the final model. The descriptor STD.DIM3 (Table 11.4) is chemically difficult to interpret, and it has very small or zero weights in all of the models. Thus the chemist elected to drop this feature. It is normally preferable to use a small number of descriptors in the final model, but this criterion is balanced by the need to retain the pertinent chemical information often carried by combinations of less important descriptors. For example, nonlinear SVMs function better with too many features rather than too few. With this consideration in mind, the remaining 12 descriptors were retained.

The starplots of 20 bootstrapped models generated by SVM regression using the 12 remaining descriptors can be seen in Figure 11.10. By comparing the starplots in Figure 11.10 with those in Figure 11.9, it can be seen that those within the dotted lines (negative contributors to retention time in this figure) are consistently significant throughout the bootstrap folds. The starplots shown within solid lines (positive contributors to retention) are also more symmetrical. These characteristics allow the model to be interpreted with confidence.

Figure 11.11 illustrates the predictive power of the final SVM model based on

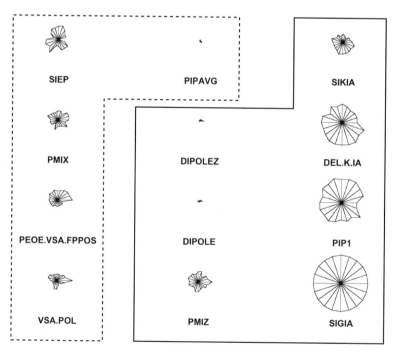

Figure 11.10   Starplots of important features after running the second SVM feature selection. There are no descriptors with flipping weights after this degree of feature selection.

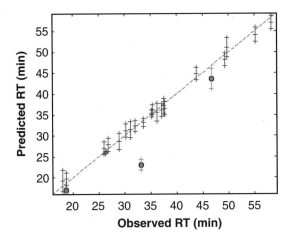

Figure 11.11   Predicted vs. actual retention times after the second feature selection.

the 12 selected features. It can be seen that the final model after feature selection is considerably more accurate than the initial model that used all of the available descriptors. It is also clear that the chemist-assisted second iteration of feature selection improved the predictive capabilities and interpretability of the final model.

11.5.3   Model Interpretation

Although the interpretation of descriptors is not a straightforward exercise, some of the most important descriptors may be indirectly attributed to specific types of non-covalent interaction. In the present example, the interaction of the Source-15S stationary phase with proteins can be described by a model with four main features: electrostatic interaction, polarity, hydrogen bonding, and protein shape.

Electrostatic interactions are represented by descriptors based on the distribution of partial charge on the van der Waals surface, and indirectly by other electron-density-derived descriptors. The MOE descriptor PEOE.VSA.FPPOS describes the fraction of solvent-accessible molecular polar-surface area that bears a positive partial charge. It is observed that when this feature increases, it causes a decrease in the protein retention time. This observation is consistent with intuition, in that the positively charged stationary-phase resin would be expected to repel positively charged molecular surface regions. It is interesting to note the consistent importance of the SIGIA descriptor in all of the models. This descriptor relates to the energy of the electron density on the surface of the molecule, and is also related to molecular size—in that sense it is related to the DELKIA descriptor. The relative electronic kinetic-energy density of the molecular surface described by these two descriptors has been associated with Lewis acidity, so this descriptor provides a unique insight into the ion-exchange interaction. It is important to note that simple molecular size and surface-area descriptors were available in the original descriptor set, but were not chosen in the feature-selection procedure. The PIP1 descriptor is also consistently important, and indicates that easily ionizable regions of electron density protein binding, most likely because they are protonated under the chromatographic conditions used in the experiment. The importance of the MOE polar surface-area descriptor VSA.POL implies that the hydrogen-bond capacity of proteins is also involved in these intermolecular interactions. The descriptors PMIX and PMIZ are shape-related descriptors that suggest that flattened proteins are retained more strongly under these conditions.

11.5.4 Summary of Case Study 1

In this Case Study, SVM regression and feature selection were introduced as a way to generate predictive models for predicting the interactions of proteins with ion-exchange resins. When this model-building procedure was used in concert with a diverse set of molecular-property descriptors, it was possible to obtain an internally consistent set of models that was able to make useful predictions of protein/resin interactions during chromatographic separations. The ability of the chemist to interact with the modeling and feature-selection process through starplot graphical representations allowed a more accurate and interpretable model to be constructed.

11.6 CASE STUDY 2: MODELING AND PREDICTION OF CACO-2 PERMEABILITY

The absorption, distribution, metabolism, and elimination/excretion (ADME) characteristics of pharmaceuticals are important in several developmental phases of therapeutic agents. A disturbing number of the compounds entering phase I–III clinical trials have failed due to issues directly related to ADME. Currently, the focus of drug discovery is not on simply achieving the best possible potency against a biological target of interest, but also on seeking favorable ADME properties as part of the design process. It has been suggested that computational models for reliable prediction of ADME properties are promising as an early screen for potential drug candidates, and can be useful in the design of combinatorial libraries, i.e., weeding out compounds with poor physicochemical properties early in the drug-discovery phase, thereby saving both time and expense.

Among these ADME properties, absorption (permeability) is of paramount importance in drug design. Currently, there is a heavy emphasis on producing drug candidates that have good oral absorption that hopefully extends to good oral bioavailability. Several in vitro test systems have been developed for measuring transport across intestinal mucosa. Of all in vitro systems that have been developed to date, the most commonly used method for determining the transport of compounds across the intestinal mucosa currently involves the use of Caco-2 cells, which were originally established by Jorgen Fogh approximately 20 years ago [36,37]. Caco-2 assays are amenable to high throughput screening, and therefore the Caco-2 cell line data have been widely accepted as a meaningful substitute for actual human intestinal absorption values.

It is generally assumed that the molecular properties relevant to intestinal absorption include lipophilicity, molecular size, charge, hydrogen bonding, and solubility [38]. Therefore, physicochemical descriptors of drug molecules have been widely applied to generate QSAR models for human intestinal permeability prediction. Lipinski et al. proposed the "rule of five" for preliminary estimation of compound absorption on the basis of molecular weight, lipophilicity, and the number of H-bond donor and acceptor atoms in the molecule [39]. The use of hydrogen bonding descriptors to quantify permeability goes back to earlier work using the HY-BOT method [40] and to estimations involving molecular polar surface area or dynamic polar surface areas [38,41,42]. Han Van de Waterbeemd et al. reported that permeability increases with increasing molecular weight and decreasing polar surface area, as shown below:

$$\text{Log } P_{app} = 0.008\% \text{ MW} - 0.043\% \text{ PSA} - 5.165. \tag{11.1}$$

(Here Log P_{app} is the logarithm of the apparent permeability (cm · s^{-1}) through the monolayer, and PSA indicates the polar surface areas of the compounds.) Quantum mechanically derived descriptors calculated by the program MolSurf were used to construct QSPR models for the drug-absorption investigation. Various complexities of computational models are applicable to this problem—they range from simple and rapid counts of atoms of fragments (e.g., Lipinski's rule of five), to time consuming approaches involving quantum mechanics calculations. Recently, in order to devise experimental protocol and computational models for the prediction of intestinal drug permeability, four different principles for molecular descriptors (atom count, molecular mechanics, fragmental, and quantum mechanics approach) were evaluated for their ability to predict intestinal membrane permeability [43].

11.6.1 Data and Descriptors

In this study, a representative Caco-2 cell permeability data set that includes 27 molecules was obtained from the literature [43]. Compounds were carefully selected from published in vivo data so that the data set contains only molecules that are predominantly absorbed through the intestinal wall by passive diffusion. Additional selection was also performed to minimize the contribution from complicating factors such as low solubility, instability in the gastrointestinal lumen, or presystemic metabolism. The structural diversity among these compounds was large enough to span a significant space of conventional drugs. The variability in lipophilicity, molecular weight and charge of the compounds is comparable to those of registered drugs intended for oral administration.

A large set of descriptors was generated by combining electron-density-based TAE descriptors, property/shape-encoded PEST descriptors, and traditional descriptors from the MOE software distributed by the Chemical Computing Group [10].

11.6.2 Modeling Results

The same feature selection and modeling procedure used in case 1 was also applied to the Caco–2 problem. The performance of the nonlinear SVM regression using all descriptors is shown in Figure 11.12. The validation data set is illustrated as large dots to show the quality of the original model, but those data were not included in the modeling process. To accomplish the modeling and subsequent feature selection, a sparse linear SVM regression model was trained using 20 different bootstraps (combinations of test and training data), and the modeling results were aggregated (averaged) to produce a final result. The combined model set was found to involve 31 descriptors with nonzero weights. These 31 features were then used within a nonlinear SVM to produce a set of nonlinear predictive models. The performance of the nonlinear SVM shown in Figure 11.13 is significantly improved, both in terms of accuracy and variance. A starplot of the 31 selected descriptors is given in Figure 11.14.

Notice that 16 of the descriptors (the ones not enclosed in boxes in Figure 11.14) have weights in the SVM regression models that changed signs in the different bootstraps. As in case 1, these 16 descriptors were removed from the descriptor field, since it is difficult to attribute chemical contributions to descriptors that change signs when used in related models. Since PEOE.VSA.FHYD has only a very small weight and other selected descriptors contain similar information, this descriptor was also removed. As illustrated in this procedure, chemistry domain knowledge can play an important part in model development to enhance the interpretability of the final model.

The remaining 15 descriptors are used to generate a final nonlinear SVM model whose results are shown in Figure 11.15. The final model still achieves a similar

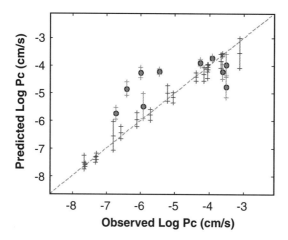

Figure 11.12 Caco-2 SVM modeling results prior to feature selection.

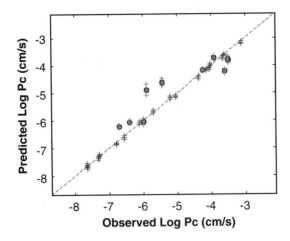

Figure 11.13 Caco-2 SVM modeling results using 31 features.

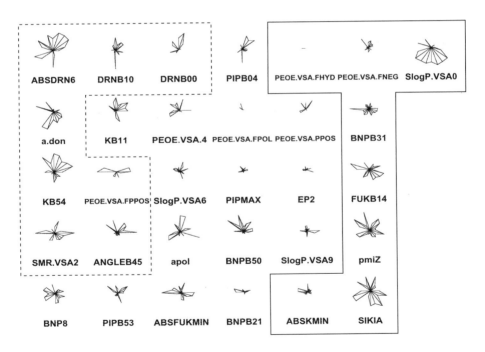

Figure 11.14 The starplots for 31 descriptors from the initial feature-selection step.

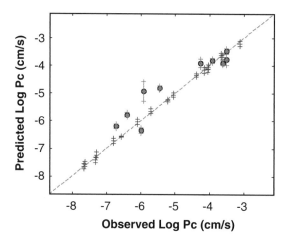

Figure 11.15 Caco-2 model of 15 descriptors after the second feature selection.

performance as the model containing 31 descriptors, however, the number of descriptors employed for model construction is reduced dramatically. A starplot of the 20 new sparse linear models based on the 15 selected descriptors are shown in Figure 11.16. The results indicate that the remaining descriptors are used consistently across the 20 models, allowing chemical interpretations to be made.

Note that descriptors in starplots surrounded by dotted lines are negatively weighted. Those surrounded by solid lines are all positively weighted. There are no descriptors with alternating positive and negative weights after this degree of feature selection.

11.6.3 Interpretation and Discussion

The absorption of therapeutic agents through the intestinal wall is a complex phenomenon that involves several different transport mechanisms. For the differentiated Caco-2 cell cultures used in HTS work to mimic small intestinal wall tissue, these mechanisms include transcellular active and passive transport, pericellular passive transport, and active transcellular efflux. Passive cellular transport phenomena are governed by a regular set of intermolecular interactions, including hydrophobicity, hydrogen bond donor/acceptor pairs, and molecular size, but are still difficult to predict with quantitative accuracy. Molecules subject to active transport or efflux are of special interest, and show up as outliers when passive diffusion models are used.

The models developed in this study are representative of a common situation that arises when quantitative accuracy is optimized at the expense of model simplicity. As a result of using complex descriptors and nonlinear modeling techniques, good quantitative predictions of Caco-2 permeability are obtained, but model interpretability is decreased. In the area of library design and virtual high-throughput screeing, this

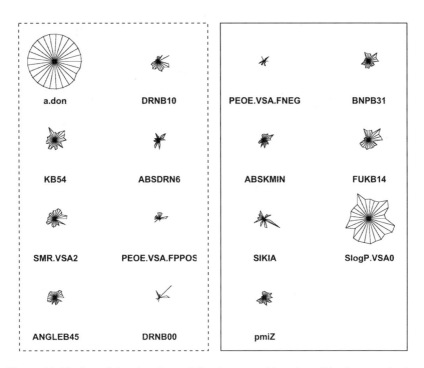

a.don DRNB10 PEOE.VSA.FNEG BNPB31

KB54 ABSDRN6 ABSKMIN FUKB14

SMR.VSA2 PEOE.VSA.FPPOS SIKIA SlogP.VSA0

ANGLEB45 DRNB00 pmiZ

Figure 11.16 Remaining descriptors following second iteration of the feature selection.

trade-off is usually favorable since the quality of the predictions is of paramount importance in these applications. When faced with an array of nonintuitive descriptors that are important in the final model, domain scientists (chemists) can still make valuble interpretations by examining the contributions of each class of molecular descriptors. For example, the most significant negative contribution to Caco-2 permeability is the number of hydrogen-bond donors as represented by the descriptor a.don. This agrees with the hypothesis that an excessive number of hydrogen-bond donor or -acceptor groups impairs molecular membrane bilayer permeability [44]. This is a nonlinear relationship in that, whereas an optimum number of H-bonding interactions enhance the ability of a drug to penetrate a cell membrane, too many of these interactions will severely curtail membrane permeability. In this example, the compensating effect is represented by the descriptor SlogP.VSA.0, which correlates with the presence of hydrophilic regions on the molecular surface.

In the model used in this example, molecular-size and -shape information are expressed by the descriptors ABSDRN6 and ANGLEB45, respectively. Neither of these descriptors is cleanly associated with specific molecular shapes or sizes, but ABSDRN6 is known to correlate with molecular surface area. The presence and negative weighting of the ANGLEB45 and KB45 descriptors suggests that large and flat hydrophobic molecules would not effectively pass through the Caco-2 cell barrier.

The nonlinear relationship between the permeability and negative charge of a compound was captured by PEOE.VSA.FNEG and PEOE.VSA.FPPOS. These descriptors suggest that a negatively charged surface area will be beneficial for permeability. Previous work is consistent with this explanation, in that it shows that peptides and pepidomimetics with only single or double anionic charges efficiently pass through Caco-2 cell membranes, while more highly charged molecules do not. Since biological membranes are negatively charged, this result is rather astonishing and interesting.

Molecular polarizability also plays a significant role in permeability as represented by the negative weighting on the SMR.VSA2 descriptor. One interpretation of this property proposed by Ulf Norinder et al. is that regions of polarizable electron density have doubly beneficial effects on Caco-2 permeability: first, they promote induced-dipole–induced-dipole interactions in the more hydrophilic phase before reaching the membrane surface; and second, they minimize electrostatic interactions in order to become as nonpolar and electrostatically "inert" as possible in the lipophilic phase that is encountered when the compound begins to penetrate the membrane [45]. The positive weighting of the FUKB14 and BNPB31 descriptors provides additional support for this interpretation.

11.6.4   Case Study Summary

In conclusion, both large-molecule and small-molecule interactions can be modeled using a combination of fundamental molecular descriptors and SVM regression. For example, Caco-2 cell permeability can be estimated simply from the structures of putative drug molecules, and protein interactions with stationary phase resins can be predicted with an unprecedented degree of accuracy.

11.7   CONCLUSIONS AND OPEN QUESTIONS

The availability of reliable QSPR and QSAR models can provide significant support to combinatorial library design and virtual high throughput screening projects. A key component in this effort is the generation of effective molecular property descriptors that capture fundamental chemical information relevant to the chemical behavior of interest. Electron-density-derived descriptors fulfill this role and enable the construction of predictive models, particularly when they are used together with hybrid shape/property PEST and traditional two- and three-dimensional MOE descriptors.

The importance of robust machine learning tools cannot be overestimated, particularly in cases where large numbers of descriptors are present in the training data and few molecules with known activities are available. In such cases, care must be taken in the modeling process to ensure that only valid models and interpretations are developed. An SVM regression methodology has been developed to address the goals of both predictive modeling and model interpretation. The inclusion of descriptive visualization tools allows the chemist to interact and guide the modeling process, thus allowing domain knowledge to be incorporated into the final model.

This chapter has investigated only one machine learning methodology, but there are many alternative methods possible within this framework. Additional new descriptors can also be incorporated to enhance specific modeling efforts.

A key component of the feature-selection process described in this chapter is the use of aggregate behavior of many sparse models to overcome the unreliability of any single model. Here a sparse SVM was used within the feature-selection process, but any sparse modeling process could be used similarly, for example, GAPLS. Potentially, nonlinear modeling methods such as GA sensitivity analysis [46] and MARKing, a heterogenous kernel modeling method [47], could be used.

The key to success of any QSPR and QSAR experiment is in the involvement of domain scientists in the modeling process. An open question is how to incorporate chemistry-domain knowledge to further guide the modeling process. The model interpretation described here is limited and nonlinear. Further work will focus on developing reliable methods to extract domain knowledge from nonlinear models.

ACKNOWLEDGEMENTS

This work was supported in part by NSF Grants IIS-9979860 and BES0079436, and by K. P. Bennett's NSF Career Award. Special thanks to Larry Lockwood and Mushin Ozdemir for valuable discussions and suggestions during the modeling process.

APPENDIX 11.A: METHODOLOGIES

In this section, we provide a more detailed discussion of the methodologies employed in the case studies just covered.

11.A.1 Support Vector Machines

SVMs are a powerful general approach for inference modeling. SVMs are based on the idea that it is not enough to just minimize empirically on training data such as is done in least-squares methods; one must balance training error with the capacity of the model used to fit the data. Through introduction of capacity control, SVMs avoid overfitting, producing models that generalize well. SVM's generalization error is not related to the input dimensionality of the problem, since the input space is implicitly mapped to a high-dimensional feature space by means of so-called kernel functions. This explains why SVMs are less sensitive to the large number of input variables than many other statistical approaches. However, reducing the dimensionality of problems can still produce lots of benefits, such as improving prediction accuracy by removing irrelevant features and emphasizing relevant features, speeding up the learning process by decreasing the size of search space, and reducing the cost of acquiring data because some descriptors or experiments may be found unnecessary. To date, SVM has been applied successfully to a wide range of problems, such as classification, regression, time series prediction, and density estimation. In this section, we focus on regression problems and present a feature-selection approach. The reader can consult recent literature [48,49] for more extensive overviews of SVM methods.

11.A.2   Support Vector Regression Models

SVM modeling was originally developed for pattern recognition, but Vapnik proposed a scheme to extend the basic properties achieved on pattern recognition to solve the regression problem. There are many variations in SVM methods. The reader should consult the literature for tutorials and full explanations of SVM [1,48,49]. The methods utilized in this work are documented in this section.

Suppose we are given training data drawn from a certain underlying distribution, our goal is to find a function $f(x)$ that has at most ε deviation from the actually obtained target y's for all the training data, and at the same time, is as flat as possible. In other words, we do not care about errors as long as they are less than ε, but will not accept any deviation larger than this. This loss function, the ε-insensitive loss function, is illustrated in Figure 11.17. In QSPR, the magnitude of ε should be roughly the experimental error in the measurement of y.

Let us start by describing the classic SVMs with linear functions, taking the form $f(x) = \langle w \cdot x \rangle + b$, with $w \in R^N$, $b \in R$ as hypothesis space where $\langle w \cdot x \rangle$ denotes the dot product in R^N. Flatness in this case means that one seeks small w. One way to ensure this is to minimize the 1-norm, i.e., $\|w\|_1 = \Sigma_{i=1}^{N}|w_i|$. Formally, this problem can be posed as as a convex optimization problem with the objective function having one term related to Euclidean norm and one term concerning the training error calculated based on ε-insensitive losses. It means that once a training data set obtains a prediction that is farther than ε away from its true response y, slack variables are added leading to the following formulation:

$$\text{minimize:} \quad \frac{1}{2}\|w\|_1 + C\sum_{i=1}^{M}(\xi_i + \xi_i^*)$$

$$\text{subject to:} \quad y_i - \langle w \cdot x_i \rangle - b \leq \varepsilon + \xi_i, \qquad \xi_i \geq 0,$$

$$\langle w \cdot x_i \rangle + b - y_i \leq \varepsilon + \xi_i^*, \qquad \xi_i^* \geq 0,$$

$$i = 1, 2, \ldots, M$$

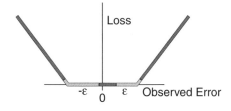

Figure 11.17 ε-Insensitive loss function.

Problem 11.A.1: Classic Sparse Linear SVM Model   The constant C determines the trade-off between the flatness of the linear function and the amount up to which deviations larger than ε are tolerated. The constant C also controls the degree of sparsity of the coefficient vector w. Minimizing the 1-norm of w forces many of the coefficients to be 0. Thus C trades off the simplicity or capacity of the function with the training set error.

The parameter ε in classic support vector regression is usually difficult to control, as one usually does not know beforehand how accurately one is able to fit the curve. This problem is partially resolved in a new algorithm, v, support vector regression, which executes automatic accuracy control [31]. In this case, parameter ε, the size of the ε tube itself, is a variable in the optimization problem. A new parameter $v \in (0,1]$ is used to control the ε. The v determines the fraction of error examples, and the fraction of support vectors.

The core aspect and advantage of v support vector regression can be captured in the following several points: assume $\varepsilon > 0$, (1) v is an upper bound on the fraction of errors; (2) v is a lower bound on the fraction of points inside the ε-insensitive tube; (3) suppose the data were generated indepedently identically distributed (i.i.d.) from a continuous distribution, with probability 1, asymptotically, v equals both the fraction of points inside the tube and the fraction of errors.

To solve the sparse linear v SVM it is convenient to reformulate the problem as a linear program. In order to formulate it as a linear program, we rewrite $w_i = \alpha_i - \alpha_i^*$, where $\alpha_i \geq 0$ and $\alpha_i^* \geq 0$, and thus $|w_i| = \alpha_i + \alpha_i^*$. Then the optimization problem can be formulated as

$$\text{minimize:} \quad \frac{1}{2} \sum_{j=1}^{N} (\alpha_j + \alpha_j^*) + C\frac{1}{M} \sum_{i=1}^{M} (\xi_i + \xi_i^*) + Cv\varepsilon$$

$$\text{subject to:} \quad y_i - \sum_{j=1}^{N} (\alpha_j - \alpha_j^*)x_{ij} - b \leq \varepsilon + \xi_i, \qquad i = 1, 2, \ldots, M$$

$$\sum_{j=1}^{N} (\alpha_j - \alpha_j^*)x_{ij} + b - y_i \leq \varepsilon + \xi_i, \qquad i = 1, 2, \ldots, M$$

$$\alpha_j, \alpha_j^*, \xi_i, \xi_i, \varepsilon \geq 0, \qquad j = 1, 2, \ldots, N, \qquad i = 1, 2, \ldots, M.$$

Problem 11.A.2: Sparse Linear SVM LP   Due to such good properties demonstrated by v support vector regression, we adopted the preceding SVM formulation into our numerical experiments on QSAR data. Solving this linear programming formulation is the vital part of our feature-selection algorithm.

For the final predictive model, nonlinear SVM are used. Nonlinear SVM regression models are constructed using the so-called "kernel" trick. The nonlinearity of the model is obtained by mapping data x into a feature space via the feature map $x \mapsto \Phi(x)$ and computing a linear regressor in feature space to acquire $f(x) = \langle w \cdot \Phi(x) \rangle + b$. The normal vector w can be expressed as $w = \sum_{i=1}^{M} \alpha_i \Phi(x_i)$, and the regression function f can be written in the form of kernel expansion as $f(x) =$

$\sum_{i=1}^{M} \alpha_i k(x_i, x) + b$, where $k(x_i, x) = \langle \Phi(x_i) \cdot \Phi(x) \rangle$. The radial basis function (RBF) kernel $k(x, x') = \exp(-\|x - x'\|^2/2\sigma^2)$ is generally adopted into our computational studies. Thus the problem is formulated as a linear program with respect to variables v, α_i, α_i^*, $i = 1, \ldots, M$ and b. Problem 11.A.3 provides the nonlinear SVM model used in this work.

$$\text{minimize:} \quad \frac{1}{2}\sum_{j=1}^{M}(\alpha_j + \alpha_j^*) + C\frac{1}{M}\sum_{i=1}^{M}(\xi_i + \xi_i^*) + Cv\varepsilon$$

$$\text{subject to:} \quad y_i - \sum_{j=1}^{M}(\alpha_j - \alpha_j^*)k(x_i, x_j) - b \leq \varepsilon + \xi_i, \qquad i = 1, 2, \ldots, M$$

$$\sum_{j=1}^{M}(\alpha_j - \alpha_j^*)k(x_i, x_j) + b - y_i \leq \varepsilon + \xi_i, \qquad i = 1, 2, \ldots, M$$

$$\alpha_j, \alpha_j^*, \xi_i, \xi_i, \varepsilon \geq 0, \qquad i, j = 1, 2, \ldots, M.$$

Problem 11.A.3: Nonlinear SVM Problem   The optimal α_i, α_i^*, $i, = 1, \ldots, M$ will have relatively few nonzero components. This allows us to obtain the solution of normal w by generating a linear combination of only a few training examples, x_i. The few training samples with nonzero weights are the support vectors. The corresponding regression function f is produced from only a few kernel functions $k(x_i, x)$. A nice property of the preceding objective function is that its solution is robust with respect to small perturbations of the training data. In other words, using linear programming regression with the ε-insensitive loss function, local movements of target values of points outside the ε-tube do not influence the regression.

11.A.4   Feature Selection Approach

To develop an approach of feature selection wrapped around this induction algorithm, a criterion measuring the quality of feature subsets has to be applied to the induction. A complicated but perhaps intuitive way is to utilize a wrapper approach utilizing some search algorithm to pick out feature subsets and then evaluate the effectiveness of each selected subset in the induction algorithm. This is in many cases computationally infeasible. So another way that may be practically easier and more efficient is to combine the two steps together by utilizing a sparse induction method such as the SVM in Problem 11.A.2. Then while performing induction, one can select and assess the quality of different feature subsets at the same time.

Therefore we decide to carry out the following scheme. Construct a series of sparse linear SVMs (Problem 11.A.2) that exhibit good generalization. Choose the subset of features having nonzero weights in the linear models. Then use this subset of features in nonlinear SVM to produce the final regression function. The method takes advantage of the fact that a linear SVM with l_1-norm regularization inherently performs feature selection as a side effect of minimizing capacity in the SVM model. In a linear l_1-norm SVM, the optimal weight vector will have relatively few nonzero weights, with the degree of sparsity depending on the SVM model parameters. The features with nonzero weights then become potential attributes to be used in the nonlinear SVM. In some sense, in SVM we trade the feature-selection problem for the model parameter-selection problem.

The method of feature selection works by assigning features different weights. Once we determine a linear regression model, i.e., a linear function $y = \langle w \cdot x \rangle + b$, we obtain a weight vector, w, with each component corresponding to each feature. Most features have zero weights, while a few features have nonzero weights. We believe these latter contribute more in the relation between attributes and response than the zero ones, especially considering the linear relationship. Why does the l_1-norm of α inherently enforce the sparseness of solution? Roughly speaking, a reason for the reduced sparseness is the fact that vectors far from the coordinate axes are larger with respect to the l_1-norm than with respect to l_p-norms with $p > 1$. For example, consider the vectors $(1,0)$ and $(1/\sqrt{2}, 1/\sqrt{2})$. For the l_2-norm, $\|(1,0)\|_2 = \|(1/\sqrt{2}, 1/\sqrt{2})\|_2 = 1$, but for the l_1-norm, $1 = \|(1,0)\|_2 < \|(1/\sqrt{2}, 1/\sqrt{2})\|_2 = 2$. For the Problem 11.A.2, the optimal solution is always a vertex solution (or can be expressed as such) and tends to be very sparse.

11.A.5 Pattern Search Model Selection

The performance of the SVM model is dependent on the selection of the model parameters. A pattern-search model selection method is utilized to search for the parmaters. The regularization parameter, C, and tube parameter, v, play important roles in SVM learning, and especially in our feature selection, since the sparsity and predictive accuracy depend on the choices of these hyperparameters. We have to try to determine the appropriate values for them before solving the optimization problems in either Problem 11.A.2 or Problem 11.A.3. This becomes another optimization problem called model selection over model parameters C and v. Parameters C, v are are optimized based on the validation set withheld during training. The tuning statistic (or validation statistic) in this work was

$$Q^2 = \frac{\sum_{i=1}^{M}(y_i - \hat{y}_i)}{\sum_{i=1}^{M}(y_i - \bar{y})},$$

where $\hat{y}_i$ is the prediction of response y_i by the learned model for sample example x_i, and $\bar{y}$ is the average of sample responses. This statistic is the mean squared error of responses normalized by the variance of response.

A regression model is produced by solving the linear-programming problem for the training data with the given C and v. To evaluate the model, the model is tested on the validation set and the predictions $\hat{y}_i$ for sample examples in the tuning set are obtained. Then Q^2 is calculated using these predictions for the given values of C and v. We formulate the model-selection problem as $\min_{C,v} Q^2$. Obviously, it is hard to express the Q^2 explicitly as a function of C and v. There is no explicit derivative, so classic methods such as the gradient-descent method and Newton's method cannot be applied. Thus a simple derivative-free pattern search strategy for SVM was utilized [50]. The method searches a range of parameters using only function evaluations. The good ranges of model parameters could be problem-specific, but we prefer to find ways general enough for most data sets, so we provide reasonably large ranges for them, such as $e^{-2} = 0.1353 \leq C \leq e^{10} = 22026$ and $0.02 \leq v \leq 0.6$.

The pattern-search algorithm for model selection is embedded in the feature-selection algorithm as a subroutine; this is depicted in Figure 11.18. The generic framework of the pattern-search method is that in each iteration it starts with an initial randomly chosen center, samples other points around the center in the search, and calculates function values of each nearest point until it finds a point with a function value less than that of the center. After that the algorithm obtains a new minimizer, so it moves the center to the new minimizer. If all the points around the center fail to decrease the function value, then the size of the search step is reduced by half. This search continues until the search step gets sufficiently small, thus ensuring convergence to a local minimizer. This searching process is called the near-

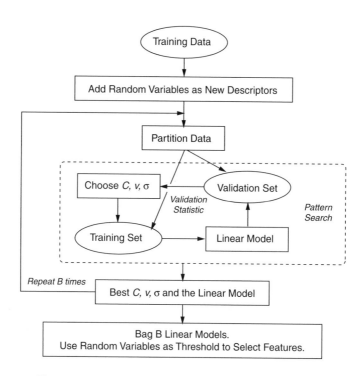

Figure 11.18   The linear SVM algorithm for feature selection.

est-neighbor search. To gain computational efficiency, we reuse function values of nearest neighbors, if appropriate, as the center moves so as to avoid evaluating the same function.

The pattern search is based on a pattern **P**. There are many ways to generate **P**. For the purpose of the two-dimensional search of C and v in this approach, we used the nearest-neighbor sampling pattern. The advantage of pattern search is that it does not require differentiability or continuity of the objective function. Furthermore, because the algorithm samples a point even if it is in an increasing direction of the objective function, pattern search can be more robust to local minima than derivative-based methods. Second, it can be applied to problems where the derivative information is not reliable or even to problems where the function is not differentiable, nor even continuous. The pattern that we use is one of the simplest patterns. It consists of only nearest neighbors. We could have chosen a different pattern, such as including next nearest neighbors or more.

11.A.6 Bootstrap Aggregation

Breiman [51] describes a technique based on the bootstrap [52,53] that can be used to augment the performance of various learning algorithms. This is called *bootstrap aggregation* or *bagging* [51]. It involves taking a number of bootstrap replicates of the training set and deriving regression predictions for the entire test set from each one and averaging these over the bootstrap replicates. Regression based on this average is usually superior to that from any of the replicates themselves, or indeed to that obtained by just using the regression model obtained from the training set itself.

To make the applied methodology more robust, **B** random partitions are performed in order to combine results produced by different partitions so that the inherent variation of the learning procedure can be reduced, since in QSAR we usually have few sample data but many descriptors. The bagging algorithm for feature selection is summarized in Figure 11.18.

There are many possible strategies that can be adopted to obtain the feature subset. A simple strategy is to select any descriptor with nonzero weights in any of the **B** linear models. Since QPSR problems can be underdetermined, allowing spurious descriptors to receive nonzero weights, the feature subset is further refined to eliminate variables that are weighted less than a random descriptor. To check this, a random descriptor with low correlation to the response vector is generated. If this random descriptor is weighted in the bagged models, then all features with average absolute weight less than this feature are eliminated from the selected subset.

11.A.7 Model Evaluation

One of the indispensable components of the QSPR processes is how to evaluate the induction performance and variable-selection approaches. Reasonably accurate estimates of predictive accuracy are needed for model selection and for estimating the validity of the final model. If the feature selection algorithm is too aggressive, the training accuracy will be low, resulting in underfitting of the model. If the search for the feature subsets is viewed as part of the induction algorithm, then overuse of the accuracy estimates may cause overfitting of the model. Both can result in poor predictive performance.

We use two schemes, one is leave-1-out cross-validation, the other one is to choose a holdout set to evaluate the predictive power of a final regression model over a selected feature subset. Experiments using cross-validation to guide the search must report the accuracy of the selected variable subset on a separate test set or on a holdout set generated by an external cross-validation loop that was never used during the variable-selection process and induction learning process.

Figure 11.6 shows the flowchart of our experimental design. Three examples in the anion dataset and nine examples in the Caco-2 dataset were selected based on references to serve as holdout test sets. They did not appear in the procedure of either feature selection or inductive learning. The test statistics are obtained by applying the resultant regression models from the whole learning process (Figure 11.6) to the holdout set.

REFERENCES

1. Vapnik, V. N. *IEEE Trans. Neural Networks,* 1999, 10: 988–999.

2. Breneman, C. M.; Rhem, M. *J. Comput. Chem.,* 1997, 18: 182–197.

3. Kewley, R. H.; Embrechts, M. J.; Breneman, C. *IEEE Trans. Neural Networks,* 2000, 11: 668–679.

4. McElroy, N. R.; Jurs, P. C. *J. Chem. Inf. Comput. Sci.,* 2001, 41: 1237–1247.

5. Hansch, C. *Acc. Chem. Res.,* 1969, 2: 232–239.

6. Famini, G. R.; Wilson, L. Y.; DeVito, S. C. *ACS Symp. Ser.,* 1994, 542: 22–36.

7. Murray, J. S.; Politzer, P.; Famini, G. R. *J. Mol. Struct. (Theochem),* 1998, 454: 299–306.

8. Mitchell, B. E.; Jurs, P. C. *J. Chem. Inf. Comput. Sci.,* 1998, 38: 489–496.

9. Breneman, C. M.; Rhem, M. O. *J. Comput. Chem.,* 1997, 18: 182–197; Sundling, M. Masters Thesis, Rensselaer Polytechnic Institute, Troy, 2001.

10. Labute, P. *J. Mol. Graph. Model.,* 2000, 18: 464–477.

11. Breneman, C. M.; Thompson, T. R.; Rhem, M.; Dung, M. *Comput. Chem.,* 1995, 19,:161–&.

12. Thompson, T. R. Ph.D. Thesis, Rensselaer Polytechnic Institute, Troy, NY, 1994.

13. Mulcahy, C. *Math. Mag.,* 1996, 69: 323–343.

14. Stollnitz, E. J.; Derose, T. D.; Salesin, D. H. *IEEE Comput. Graphics Appl.,* 1995, 15(3): 76–84.

15. Stollnitz, E. J.; Derose, T. D.; Salesin, D. H. *IEEE Comput. Graphics Appl.,* 1995, 15(4): 75–85.

16. Breneman, C. M.; Sukumar, N.; Bennett, K. P.; Embrechts, M. J.; Sundling, M.; Lockwood, L. *Abstr. Pap. Am. Chem. Soc.,* 2000, 220: U278.

17. Lockwood, L.; Breneman, C. M.; Embrechts, M.; Bennett, K. P.; Arciniegas, F. *Abstr. Pap. Am. Chem. Soc.,* 2000, 220: U296–U297.

18. Embrechts, M. J.; Breneman, C. *Abstr. Pap. Am. Chem. Soc.,* 1998, 215: U511–U512.

19. Sundling, C. M.; Breneman, C. M. *Abstr. Pap. Am. Chem. Soc.,* 2001, 221: U437.

20. Zauhar, R. J.; Welsh, W. J. *Abstr. Pap. Am. Chem. Soc.,* 2000, 220: U287–U287.

21. Cecchetti, V.; Filipponi, E.; Fravolini, A.; Tabarrini, O.; Bonelli, D.; Clemeti, M.; Cruciani, G.; Clementi, S. *J. Med. Chem.,* 1997, 40: 1698.

22. Hall, L. H.; Kier, L. B. *J. Chem. Inf. Comput. Sci.,* 2000, 40: 784–791.

23. Heinzer, V. E. F.; Yunes, R.A. *J. Chromatogr. A,* 1996, 719: 462–467.

24. Pompe, M.; Razinger, M.; Novic, M.; Veber, M. *Anal. Chim. Acta,* 1997, 348: 215–221.

25. Sutter, J. M. P.; Peterson, T. A.; Jurs, P. C. *Anal. Chim. Acta,* 1997, 342: 113–122.

26. M. Balcan, T. C.; Forgacs, E.; Anghel, D. F. *Biomed. Chromatogr.,* 1999, 13: 225.

27. Loukas, Y. L. *J. Chromatogr. A,* 2000, 904: 119–129.

28. Rhem, M. Ph.D. Thesis, Rensselaer Polytechnic Institute, Troy, NY, 1996.

29. Whitehead, C. E.; Breneman, C.; Ryan, D. *Abstr. Pap. Am. Chem. Soc.,* 1999, 217: U656–U656.

30. Mazza, C. B.; Sukumar, N.; Breneman, C. M.; Cramer, S. M. *Analyt. Chem.,* 2001, 73: 5457–5461.

31. Schoelkopf, B.; Bartlett, P.; Smola, A.; Williamson , R. In *Advances in Neural Information Processing Systems,* Kearns, M. S., Solla, S. A.; Cohn, D.A., Eds., MIT Press, Cambridge, MA, 1999, 11.

32. Song, M.; Breneman, C. M.; Bi, J.: Sukumar, N.; Bennett, K. P.; Cramer, S.; Tugcu, N. *J. Chem. Inf. Comp. Sci.*(in press).

33. http://www.rcsb.org/pdb

34. Clark, M.; Cramer III, R. D.; Opdenbosch, N. V. *J. Comput. Chem.,* 1989, 10: 982–1012.

35. Gasteiger, J., Marsali, M. *Tetrahedron,* 1980, 36: 3219.

36. Fogh, J.; Trempe, G. In *Human Tumor Cell Lines in Vitro,* Fogh, J., Ed., Plenum, New York, 1975, 115–141.

37. Fogh, J.; Fogh, J. M.; Orfedo, T. *J. Natl. Cancer Inst.,* 1977, 59: 221–226.

38. Han van de Waterbeemd, G. C.; Folkers, G.; Raevsky, O. A. *Quant. Struc.-Act. Relat.,* 1996, 15: 480–490.

39. Lipinski, C. A.; Lombardo, F.; Dominy, B. W.; Feeney, P. J. *Adv. Drug Delivery Rev.,* 1997, 23: 3–25.

40. Raevsky, O. A.; Trepalina, E. P.; McFarland, J. W.; Schaper, K.-J. *Quant. Estimation Drug Absorpt. Hum. Passively Transp. Compd. Basis Phys.-Chem. Parameters,* 2000, 19: 366–374.

41. Han van de Waterbeemd, K. M. *Chimia,* 1992, 46: 299–303.

42. Palm, K. L.; Ungell, A.; Strandlund, G.; Artursson, P. *J. Pharm. Sci.,* 1996, 85: 32–39.

43. Stenberg, P. U. N., Luthman, K.; Artursson, P. *J. Med. Chem.,* 2001, 44: 1927–1937.

44. Abraham, M. H.; Chadha, H. S.; Mitchell, R. C. *J. Pharm. Sci.,* 1994, 83: 1257–1268.

45. Norinder, U., Osterberg, T.; Artursson, P. *Pharm. Res.,* 1997, 14: 1786–1791.

46. Embrechts, M.; Breneman, C. M.; Arciniegas, F.; Ozdemir, M.; Bennett, K. In *Intelligent Engineering Systems through Artificial Neural Networks: Smart Engineering System Design,* Vol. 11, Dagli, C. H., Ed., ASME Press, New York, 2001, 345–350.

47. Demiriz, A.; Bennett, K.; Embrechts, M. *Proc. Interface 2002,* Interface Foundation, Fairfax, VA, 2002.

48. Bennett, K.; Campbell, C. *SIGKDD Explor.,* 2000, 2: 1–13.

49. Cristianini, N.; Shawe-Taylor, J. *An Introduction to Support Vector Machines,* Cambridge University Press, Cambridge, 2000.

50. Momma, M.; *Proc. SIAM Int. Conf. Data Mining,* 2002.

51. Breiman, L. *Mach. Learn.,* 1996, 24: 123–140.

52. Tibshirani, R. *Biometrika,* 1988, 75: 433–444.

53. Efron, B.; Tibshirani, R. *J. Am. Stat. Assoc.,* 1997, 92: 548–560.

CHAPTER 12

THE MONTE CARLO APPROACH TO LIBRARY DESIGN AND REDESIGN

MICHAEL W. DEEM
Rice University

12.1 INTRODUCTION

Once a high throughput laboratory has the ability to synthesize and screen for figure of merit large numbers of compounds, the question naturally arises as to what compounds should be made. This question exposes the basic feature of HTE, which is that only an infinitesimally small fraction of the space of possible chemical compounds can actually be explored. To search a system with five mole-fraction variables and two noncomposition variables at a resolution of 1% would take, for example, a library of 9×10^{10} compounds. Similarly, to search a small molecule system with 15 possible templates, each with 6 attachment sites, and 1000 possible ligands would take a library of 1.5×10^{19} compounds. Even more dramatically, it would take a library of $20^{100} \approx 10^{130}$ members to search exhaustively the space of all 100 amino-acid-long proteins.

The task of deciding what to make, that is, the task of deciding how to search the chemical and experimental variable spaces, is central to the design of high-throughput experiments. There is a striking analogy between searching this variable space and searching configuration space in Monte Carlo computer simulation. Searching configuration space using Monte Carlo methods is also an apparently challenging problem, as very often systems with many thousands of continuous degrees of freedom are examined, with effective complexities in excess of 10^{2000}. To date, HTE has not been successful at searching spaces of anywhere near this complexity. This chapter pursues the analogy between searching composition and noncomposition space in HTE and searching configuration space using Monte Carlo methods.

Monte Carlo methods are based upon the rigorous mathematical framework of Markov sampling. This framework ensures that points are sampled in a certain statistical way in a Monte Carlo protocol. In particular, the probability of visiting any given set of composition and noncomposition variables is given by $p = \exp(-\beta E)$, where E is the negative of the figure of merit, and β is a control parameter that de-

Experimental Design for Combinatorial and High Throughput Materials Development, Edited by James N. Cawse.
ISBN 0-471-20343-2 © 2003 John Wiley & Sons, Inc.

termines the degree to which favorable figures of merit are sampled. In some cases, E may be a function of the value returned by the experimental screen. If chemical binding energies were to be optimized, for example, E would typically be the logarithm of the measured binding constant. If a Monte Carlo protocol for library design is to be valid, it must possess several properties [1]. First, the protocol must be Markovian. That is, the library design for the next round can depend only on the current compositions, not on compositions in previous rounds. Second, the library design must lead to effective sampling of the entire variable space, or what is known as regular sampling. Third, the library design must satisfy what is known as the *balance condition*. This last condition ensures that the figure of merit will be properly sampled. An easier and sufficient condition of detailed balance is often imposed rather than balance. Detailed balance basically requires that the forward and backward fluxes to go from one state to another by a specific type of move are equal. This condition, which is overly strict and can be relaxed, is one way to ensure that equilibrium is reached. The Monte Carlo methods presented in this chapter satisfy these three conditions of validity.

In practice, a Monte Carlo method modifies the variables in the system in a random, yet statistical way. In the simplest possible method, the current value of the variables is modified by a small random amount. If the figure of merit increases by this change, the new value of the variables is used. If the figure of merit decreases, the new value of the variables is occasionally used. The probability with which the new value is used in this case is given by $\exp[-\beta(E_{\text{proposed}} - E_{\text{current}})]$. This simple procedure is called *Metropolis Monte Carlo*.

A significant advantage of Monte Carlo methods is that they are guaranteed to sample the figure-of-merit space. This is in contrast to most other optimization methods, in which a single global optimum is sought. Of course, Monte Carlo methods are themselves effective techniques for global optimization. However, the sampling that Monte Carlo provides is an additional, significant advantage. For example, simple global optimization may be misleading, since concerns such as ease of synthesis, patentability, and cost of materials are not usually included in the experimental figure of merit. Another way of understanding this issue is to realize that the primary screen, or the screen most easily performed in the laboratory, is usually correlated only roughly with the true figure of merit. That is, after materials are found that look promising based upon the primary screen, secondary and tertiary screens are usually performed to identify that material that is truly optimal. Finally, it may be that the figure of merit is only roughly known or that it has not been fully defined by the experimenters or management or that several figures of merit are of interest. In this case, sampling on the roughly defined figure of merit would lead to a set of possible compounds, some of which may be found to be suitable when later examined with a more refined figure of merit. Global optimization on the rough figure of merit, on the other hand, would produce a single molecule, which would have a high chance of being rejected when the more refined figure of merit is identified and applied.

The ultimate test of new, theoretically motivated protocols for HTE is, of course, experimental. So as to motivate such experimentation, the effectiveness of Monte

Carlo protocols will be demonstrated in this chapter by high throughput experiments where the experimental screening step is replaced by a figure of merit returned by a model. That is, to test the proposed protocols in an efficient fashion, a model will be used in lieu of the real experimental screening process. The model is not fundamental to the protocols; it is introduced as a simple way to test, parameterize, and validate the various library design methods. Nonetheless, the validity of the model is important if discrimination between the various searching protocols is to be achieved. The particular type of model used is known in the statistical physics literature as a random-energy model. This type of model cannot be used to design a compound on a computer, but it can be used to determine how well various protocols would work on a variety of possible compounds or a variety of possible figures of merit. A random-energy model, in particular, can mimic the generic features of an experimental figure of merit. For example, the NK model is used to model combinatorial chemistry experiments on peptides [2], the block NK [3] and generalized NK [4] models are used to model protein molecular evolution experiments, the Random Phase Volume Model is used to model materials discovery [5], and a random-energy model is used to model small molecule design [6]. These random-energy models are rather subtle aspects of statistical mechanics. The reader interested mainly in the Monte Carlo protocols themselves may skip the subsections that describe the details of these models, simply assuming that the figure of merit has been measured experimentally.

This chapter explores the Monte Carlo approach to library design and redesign in HTE of both materials and molecular systems, summarizing our recent work [5,6,7]. The task of library design for materials discovery is addressed first in Section 12.2. The space of variables is identified, and how to search this space by a Monte Carlo method is discussed. A Random Phase Volume Model is introduced as a surrogate for the experimental screen, and the effectiveness of the Monte Carlo protocols is judged. Attention is then turned to library design for small molecule discovery in Section 12.3. The space of variables is identified, and a random-energy model for the experimental screen is introduced. Several Monte Carlo strategies are described, and their performance on both figure of merit and diversity is compared with genetic algorithms. Finally, the interesting case of templated materials synthesis, where there are both molecular and solid-state variables, is considered in Section 12.4. Concepts from the rest of the chapter are brought together to show how Monte Carlo methods can be applied to this final case of library design and redesign. Some final thoughts are offered in Section 12.5.

12.2 MATERIALS DISCOVERY

The goal of high throughput materials discovery is to find compositions of matter that maximize a specific material property [8,9,10]. The property may be, for example, superconductivity [11], magnetoresistance [12], luminescence [13,14,15], ligand specificity [16], sensor response [17], or catalytic activity [9,18,19,20,21,22,23]. The task of finding desirable materials can be reformulated as one of searching a multidimensional space, with the material composition, impurity levels, and synthesis conditions as variables. The property to be optimized, the figure of merit, is generally an unknown function of the variables. Indeed, if how the figure of merit varied with composition were known, desirable materials could be determined purely by theoretical means. Generally this will not be the case, and the figure of merit will be measured only experimentally. It is under these conditions that HTE is useful.

Approaches to high throughput materials discovery to date almost always perform a grid search in composition space, followed by a gradient-type optimization of the figure of merit. This approach becomes inefficient in high-dimensional spaces or when the figure of merit does not vary smoothly with the variables, and use of the grid search has limited most current combinatorial chemistry experiments to ternary or quaternary compositions.

12.2.1 The Space of Variables

There is a large chemical and experimental space to search when seeking the material with the optimal figure of merit. Material composition is clearly a variable. In addition, there is a variety of noncomposition variables. Film thickness [24] and deposition method [25], for example, are variables for materials made in thin-film form. The processing history, such as pH, pressure, temperature, and atmospheric composition, is a variable. The impurity levels or guest compositions can greatly affect the figure of merit [23]. The crystal habit or "crystallinity" of the material can affect the observed figure of merit [24]. Finally, the method of synthesis or nucleation may affect the phase or morphology of the material and thereby affect the figure of merit [26].

There are several mechanisms by which these variables can affect the figure of merit, often in unexpected ways. First, a small impurity composition can cause a large change in the figure of merit, as seen by the rapid variation of activity with Cu content in a Cu/Rh oxidation catalyst [23]. Second, the phases formed in a thin-film synthesis are not necessarily the same as those formed in bulk, as seen in the case of a thin-film dielectric, where the optimal material was found outside the region where the bulk phase forms [24]. Third, the crystallinity of the material can affect the observed figure of merit, again as seen in the thin-film dielectric example [24].

12.2.2 Library Design and Redesign

HTE differs from usual Monte Carlo simulation in that multiple searches of the variable space are simultaneously carried out. That is, in a typical high throughput experiment, many samples, e.g., 10,000, are synthesized and screened for figure of merit at one time. With the results of this first round of experimentation, a new set of samples may then be synthesized and screened. This procedure may be repeated for several rounds, although current materials discovery experiments have not systematically made use of this feature.

Using the analogy with Monte Carlo, each round of combinatorial chemistry corresponds to a move in a Monte Carlo algorithm. Instead of tracking a single system with many configurational degrees of freedom, however, many samples are synthesized and screened, each with several composition and noncomposition variables. Modern robotic technology is what allows for the cost-effective synthesis and screening of these multiple-sample compositions.

The development of robotic technology for materials discovery continues, and future progress can and should be influenced by theoretical considerations. In this spirit, the composition and noncomposition variables of each sample are assumed to be independently adjustable, as in spatially addressable libraries [8,21]. This flexibility is significant, because it allows great latitude in how the variable space can be searched with a limited experimental budget. In fact, the term *high throughput experimentation* is to be preferred to *combinatorial chemistry* when spatially addressable libraries are used. Any constraints that exist on how the variables can be changed in an experiment can easily be accommodated in a Monte Carlo protocol, so the approach is generally applicable.

Current materials discovery experiments uniformly tend to perform a grid search on the composition and noncomposition variables. As will be shown in this chapter, however, it is preferable to choose the variables randomly and statistically from the allowed values. It is also possible to consider choosing the variables so as to maximize the amount of information gained from the limited number of samples screened, via a quasi-random, low-discrepancy sequence [27,28]. These sequences attempt to eliminate the gaps and the redundancy that naturally occur when a space is searched randomly, and they have several desirable theoretical properties. Figure 12.1 depicts these three approaches to materials discovery library design.

As a high throughput experiment proceeds, information is gathered about the figure-of-merit landscape, and this information can be incorporated by multiple rounds of screening. The Monte Carlo protocol provides one convenient method to

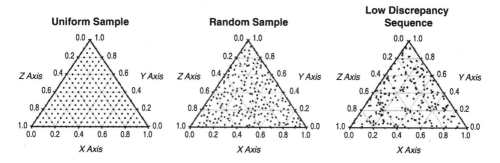

Figure 12.1 The grid, random, and low-discrepancy sequence protocols for designing the first library in a materials discovery experiment with three compositional variables. The random approach is freed from the regular pattern of the grid search , and the low-discrepancy sequence approach avoids gaps and overlapping points that may arise in the random approach. (Adapted, with permission from Deem, M. W., *Advances in Chemical Engineering,* Vol. 28, A. Chakraborty, Ed., 81–121; copyright © 2001. Academic Press, San Diego.)

incorporate this feedback in multiple rounds of experimentation. This approach leads to sampling the experimental figure of merit, E, proportional to $\exp(-\beta E)$. If the control parameter β is large, the Monte Carlo procedure will seek out values of the composition and noncomposition variables that optimize the figure of merit. If β is too large, however, the procedure will become trapped in relatively low-lying local optima. The first round is initiated by choosing the composition and noncomposition variables either from a grid search or statistically from the allowed values. The variables are changed in succeeding rounds as dictated by the Monte Carlo procedure.

12.2.3 Searching the Variable Space by a Monte Carlo Protocol

The Monte Carlo approach to HTE redesigns the library for the next round based upon the results of the screen of the library for the current round. In particular, the composition and noncomposition variables of the current round are perturbed in some fashion to produce the samples to be examined in the next round. Two ways of changing the variables are considered, either a random change of the variables of a randomly chosen sample or a swap of a subset of the variables between two randomly chosen samples. Swapping is productive when there is a hierarchical structure to the variables. The swapping event allows for the combination of beneficial subsets of variables between different samples. Swapping is, in fact, the same as the crossover move from genetic algorithms. A swap move might, for example, combine a good set of composition variables with a particularly good impurity composition. Alternatively, a good set of processing variables might be combined with a good set of composition variables. These moves are performed on the samples until all the samples have been modified. The new samples are then screened for figure of merit. Whether the newly proposed samples will be accepted or the current samples will be kept for the next round is decided according to the detailed balance acceptance criterion. For a random change of one sample, the Metropolis acceptance probability is applied:

$$p_{acc}(c \to p) = \min\{1, \exp[-\beta(E_{proposed} - E_{current})]\}. \tag{12.1}$$

Proposed samples that increase the figure of merit are always accepted. Proposed samples that decrease the figure of merit are accepted with the Metropolis probability. By allowing the figure of merit to decrease occasionally, the protocol is able to escape from local optima. Periodic boundary conditions are used on the noncomposition variables. As discussed in Appendix 12.B, reflecting boundary conditions are used on the composition variables.

A similar acceptance criterion is applied when the swapping move is used. In particular, if a swap is made between samples i and j, the acceptance probability is

$$p_{acc}(c \to p) = \min\{1, \exp[-\beta(E^i_{proposed} + E^j_{proposed}$$
$$- E^i_{current} - E^j_{current})]\}. \tag{12.2}$$

One round of the Monte Carlo procedure is shown in Figure 12.2a. The parameter β is not related to the real temperature of the system in the experiment and should be optimized for best efficiency. The characteristic sizes of the random changes in the composition and noncomposition variables are also parameters that should be optimized.

When the number of composition and noncomposition variables is too large, or when the figure of merit changes too roughly with the variables, normal Monte Carlo will not succeed in achieving effective sampling. Parallel tempering is a Monte Carlo method that is used to study statistical [29], spin glass [30,31], and molecular [32] systems with rough energy landscapes. The most powerful protocol for materials discovery incorporates parallel tempering for changing the system variables. In the parallel tempering approach, some of the samples are updated by Monte Carlo with parameter β_1, some by Monte Carlo with parameter β_2, and so on. After a round of synthesis and screening, samples are randomly exchanged between groups with different β's, as illustrated in Figure 12.2b. The acceptance probability for such an exchange of two samples is

$$p_{acc}(c \rightarrow p) = \min\{1, \exp[\Delta\beta\Delta E]\}. \tag{12.3}$$

Here $\Delta\beta = \beta_i - \beta_j$, and ΔE is the difference, before the exchange is made, of the figures of merit between the sample in group i and the sample in group j. This parallel tempering exchange step involves no extra screening, and is, therefore, free in terms

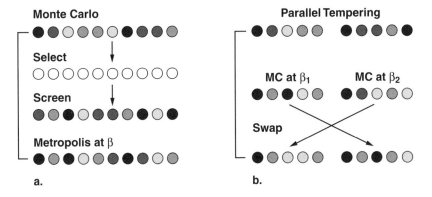

Figure 12.2 Schematic of the Monte Carlo library design and redesign strategy. (a) One Monte Carlo round with 10 samples. Shown are an initial set of samples, modification of the samples, and the Metropolis criterion for acceptance or rejection of the new samples. (b) One parallel tempering round with 5 samples at β_1 and 5 samples at β_2. In parallel tempering, several Monte Carlo simulations are performed at different temperatures, with the additional possibility of sample exchange between the simulations at different temperatures. (Reproduced, with permission, from Falcioni, M., and Deem, M.W., *Physics Review E,* 2000, 61: 5948–5952; copyright © 2000. American Physical Society, College Park, MD.)

of experimental costs. This step, however, can be, effective at helping the protocol to escape from local maxima. The number of different systems and the temperatures of each system are parameters that must be optimized for best experimental performance.

12.2.4   The Random Phase Volume Model

The effectiveness of these protocols is demonstrated by combinatorial chemistry experiments as simulated by the Random Phase Volume Model. The Random Phase Volume Model is not fundamental to the protocols; it is introduced as a simple way to test, parameterize, and validate the various searching methods. The model relates the figure of merit to the composition and noncomposition variables in a statistical way. The model is fast enough to allow for validation of the proposed searching methods on an enormous number of samples, yet possesses the correct statistics for the figure-of-merit landscape. The d-dimensional vector of composition mole fractions is denoted by $\mathbf{x}$. Composition mole fractions are nonnegative and sum to unity; so the allowed compositions are constrained to lie within a simplex in $d - 1$ dimensions. For the familiar ternary system, this simplex is an equilateral triangle. The composition variables are grouped into phases centered around N_x points $\mathbf{x}_\alpha$ randomly placed within the allowed composition range (the phases form a Voronoi diagram [33]; see Figure 12.3). The model is defined for any number of composition variables, and the number of phase points is defined by requiring the average spacing between phase points to be $\xi = 0.25$. To avoid edge effects, additional points are added in a belt of width 2ξ around the simplex of allowed compositions.

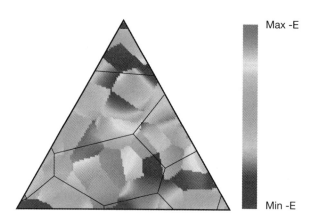

Figure 12.3   The Random Phase Volume Model. The model is shown for the case of three composition variables and one noncomposition variable. The boundaries of the $\mathbf{x}$ phases are evident by the sharp discontinuities in the figure of merit, $-E$. To generate this figure, the $\mathbf{z}$ variable was held constant. The boundaries of the $\mathbf{z}$ phases are shown as thin dark lines. (Reproduced, with permission, from Falcioni, M., and Deem, M. W., *Physics Review E,* 2000, 61: 5948–5952; copyright © 2000. American Physical Society, College Park, MD.)

The figure of merit should change dramatically between composition phases. Moreover, within each phase, α, the figure of merit should also vary with $\mathbf{y} = \mathbf{x} - \mathbf{x}_\alpha$ due to crystallinity effects such as crystallite size, intergrowths, defects, and faulting [24]. In addition, the noncomposition variables should also affect the measured figure of merit. The noncomposition variables are denoted by the b-dimensional vector $\mathbf{z}$, with each component scaled so as to fall within the range $[-1,1]$ without loss of generality. There can be any number of noncomposition variables. The figure of merit depends on the composition and noncomposition variables in a correlated fashion, and so the noncomposition variables also fall within N_z "z-phases" defined in the space of composition variables. There are a factor of 10 fewer noncomposition phases than composition phases. The functional form of the model when $\mathbf{x}$ is in composition phase α and noncomposition phase γ is

$$
E(\mathbf{x},\mathbf{z}) = U_\alpha + \sigma_x \sum_{k=1}^{q} \sum_{\substack{i_1 + \cdots + i_d = k \\ i_1, \ldots, i_d \geq 0}} f_{i_1 \ldots i_d; k} \xi_x^{-k} A_{i_1, \ldots, i_d}^{(\alpha k)} y_1^{i_1}, \ldots, y_d^{i_d}
$$

$$
+ \frac{1}{2}\left(W_\gamma + \sigma_z \sum_{k=1}^{q} \sum_{\substack{i_1 + \cdots + i_b = k \\ i_1, \ldots, i_b \geq 0}} f_{i_1, \ldots, i_b; k} \xi_z^{-k} B_{i_1, \ldots, i_b}^{(\gamma k)} z_1^{i_1}, \ldots, z_b^{i_b} \right), \quad (12.4)
$$

where $f_{i_1, \ldots, i_n; k}$ is a constant symmetry factor, ξ_x and ξ_z are constant scale factors, and U_α, W_γ, $A_{i_1, \ldots, i_d}^{(\alpha k)}$, and $B_{i_1, \ldots, i_b}^{(\gamma k)}$ are random Gaussian variables with zero mean and unit variance. In more detail, the symmetry factor is given by

$$
f_{i_1, \ldots, i_n; k} = \frac{k!}{i_1!, \ldots, i_n!}. \quad (12.5)
$$

The scale factors are chosen so that each term in the multinomial contributes roughly the same amount: $\xi_x = \xi/2$ and $\xi_z = (\langle z^6 \rangle / \langle z^2 \rangle)^{1/4} = (3/7)^{1/4}$. The σ_x and σ_z are chosen so that the multinomial, crystallinity terms contribute 40% as much as the constant, phase terms in the root-mean-square sense. For both multinomials $q = 6$. As Figure 12.3 shows, the Random Phase Volume Model describes a rugged figure of merit landscape, with subtle variations, local maxima, and discontinuous boundaries.

12.2.5 Several Monte Carlo Protocols

Six different search protocols are tested with increasing numbers of composition and noncomposition variables. The total number of samples whose figure of merit will be measured is fixed at $M = 100{,}000$, so that all protocols have the same experimental cost. The single-pass protocols grid, random, and low-discrepancy sequence (LDS) are considered. For the grid method, the constants $M_x = M^{(d-1)/(d-1+b)}$ and $M_z = M^{b/(d-1+b)}$ are defined. The grid spacing of the composition variables is $\zeta_x = (V_d/M_x)^{1/(d-1)}$, where

$$
V_d = \frac{\sqrt{d}}{(d-1)!} \quad (12.6)
$$

is the volume of the allowed composition simplex. Note that the distance from the centroid of the simplex to the closest point on the boundary of the simplex is

$$R_d = \frac{1}{[d(d-1)]^{1/2}}. \qquad (12.7)$$

The spacing for each component of the noncomposition variables is $\zeta_z = 2/M_z^{1/b}$. For the LDS method, different quasi-random sequences are used for the composition and noncomposition variables. The feedback Monte Carlo protocols, Monte Carlo with swap, and parallel tempering are considered. The Monte Carlo parameters were optimized on test cases. It was optimal to perform 100 rounds of 1000 samples with $\beta = 2$ for $d = 3$ and $\beta = 1$ for $d = 4$ or 5, and $\Delta x = 0.1 \ R_d$ and $\Delta z = 0.12$ for the maximum random displacement in each component. The swapping move consisted of an attempt to swap all of the noncomposition values between the two chosen samples, and it was optimal to use $P_{\text{swap}} \simeq 0.1$ for the probability of a swap versus a regular random displacement. For parallel tempering it was optimal to perform 100 rounds with 1000 samples, divided into three subsets: 50 samples at $\beta_1 = 50$, 500 samples at $\beta_2 = 10$, and 450 samples at $\beta_3 = 1$. The 50 samples at large β essentially perform a "steepest-ascent" optimization and have smaller $\Delta x = 0.01 \ R_d$ and $\Delta z = 0.012$.

12.2.6 Effectiveness of the Monte Carlo Strategies

The figures of merit found by the protocols are shown in Figure 12.4. The random and LDS protocols find better solutions than does grid in one round of experiment. More importantly, the Monte Carlo methods have a tremendous advantage over one pass methods, especially as the number of variables increases, with parallel tempering the best method. The Monte Carlo methods, in essence, gather more information about how best to search the variable space with each succeeding round. This feedback mechanism proves to be effective even for the relatively small total sample size of 100,000 considered here. The advantage of the Monte Carlo methods will become even greater for larger sample sizes. Note that in cases such as catalytic activity, sensor response, or ligand specificity [4] the physical property depends on the exponential of the energy. In these cases, the experimental figure of merit would be exponential in the values shown in Figure 12.4, so that the success of the Monte Carlo methods would be even more dramatic.

The question of how to design and redesign materials discovery experiments has been addressed by analogy with Monte Carlo computer simulation. The Random Phase Volume Model has been used to compare various strategies. The multiple-round Monte Carlo protocols are found to be especially effective on the more difficult systems with larger numbers of composition and noncomposition variables.

An efficient implementation of the search strategy is feasible with existing library-creation technology. Moreover "closing the loop" between library design and redesign is achievable with the same database technology currently used to track and record the data from combinatorial chemistry experiments. These multiple-

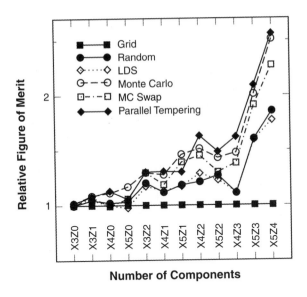

Figure 12.4 The maximum figure of merit found with different protocols on systems with a different number of composition (**x**) and noncomposition (**z**) variables. The results are scaled to the maximum found by the grid searching method. Each value is averaged over scaled results on 10 different instances of the Random Phase Volume Model with different random phases. The Monte Carlo methods are especially effective on the systems with a larger number of variables, where the maximal figures of merit are more difficult to locate. (Reproduced, with permission, from Falcioni, M., and Deem, M. W., *Physics Review E,* 2000, 61: 5948–5952; copyright © 2000. American Physical Society, College Park, MD.)

round protocols, when combined with appropriate robotic controls, should allow the practical application of high throughput experimentation to more complex and interesting systems.

12.3 SMALL-MOLECULE-DISCOVERY[1]

High throughput synthesis is established as one of the methods for the discovery of small molecules, such as drugs or homogeneous catalysts. High throughput or combinatorial methods allow for simultaneous creation of a large number of structurally diverse and complex compounds [34–36], generalizing the traditional techniques of single-compound synthesis. Parallel synthesis and split/pool synthesis [37,38] on solid phase, for example, are two commonly used methods for combinatorial synthesis. Among the high-throughput methodologies, small-molecule combinatorial

---

[1]Portions of this section are reprinted with permission from Chen, L., and Deem, M. W., *Journal of Chemical Information and Computer Sciences,* 2001, 41: 950–957. Copyright © 2001. American Chemical Society, Washington, DC.

chemistry is the most developed and has been applied successfully in areas such as transition metal complexation [39], chemical genetic screening [40], catalysis [41], and drug discovery [42].

The parallel synthesis and split/pool synthesis methods search the composition space in a regular, gridlike fashion. As the complexity of the molecular library grows, the number of dimensions in the composition variable space grows, and with a gridlike method, the number of compounds that must be synthesized to search the space grows exponentially. Synthesis and screening of mixtures of compounds can partially alleviate the dimensional curse [43]. However, a mixture approach raises the question of how to deconvolute and interpret the results. The greater the degree of mixing, the stronger the synergistic effects can be in the mixture, and the more difficult it is to identify individual compounds responsible for the activity [44].

The challenge of searching the composition space in an efficient way has led to extensive efforts in the rational design of combinatorial, or high throughput, libraries. A basic assumption in library design is that structurally similar compounds tend to display similar activity profiles. By designing libraries with maximum structural diversity, the potential for finding active compounds in the high throughput screenings can be enhanced. This design approach requires a quantitative account of the structural and functional diversity of the library, and many descriptors have been developed [45]. Optimization of a library to maximum diversity is then driven by a reliable statistical method. Several structurally diverse libraries have been successfully designed along these lines [46,47]. For example, strategies have been presented to optimize the structural diversity of libraries of potential ligands and molecules by using stochastic optimization of diversity functions and a point mutation Monte Carlo technique [48]. Peptide libraries have been designed by using topological descriptors and quantitative structure–activity relationships combined with a genetic algorithm and simulated annealing [49–51]. Diverse libraries of synthetic biodegradable polymers have been designed by using molecular topology descriptors and a genetic algorithm [52]. Similarly, peptoid libraries have been designed by using multivariate quantitative structure–activity relationships and statistical experimental design [53].

The question of how an initial library should be redesigned for subsequent rounds of HTE in light of the results of the first round of screening has remained largely unanswered. In this chapter, it is suggested that Monte Carlo methods provide a natural means for library redesign in HTE. There is a striking analogy between searching configuration space for regions of low free energy in a Monte Carlo simulation and searching composition space for regions of high figure of merit in a high throughput experiment. Importantly, Monte Carlo methods do not suffer the curse of dimensionality. A Monte Carlo approach should, therefore, be exponentially more efficient than a regular, gridlike method for libraries of complex molecules.

GAs are the computational analog of Darwinian evolution. Typically, a GA consists of three basic processes: crossover, mutation, and selection. In the crossover step, new compounds are generated by mixing the compositions of parent compounds. In the mutation step, individual molecules are changed at random. In the selection step, the best molecules are identified for the next round. The application of GAs to combinatorial synthesis and library design has achieved considerable success [54–57]. Nonetheless, unlike Monte Carlo algorithms, GAs do not satisfy detailed balance. Because of this, GAs cannot be guaranteed to sample the variable space in a defined statistical way or to locate optimal molecules. Furthermore, in most experiments, one wants to identify several initially promising molecules in the hope that, among them, a few can survive further stringent screenings, such as patentability or lack of side effects [58]. In the genetic approach, however, all the molecules in the library tend to become similar to each other due to the crossover step. While diversity can be encouraged in a genetic approach [59,60], diversity can never be guaranteed. The Monte Carlo approach, on the other hand, can maintain or even increase the diversity of a molecular library, due to the satisfaction of detailed balance.

Several strategies for small-molecule HTE will be derived by analogy with Monte Carlo methods. These Monte Carlo protocols will be compared to the GA approach. In order to make this comparison and to demonstrate the effectiveness of the Monte Carlo approach, simulated high-throughput experiments are performed. A random-energy model is introduced and used as a surrogate for experimental measurement of the figure of merit. The random-energy model is not fundamental to the protocols; it is introduced as a simple way to test, parameterize, and validate the various searching methods. In an experimental implementation, the random-energy model would be replaced by the value returned by the screen. Various Monte Carlo protocols are introduced, and a means to calculate the diversity of a library is provided. The effectiveness of the protocols is gauged, and some discussion of the results is presented.

12.3.1 The Space of Chemical Variables

The molecules in a high throughput library are uniquely characterized by their composition, such as the identity of the template and ligands. For specificity, the figure of merit of interest will be considered to be a binding constant, but the results are generically valid. A schematic view of the model is presented in Figure 12.5. For simplicity, the small molecule is considered to consist of one template, drawn from a library of templates, and six binding ligands, each drawn from a single library of ligands [61]. Numerous energetic interactions could exist between this molecule and the substrate. It is commonly believed that descriptors can be directly related to compound performance. A large class of descriptors, such as one-dimensional, two-dimensional, three-dimensional, and BCUT descriptors, has been used to measure the diversity between ligands, templates, and molecules in the literature [45–47]. To simplify, consideration is limited to a set of six weakly correlated descriptors for each ligand and template. For example, the descriptors could be hydrogen-bond donors, hydrogen-bond acceptors, flexibility, an electrotopological calculation, clog$P$, and aromatic density [47]. These descriptors are needed only for the definition of the random-energy model. In a real experiment, the figure of merit would be measured directly, and the descriptors would likely not be needed.

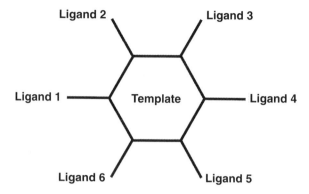

Figure 12.5   Schematic view of the small molecule model. (After [6].)

12.3.2   A Random-Energy Model for the Experimental Screen

To carry out simulated experiments, a figure-of-merit function that mimics the experimental step of measuring the figure of merit is needed. Once constructed or synthesized, the molecules are scored by the model, which takes the composition or molecular descriptors as input.

The basic building block for the random-energy model is a random polynomial of n descriptors, $w_1, \ldots, w_n$:

$$F(w_1, \ldots, w_n, \{G\}) = \sum_{k=0}^{q} \sum_{\substack{i_1 + \cdots + i_n = k \\ i_1, \ldots, i_n \geq 0}} f_{i_1, \ldots, i_n; k}^{1/2} \xi^{-k} G_{i_1, \ldots, i_n} w_1^{i_1}, \ldots, w_n^{i_n} \quad (12.8)$$

Here q is the degree of the polynomial, $G_{i_1, \ldots, i_n}$ are the fixed coefficients of the polynomial, and $f_{i_1, \ldots, i_n; k}$ are constant symmetry factors given by equation (12.5). The square root of f is taken here because each term of the unsymmetrized polynomial is considered to be random. The scale factor, ξ, is used to equalize roughly the contributions from each term of the polynomial. Since w_i will be drawn from a Gaussian random distribution of zero mean and unit variance, the value

$$\xi = \left(\frac{\langle x^q \rangle}{\langle x^2 \rangle} \right)^{1/(q-2)} = \left(\frac{q!}{(q/2)! 2^{q/2}} \right)^{1/(q-2)} \quad (12.9)$$

is appropriate. The values $q = 6$ and $n = 6$ are used in the random-energy model.

The random-energy model accounts for contributions to substrate binding arising from interactions between the substrate and the template and from interactions between the substrate and each of the ligands. In addition, synergistic effects between the ligands and template are incorporated. Consider, for example, a molecule made from template number m from the template library and ligands number $l_1, \ldots, l_6$ from the ligand library. The template is characterized by six descriptors, $D_1^{(m)}, \ldots,$ $D_6^{(m)}$. Similarly, each ligand is characterized by six descriptors, $d_1^{(li)}, \ldots, d_6^{(li)}$. The template contribution to binding is denoted by E_T and the ligand contributions by E_L. The contribution due to synergistic ligand–ligand interactions is denoted by E_{LL} and the contribution due to synergistic template–ligand interactions by E_{TL}. The total contribution to the figure of merit is, then,

$$E = E_L + E_T + E_{LL} + E_{TL}. \tag{12.10}$$

Each of these factors is given in terms of the random polynomial:

$$E_L = \alpha_1 \sum_{i=1}^{6} F(d_1^{(li)}, \ldots, d_6^{(li)}, \{G_L\}) \tag{12.11}$$

$$E_T = \alpha_2 F(D_1^{(m)}, \ldots, D_6^{(m)}, \{G_T\}) \tag{12.12}$$

$$E_{LL} = \alpha_3 \sum_{i=1}^{6} h_i F(d_{j_1}^{(li)}, d_{j_2}^{(li)}, d_{j_3}^{(li)}, d_{j_4}^{(li+1)}, d_{j_5}^{(li+1)}, d_{j_6}^{(li+1)}, \{G_{LL}^i\}) \tag{12.13}$$

$$E_{TL} = \alpha_4 \sum_{i=1}^{6} h_i F(d_{k_1}^{(li)}, d_{k_2}^{(li)}, d_{k_3}^{(li)}, D_{k_4}^{(m)}, D_{k_5}^{(m)}, D_{k_6}^{(m)}, \{G_{TL}^i\}), \tag{12.14}$$

where the $\{G_L\}$, $\{G_T\}$, $\{G_{LL}^i\}$, and $\{G_{TL}^i\}$ are sets of fixed random Gaussian variables with zero mean and unit variance. The α_i are constants to be adjusted so that the synergistic terms will contribute in desired percentages, and h_i is a structural constant indicating the strength of the interaction at binding site, i. The interaction strengths, h_i, are chosen from a Gaussian distribution of zero mean and unit variance for each site on each template. Only synergistic interactions between neighboring ligands are considered in E_{LL}, and it is understood that l_7 refers to l_1 in equation (12.13). In principle, the polynomial in equation (12.13) could be a function of all 12 descriptors of both ligands. It is reasonable to assume, however, that important contributions come from interactions among three randomly chosen distinct descriptors of ligand l_i, $d_{j_1}^{(li)}$, $d_{j_2}^{(li)}$, and $d_{j_3}^{(li)}$, and another three randomly chosen distinct descriptors of ligand l_{i+1}, $d_{j_4}^{(li+1)}$, $d_{j_5}^{(li+1)}$, and $d_{j_6}^{(li+1)}$. Similarly, it is reasonable to assume that template–ligand contributions come from interactions between three randomly chosen distinct descriptors of the ligand, $d_{k_1}^{(li)}$, $d_{k_2}^{(li)}$, and $d_{k_3}^{(li)}$, and another three randomly chosen distinct descriptors of the template, $D_{k_4}^{(m)}$, $D_{k_5}^{(m)}$, and $D_{k_6}^{(m)}$. Both j_i and k_i are descriptor indices ranging from 1 to 6. Assuming that the degrees of freedom of the substrate have been integrated out, these indices depend only on the template.

The parameters in the random-energy model are chosen to mimic the complicated interactions between a small molecule and a substrate. Focus is directed to the case where these interactions are unpredictable, which is typical. That is, in a typical experiment, it would not be possible to predict the value of the screen in terms of molecular descriptors. Indeed, when rational design fails, an intelligent use of HTE is called for. This task of library design and redesign, rather than single-molecule design, is addressed in the next subsection.

12.3.3 Monte Carlo Strategies that Include A Priori Knowledge

To initiate the Monte Carlo protocol, the first template and ligand libraries are built. The size of the template library is denoted by N_T and the size of the ligand library by N_L. In a real experiment, in all likelihood, the descriptors for each ligand and template would not even be calculated, as the figure of merit would be directly measured in the screen. In the simulated experiment, however, the model figure of merit is a function of the descriptors, and the values for the six descriptors for each ligand and template are taken from a random Gaussian distribution with zero mean and unit variance. In the simulated experiment, two sets of random interaction descriptor indices are also associated to each template for the interaction terms in equations (12.13) and (12.14).

To give a baseline for comparison, the library is first designed using a random construction. New molecules are constructed by random selection of one template and six ligands from the libraries. Since the properties of each ligand and template are assigned randomly, this first library should be reasonably diverse and comparable to examples in the literature.

For the Monte Carlo schemes, the initial molecular configurations are still assigned randomly. The library is modified in subsequent rounds of HTE, however, by the Monte Carlo protocol. Two kinds of move are possible for each molecule in the library: template changes and ligand changes. Either the template is changed with probability p_{template}, or one of the six ligands is picked randomly to change with probability $1 - p_{\text{template}}$. The probability of changing from template m to m' is denoted by by $T(m \rightarrow m')$ and from ligand i to i' by $t(i \rightarrow i')$. The new configurations are updated according to the acceptance rule at β, the inverse of the protocol temperature. All the samples are sequentially updated in one Monte Carlo round.

For the simple Metropolis method, the transition matrices are

$$T(m \rightarrow m') = 1/N_T \qquad (12.15)$$

$$t(i \rightarrow i') = 1/N_L, \qquad (12.17)$$

and the acceptance rule is

$$p_{\text{acc}}(c \rightarrow p) = \min[1, \exp(-\beta \Delta E)]. \qquad (12.17)$$

To make use of the idea that smaller moves are accepted more often, one could try to choose a modified ligand or template that is similar to the current one, that is, one could use a transition matrix weighted toward the proposed ligands or templates close to the current one in the six-dimensional descriptor space. Interestingly, this refinement turns out not to work any better than does the simple random move. It seems that even a small move in the descriptor space is already much larger than the typical distance between peaks on the figure-of-merit landscape.

Biased Monte Carlo methods have been shown to improve the sampling of complex molecular systems by many orders of magnitude [62]. In contrast to conventional Metropolis Monte Carlo, trial moves in biased schemes are no longer chosen completely at random. By generating trial configurations with a probability that depends on a priori} knowledge, the moves are more likely to be favorable and more likely to be accepted. Since small-molecule discovery deals with a discrete configurational space, the implementation of biased Monte Carlo in this case is relatively simple. First, a biasing term is needed for both ligands and templates. Since the form of this term is not unique, one can proceed in several different ways. One strategy is to bias the choice of template and ligand on the individual contributions of the templates and ligands to the figure of merit. One might know, or be able to estimate, these contributions from theory. For the random-energy model, for example,

$$e^{(i)} = \alpha_1 F(d_1^{(i)}, \ldots, d_6^{(i)}, \{G_L\}) \tag{12.18}$$

$$E^{(m)} = \alpha_2 F(D_1^{(m)}, \ldots, D_6^{(m)}, \{G_T\}), \tag{12.19}$$

where $e^{(i)}$ is the bias energy to the ligand i in the library, and $E^{(m)}$ is the bias energy of template m in the library. Alternatively, one can estimate the contribution of each ligand or template to the figure of merit experimentally [63–65]. An electrospray ionization source coupled to a mass spectrometer, for example, can serve this purpose [66]. To measure the contributions experimentally, a preexperiment is performed on 10,000 randomly constructed molecules. This number of compounds will give on average each ligand 60 hits and each template 667 hits. By averaging the figure of merit of the molecules containing a particular ligand or template over the total number of hits, experimental estimates of $e^{(i)}$ and $E^{(m)}$ can be obtained. Using these two methods of bias, three different types of biased Monte Carlo schemes are considered: theoretical biased move, experimental biased move, and mixed biased move. In theoretical bias, both $e^{(i)}$ and $E^{(m)}$ are from the random-energy model. In experimental bias, both $e^{(i)}$ and $E^{(m)}$ are calculated from the preexperiment. In mixed bias, $e^{(i)}$ comes from the random-energy model, while $E^{(m)}$ comes from the preexperiment.

These biases tend to exhibit a large gap between a few dominant templates and ligands and the rest. To ensure the participation of more ligands and templates in the strategy, cutoff energies are introduced for the ligand and template, e_c and E_c. These cutoffs are chosen so that e_c is the 21st lowest ligand energy and E_c is the fourth lowest template energy. The biased energy, $e_b^{(i)}$, for the ith ligand is

$$e_b^{(i)} = \begin{cases} e^{(i)} & \text{if } e^{(i)} > e_c \\ e_c & \text{otherwise,} \end{cases} \tag{12.20}$$

and the biased energy, $E_b^{(m)}$, for the mth template is

$$E_b^{(m)} = \begin{cases} E^{(m)} & \text{if } E^{(m)} > E_c \\ E_c & \text{otherwise.} \end{cases} \quad (12.21)$$

To correct for this bias, Rosenbluth factors are introduced [62]. Since the transition probabilities are the same at each Monte Carlo step, the Rosenbluth factor for the ligand is constant:

$$w(p) = w(c) = \sum_{i=1}^{N_L} \exp(-\beta e_b^{(i)}). \quad (12.22)$$

The probability of transition from ligand i to i' is

$$t(i \rightarrow i') = \frac{\exp(-\beta e_b^{(i')})}{w(p)}. \quad (12.23)$$

In the same way, the Rosenbluth factor for the template is

$$W(p) = W(c) = \sum_{m=1}^{N_T} \exp(-\beta E_b^{(m)}). \quad (12.24)$$

The probability of transition from template m to m' is

$$T(m \rightarrow m') = \frac{\exp(-\beta E_b^{(m')})}{W(p)}. \quad (12.25)$$

Finally, the remaining, nonbiased part of the figure of merit is defined to be

$$E_b = E - E_b^{(m)} - \sum_{i=1}^{6} e_b^{(li)}. \quad (12.26)$$

To satisfy the detail balance, the acceptance rule becomes

$$p_{\text{acc}}(c \rightarrow p) = \min[1, \exp(-\beta \Delta E_b)]. \quad (12.27)$$

The idea of a swapping move that attempts to exchange fragments between two molecules is also used. The probability of attempting a swap instead of a single-molecule move is denoted by p_{swap}. In a swap move, the templates or a pair of ligands can be swapped between two randomly selected molecules. The probability of switching the template or ligand at the same position is given by p_{swap_T} and p_{swap_L}, respectively. The crossover event from genetic algorithms could also be introduced in the swap moves, but this additional move did not improve the results. The acceptance rule for swapping is $p_{acc}(c \rightarrow p) = \min[1, \exp(-\beta \Delta E)]$.

Parallel tempering is also used in this redesign approach. The acceptance rule for this move is equation (12.3). The number of groups, the number of samples in each group, the value of β_i, and the exchange probability, p_{ex}, are experimental parameters to be tuned.

For comparison, these Monte Carlo protocols are compared to a standard GA approach [54–57]. In the GA, as in the Monte Carlo strategy, multiple rounds of experimentation are performed on a large set of compounds. The difference between the Monte Carlo and the GA lies in how the library is redesigned, that is, how the compounds are modified in each round. In the GA, first two parents are randomly selected. Then the explicit composition of each molecule is listed, i.e., template, ligand 1, . . . , ligand 6. After aligning the sequences from the two parents, a random cut is made and the part of the sequences before the cut is exchanged. Random changes, or mutations, are also performed on the templates and ligands of the offsprings. Finally, since the population is doubled by crossover, the better half of the molecules is selected to survive this procedure and continue on to the next round.

12.3.4   A Measurement of Library Diversity

The diversity of the library as it passes through the rounds of HTE is an important quantity. The diversity, $\mathcal{D}$, is calculated as the standard deviation of the library in the 42-dimensional descriptor space:

$$\mathcal{D}^2 = \frac{1}{N} \sum_{i=1}^{N} \left[\sum_{j=1}^{6} (D_j^{(m(i))} - \langle D_j \rangle)^2 + \sum_{j=1}^{6} \sum_{k=1}^{6} (d_j^{(sk(i))} - \langle d_j^{(sk)} \rangle)^2 \right], \qquad (12.28)$$

where $m(i)$ is the index of the template of molecule i, $s_k(i)$ is the index of ligand k of molecule i, and j is the index for the descriptor. The average value in each descriptor dimension is given by $\langle D_j \rangle = N^{-1} \sum_{i=1}^{N} D_j^{(m(i))}$ and $\langle d_j^{(sk)} \rangle = N^{-1} \sum_{i=1}^{N} d_j^{(sk(i))}$. The diversity of the library will change as the library changes. A larger library will generally possess a higher absolute diversity simply due to the increased number of compounds. This important, but trivial, contribution to the diversity is scaled out by the factor of $1/N$ in equation (12.28).

12.3.5   Effectiveness of the Monte Carlo Strategies

To gauge how the synergistic terms in the figure of merit affect the efficiency of the Monte Carlo protocols, three models with increasingly important synergistic effects are considered. This is done by adjusting the α_i in equations (12.11)–(12.14), so that the root-mean-square values of the terms are in the ratio $E_L : E_T : E_{LL} : E_{TL} = 1 : 1 : 0.5 : 0.3$ in model I, $1 : 1 : 1 : 0.6$ in model II, and $1 : 1 : 2 : 1.2$ in model III. Finally, $\alpha_1 = 0.01$. To maintain roughly the same average magnitude of the total energy in each model, $\alpha_1 = 0.01 \times (2.8/3.6) = 0.00778$ in model II, and $\alpha_1 = 0.01 \times (2.8/5.2) = 0.00538$ in model III. It turns out that the results are very similar for all three models, and so the results for the most challenging model III only are presented.

The size of the library is fixed at $N_T = 15$ and $N_L = 1000$. The compositional space of this model has 15×1000^6 distinct molecules. Clearly, it is impossible to search exhaustively even this modestly complex space. The total number of molecules to be synthesized is fixed at 100,000, that is, all protocols will have roughly the same experimental cost. Specifically, 100,000 molecules will be made in the random library design protocol, while in the case of the Monte Carlo or genetic protocols, the number of molecules times the number of simulation rounds is kept fixed at 100,000.

To locate optimal parameters for the protocols, a few short preexperiments were performed. First the energy coefficients in the energy function and the descriptors of the ligand and template libraries were fixed. For simple Metropolis, it is optimal to use 10 samples with 10,000 rounds, suggesting that the system is still far from equilibrium at the random initial configuration. With the biased Monte Carlo method, 100 samples and 1000 rounds is optimal. Focus is restricted on systems with 1000 or 100 rounds, since a minimal number of rounds is typically preferred in experiments. It is more difficult to achieve effective sampling in the system with 100 rounds, and so this system is used when setting optimal parameter values. For parallel tempering, it was optimal to have the samples divided into three subsets, with 30% of the population at β_1, 40% at β_2, and 30% at β_3. The optimal parameters are listed in Table 12.1 for each model. Determination of these parameter values corresponds experimentally to gaining familiarity with the protocol on a new system.

The various Monte Carlo schemes are compared with the random-selection method and the GA. Once the optimal parameters are chosen, the coefficients of the energy function and the descriptor values of the ligand and template libraries are generated differently in each simulated experiment. The simulation results are shown in Figure 12.6. Each data point in the figure is an average over 20 independent runs. This averaging is intended to give a representative performance of the protocols on various figures of merit of experimental interest. Since there is much randomness in the results, the standard deviation of the average is shown as well.

12.3.6   Discussion of the Small Molecule Approach

It is clear that for all systems, the Metropolis methods perform better than does random selection. The system with 1000 molecules and 100 rounds is not well equili-

Table 12.1   Optimal Parameters Used in Simulations for the Three Random-Energy Models

Model	β	p_{template}	p_{swap}	p_{swap_T}	p_{swap_L}	β_1	β_2	β_3	p_{ex}
I	30	0.02	0.1	0.05	0.2	5	30	200	0.1
II	30	0.02	0.2	0.2	0.3	5	30	500	0.1
III	50	0.02	0.4	0.2	0.2	5	50	500	0.1

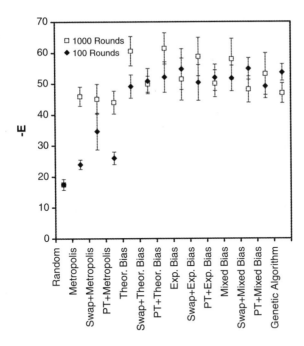

Figure 12.6 Comparison of different Monte Carlo schemes with random and genetic schemes for model III ($E_L : E_T : E_{LL} : E_{TL} = 1 : 1 : 1 : 1.2$). Data from two cases are shown, one with 1000 molecules and 100 rounds (filled diamonds) and one with 100 molecules and 1000 rounds (unfilled squares). (Reproduced, with permission, from Chen, L. and Deem, M. W. *Journal of Chemical Information and Computer Sciences,* 2001, 41: 950–957; copyright © 2001. American Chemical Society, Washington, DC.)

brated by the Metropolis schemes, and an experiment with 100 molecules and 1000 rounds significantly improves the optimal compounds identified. However, by incorporating a priori knowledge, the biased Monte Carlo schemes are able to equilibrate the experiment with either 1000 or 100 rounds. Interestingly, the theoretical bias and experimental bias methods yield similar results. This result strongly suggests that a minimal number of preexperiments can be very useful, both for the understanding of the structure of the figure-of-merit landscape and for improving the performance in future rounds.

The results produced with the composite moves including swap and parallel tempering are slightly improved relative to those from the plain Monte Carlo schemes. Typically, however, these composite moves significantly improve the sampling of a rough landscape. Indeed, swapping and crossover moves are very effective in protein molecular evolution, where the variable space is extremely large [4]. Perhaps the variable space is not so large in small molecule HTE that these composite moves are required. Alternatively, the random energy model may underestimate the ruggedness of the landscape. The landscape for RNA ligands, for example, is estimated to be extremely rough [66], and composite moves may prove more important in this case.

The GA is relatively easy to use. It does not satisfy detailed balance, however, so there is no theoretical guarantee of the outcome. The optimal figures of merit identified are, nonetheless, comparable to those from the better Monte Carlo methods for all three models. However, due to the crossover and selection steps in the GA, the molecules in the library tend to become similar to each other, which prevents this scheme from sampling the whole variable space. To help elucidate this point, diversity measurements for model I are shown in Figure 12.7. It is clear that the GA has reduced the diversity of the library by 60% relative to the biased schemes. Interestingly, the Monte Carlo simulations actually increase the diversity from the initial random configurations. The biased schemes tend to bring the system to equilibrium relatively quickly, and the diversity measurements are similar for the 100- and

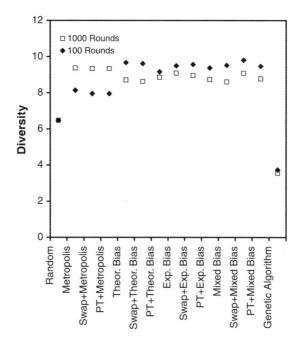

Figure 12.7 Diversity measurement of the final configurations for model I. Data from two cases are shown, one with 1000 molecules and 100 rounds (filled diamonds) and one with 100 molecules and 1000 rounds (unfilled squares). The error bars are negligible. As in equation (12.28), the contribution to the absolute diversity that scales as the square root of the number of molecules per round has been scaled out in this figure. (Reproduced, with permission, from Chen, L. and Deem, M. W. *Journal of Chemical Information and Computer Sciences,* 2001, 41: 950–957; copyright © 2001. American Chemical Society, Washington, DC.)

1000-round experiments. For the Metropolis method, on the other hand, an experiment with 100 rounds is less diverse than an experiment with 1000 rounds. The GA approach finds less favorable figure-of-merit values in the 100-compound 1000-round experiment, presumably due to a greater sensitivity to the $\sqrt{10}$ reduction in the absolute diversity relative to the already small absolute diversity in the 1000-compound 100-round experiment.

The greater the number of potentially favorable molecules in the library space, the greater the diversity of the experimental library will be for the Monte Carlo methods. The GA, on the other hand, will tend to produce a library that contains many copies of a single favorable molecule. A key distinction, then, is that a Monte Carlo strategy will sample many compounds from the figure-of-merit landscape, whereas a GA will tend to produce a single molecule with a favorable figure-of-merit value. How strongly the compounds with high figures of merit are favored in the Monte Carlo strategy is determined by the protocol temperature, since the probability of observing a compound with figure of merit $-E$ is proportional to $\exp(-\beta E)$. The sampling achieved by the Monte Carlo methods is important not only because it ensures that the composition space is thoroughly sampled, but also because it ensures that the library of final hits will be as diverse as possible.

The Monte Carlo methods perform equally well on all three models. The three models were introduced to gauge the impact of unpredictable, synergistic effects in the experimental figure of merit. It might be expected that the a priori bias methods would perform less well as the synergistic effects become more pronounced. That the biased methods perform well even in model III suggests that the bias Monte Carlo approach can be rather robust. In other words, even a limited amount of a priori information is useful in the Monte Carlo approach to library redesign.

Monte Carlo appears to be a fruitful paradigm for experimental design of multiround HTEs. A criticism of HTE has been its mechanical structure and lack of incorporation of a priori knowledge. As shown here, a biased Monte Carlo approach handily allows the incorporation of a priori knowledge. Indeed, the results reveal that biased Monte Carlo schemes greatly improve the chances of locating optimal compounds. For the moderately complex libraries considered here, the bias can be determined equally well by experimental or theoretical means. Although the compounds identified from a traditional GA are comparable to those from the better Monte Carlo schemes, the diversity of identified molecules is dramatically decreased in the genetic approach. GAs, therefore, are less suitable when the list of good molecules is further winnowed by a secondary screen, a tertiary screen, patentability considerations, lack of side effects, or other concerns. Interestingly, composite Monte Carlo moves, such as swap or parallel tempering, bring only a slight improvement to the plain biased Monte Carlo protocols, possibly due to the relatively small size of the composition space in small-molecule HTE. Presumably, as the complexity of the library is increased, these composite moves will prove more useful for the more challenging figures of merit. Although for simplicity the initial library configurations are chosen at random, the sophisticated initial library design strategies available in the literature can be used, and they would complement the multiround library redesign strategies presented here.

12.4 TEMPLATED MATERIALS SYNTHESIS

As a last example of the Monte Carlo approach to library design and redesign, ideas from Sections 12.2 and 12.3 are combined to address the topic of high throughput, templated materials synthesis.

A canonical example of templated materials synthesis is that of zeolite synthesis. Zeolites are widely used in industrial applications such as catalysis, molecular sieving, gas separation, and ion exchange. The catalytic activity and selectivity of these materials are attributed to the highly distributed active sites and the large internal surface area accessible through uniformly sized pores. The present set of zeolite frameworks is incomplete, and additional geometries and chemistry are desired. In addition, even the present zeolite geometries are often imperfect, and there is strong motivation to reduce the defects and faults in these materials. Syntheses of novel zeolites with new frameworks and chemical compositions has therefore received heavy research attention. To date, roughly 150 framework structures are known. Roughly six new structures are added per year. The use of organocation template molecules to induce structure direction is a major tool in the quest for new structure [26,67]. The addition of organic templates such as alkylammonium ions or amines to zeolite synthesis gels affects the rate at which a particular structure is nucleated and grown, and this control can be used to make new structures or framework chemical compositions accessible. Indeed, the nature and extent of interaction between the organic templates and the inorganic components of the zeolite synthesis gel are important factors that influence the final zeolite pore architecture [68,69].

As shown in this chapter, Monte Carlo protocols are efficient methods for library design and redesign in both material discovery and small-molecule design. Material discovery deals with the continuous variables of composition and noncomposition. Small-molecule design deals with the discrete variables of the template and ligand identities. For templated zeolite synthesis, both the continuous variables and the discrete variables are present. All these variables affect the function of the final zeolite material in a correlated way.

Several strategies are described here for templated, high throughput zeolite synthesis. A random energy model is built from those previously introduced. This model will serve as a surrogate for experimental measurement of the figure of merit in simulated experiments. Again, the random energy model is not fundamental to the protocols; it is introduced as a simple way to test, parameterize, and validate the various searching methods. In an experimental implementation, the random-energy model would be replaced by the value returned by the experimental primary screen. Details of the implementation of the Monte Carlo protocols are provided. The effectiveness of the protocols is gauged, and some implications are discussed.

12.4.1 The Random-Energy Model

For demonstration of the proposed protocols in an efficient fashion, a model is necessary in lieu of the real experimental screening process, that is, a model that relates the chemical composition of the zeolite and structure-directing agent to the figure of merit of the material. Again, this model is not essential to the protocols; it is simply a cheap and fast means to evaluate proposed protocols. This model captures the essence of the physical system and provides validation for the protocols.

To consider template-assisted zeolite synthesis, a model that combines ideas from Sections 12.2 and 12.3 is used. The figure of merit is naturally given by a sum of a zeolite energy and an organic molecular energy:

$$E = E_{\text{zeolite}} + E_{\text{molecule}}. \tag{12.29}$$

The d zeolite framework composition variables x_i are certainly key variables in E_{zeolite}. Typical elemental compositions in zeolite frameworks include silicon, aluminum, oxygen, phosphorus, germanium, and boron, among others. The Random Phase Volume Model from Section 12.2 is used to capture the dependence of the figure of merit on these composition variables. Noncomposition variables also affect the measured figure of merit, and they are accounted for by the Random Phase Volume Model. When the zeolite is in composition phase α and noncomposition phase γ, the contribution to the figure of merit from the zeolite is given by

$$E_{\text{zeolite}}(\mathbf{x}, \mathbf{z}) = H(\mathbf{x} - \mathbf{x}^{\alpha}, \{G^{\alpha}\}) + \tfrac{1}{2} H(\mathbf{z}, \{G^{\gamma}\}). \tag{12.30}$$

Here $\{G^{\alpha}\}$ and $\{G^{\gamma}\}$ are phase-dependent, Gaussian random variables with zero mean and unit variance. The parameters $\{G^{\alpha}\}$ are different in each composition phase, and the parameters $\{G^{\gamma}\}$ are different in each noncomposition phase. The H function is the familiar random polynomial:

$$H(w_1, \ldots, w_n, \{G\}) = G_0 + \sigma_H \sum_{k=1}^{q} \sum_{\substack{i_1 + \cdots + i_n = k \\ i_1, \ldots, i_n \geq 0}} f_{i_1 \ldots i_n; k} \, \xi_H^{-k} G_{i_1, \ldots, i_n} w_1^{i_1} \ldots w_n^{i_n} \tag{12.31}$$

The degree $q = 6$ is chosen for the polynomial. The coefficients G_0 and $G_{i_1, \ldots, i_n}$ are denoted in compact notation by $\{G\}$. The symmetry factors $f_{i_1 \ldots i_n; k}$ are given by equation (12.5). The scale factor ξ_H is chosen so that each term in the multinomial contributes roughly the same amount in the root-mean-square sense. For the composition variables $\xi_H = \xi/2$, and for the noncomposition variables $\xi_H = (\langle z^6 \rangle / \langle z^2 \rangle)^{1/4} = (3/7)^{1/4}$. The σ_H are chosen so that the multinomial terms contribute 40% as much as the corresponding constant, phase terms, G_0 in Equation (12.31), in the root-mean-square sense.

The organic structure-directing molecules are characterized by composition, such as the identity of the ligands and template, as shown in Figure 12.8. As in Section 12.3, the small molecule is considered to consist of one template and six binding ligands. There is a template library and a ligand library. Six weakly correlated descriptors are used to describe the characteristics of the molecule, as in Section 12.3.

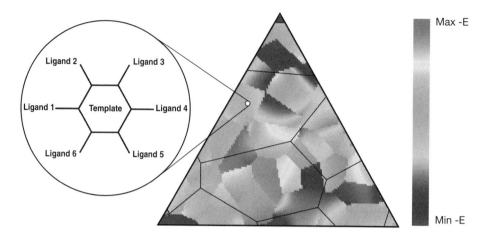

Figure 12.8   Schematic of the random-energy model for templated materials synthesis. (Adapted, with permission, from Chen, L., and Deem, W. *Mol. Phys.*, 2002, 100: 2175–2181, copyright © 2002, Taylor & Francis, Ltd., London.)

As in Section 12.3, a random-energy model is used to account for the contributions to structure direction arising from interactions between the zeolite, template, and ligands. The total molecular contribution to the figure of merit is, then,

$$E_{\text{molecule}} = E_L + E_T + E_{LL} + E_{TL}. \tag{12.32}$$

Each of these terms is given in the form of a random polynomial, where the coefficients in the random polynomial now depend on which zeolite phase is being made:

$$E_L(\mathbf{x}, \mathbf{z}) = \sum_{i=1}^{6} F[d_1^{(li)}, \ldots, d_6^{(li)}, \{G_L(\mathbf{x}, \mathbf{z})\}], \quad G_L(\mathbf{x}, \mathbf{z}) = \lambda_{x,1} G_L^\alpha + \tfrac{1}{2} \lambda_{z,1} G_L^\gamma \tag{12.33}$$

$$E_T(\mathbf{x}, \mathbf{z}) = F[D_1^{(m)}, \ldots, D_6^{(m)}, \{G_T(\mathbf{x}, \mathbf{z})\}], \quad G_T(\mathbf{x}, \mathbf{z}) = \lambda_{x,2} G_T^\alpha + \tfrac{1}{2} \lambda_{z,2} G_T^\gamma \tag{12.34}$$

$$E_{LL}(\mathbf{x}, \mathbf{z}) = \sum_{i=1}^{6} h_i(\mathbf{x}, \mathbf{z}) F[d_{j1}^{(li)}, d_{j2}^{(li)}, d_{j3}^{(li)}, d_{j4}^{(li+1)}, d_{j5}^{(li+1)}, d_{j6}^{(li+1)}, \{G_{LL}(\mathbf{x}, \mathbf{z})\}],$$

$$G_{LL}(\mathbf{x}, \mathbf{z}) = \lambda_{x,3} G_{LL}^{i,\alpha} + \tfrac{1}{2} \lambda_{z,3} G_{LL}^{i,\gamma} \tag{12.35}$$

$$E_{TL}(\mathbf{x}, \mathbf{z}) = \sum_{i=1}^{6} h_i(\mathbf{x}, \mathbf{z}) F[d_{k1}^{(li)}, d_{k2}^{(li)}, d_{k3}^{(li)}, D_{k4}^{(m)}, D_{k5}^{(m)}, D_{k6}^{(m)}, \{G_{TL}(\mathbf{x}, \mathbf{z})\}],$$

$$G_{TL}(\mathbf{x}, \mathbf{z}) = \lambda_{x,4} G_{TL}^{i,\alpha} + \tfrac{1}{2} \lambda_{z,4} G_{TL}^{i,\gamma}. \tag{12.36}$$

The building block for our random-energy model as a function of those descriptors is equation (12.8). The $\{G\}$, again, will be composed of a composition and a non-composition piece, as indicated in equations (12.33)–(12.36). The orders of the polynomials are $q = 6$ and $n = 6$. The scaling factors are as in equation (12.9). In general, the polynomial coefficients, $\{G\}$ in equations (12.33)–(12.36), are functions of the zeolite composition and noncomposition variables $\mathbf{x}$ and $\mathbf{z}$. However, for simplicity, it is assumed that the coefficients are phase-dependent only. The coefficients $\{G_L^\alpha\}$, $\{G_L^\gamma\}$, $\{G_T^\alpha\}$, $\{G_T^\gamma\}$, $\{G_{LL}^{i,\alpha}\}$, $\{G_{LL}^{i,\gamma}$, $\{G_{TL}^{i,\alpha}\}$, and $\{G_{TL}^{i,\gamma}\}$ are chosen from random Gaussian distributions with zero mean and unit variance. These coefficients are different in each phase. The strength of the interaction at structure directing site i is given by the structural constant h_i, which depends on the zeolite variables $\mathbf{x}$ and $\mathbf{z}$ as

$$h_i(\mathbf{x}, \mathbf{z}) = H(\mathbf{x} - \mathbf{x}^\alpha, \{G_i^\alpha\}) + \tfrac{1}{2} H(\mathbf{z}, \{G_i^\gamma\}). \tag{12.37}$$

The coefficients $\{G_i^\alpha\}$ and $\{G_i^\gamma\}$ are sets of phase-dependent fixed Gaussian random variables with zero mean and unit variance. The parameter σ_H is adjusted so that the multinomial terms contribute in the mean-square sense roughly the same as the constant, phase terms. It is assumed that the interaction indices, j_m and k_n, in equations (12.35) and (12.36) depend only on the template, since the molecular details of the zeolite have been suppressed. The constants $\lambda_{x,i}$ are adjusted so that the terms depending on $\mathbf{x}$ contribute as $E_L : E_T : E_{LL} : E_{TL} = 1 : 1 : 2 : 1.2$ in the root-mean-square sense. Furthermore, the $\lambda_{x,i}$ are adjusted so that the total molecular contribution from the terms depending on $\mathbf{x}$ account for roughly 15% of the constant, compositional phase term in E_{zeolite} in the root-mean-square sense. The values of $\lambda_{z,i}$ are similarly adjusted so that the contribution from the terms depending on $\mathbf{z}$ is 15% of the constant, noncomposition phase term in E_{zeolite} in the root-mean-square sense.

As always, the energy returned by the model is minimized. That is, the figure of merit, $-E$, is sampled by $\exp(-\beta E)$.

12.4.2 Multiround Monte Carlo Strategies

As in the small-molecule case, the template and ligand libraries are built first. The size of the template library is denoted by $N_T = 15$, and the size of the ligand library by $N_L = 1000$. As in Section 12.3, the values of the six descriptors of each ligand and template are extracted from a Gaussian random distribution with zero mean and unit variance. Two sets of random interaction descriptor indices are also associated to each template for the interaction terms in equations (12.35) and (12.36).

As with previous examples, the total number of samples to be synthesized is kept fixed at 100,000 for all protocols. This condition ensures that the total experimental cost is kept roughly constant.

Both grid and random single-pass protocols are considered. Most high throughput experiments tend to perform a single-pass grid search on all continuous variables. To mimic this procedure, a grid search is used on the composition and noncomposition variables. For the discrete molecular variables, however, a random approach is used to pick the template and ligands from the organocation libraries. In the random single-pass protocol, all variables are searched at random, i.e., the composition variables and noncomposition variables are chosen at random as well.

Several multipass Monte Carlo protocols are considered. Unlike single-pass protocols, multipass protocols allow one to learn about the characteristics of the system as the experiment proceeds. The multiple samples considered per round allow for a rather diverse population and increase the opportunity for several zeolite samples to survive more elaborate tests for application performance, tests that are only roughly correlated with the primary screen. A Monte Carlo protocol is used so as to maintain diversity in the samples. To be consistent with current experimental technology, 1000 samples are synthesized for a total of 100 rounds. The initial sample configurations are assigned by the random protocol.

Both the zeolite variables and the components of the structure-directing agent are changed at each round of the Monte Carlo protocol. In the simplest approach, the composition and the noncomposition variables are perturbed about their original values using the traditional Metropolis-type method. For the noncomposition variables, periodic boundary conditions are used. For the composition variables, reflecting boundary conditions are used, as discussed in Appendix 12.B. For the organocation variables, either the template is changed with probability p_{template}, or one of the six ligands is randomly changed with probability $1 - p_{\text{template}}$ in each Monte Carlo move. The proposed samples are accepted or rejected according to the acceptance rule at β, the inverse of the protocol temperature. In one Monte Carlo round, all the samples are sequentially updated. Either a simple Metropolis method or a biased Monte Carlo method can be incorporated for the organocation variables. In the simple Metropolis method, the proposed new template or ligand is selected randomly with uniform probability from the library. After the zeolite and organocation variables have been modified in an unbiased protocol, the acceptance probability is

$$p_{\text{acc}}(c \to p) = \min[1, \exp(-\beta\Delta E)]. \tag{12.38}$$

Biased Monte Carlo schemes allow the generation of new samples with a probability that depends on the figure of merit of the new sample. In the small-molecule high throughput experiment design of Section 12.3, biased energy forms were constructed from either theory or preexperiments. Since the total figure of merit is a function of both the organocation structure and the zeolite composition and noncomposition variables, it is not feasible to construct the bias from preexperiments. The theoretical bias is therefore used. The bias energy for template m for a zeolite at compositional phase α and noncompositional phase γ is

$$E^{(m)}(\mathbf{x}, \mathbf{z}) = F[D_1^{(m)}, \ldots, D_6^{(m)}, \{G_T(\mathbf{x}, \mathbf{z})\}], \quad G_T(\mathbf{x}, \mathbf{z}) = \lambda_{x,2} G_T^\alpha + \tfrac{1}{2}\lambda_{z,2} G_T^\gamma. \tag{12.39}$$

The bias for ligand i is

$$e^{(i)}(\mathbf{x}, \mathbf{z}) = F[d_1^{(i)}, \ldots, d_6^{(i)}, \{G_L(\mathbf{x}, \mathbf{z})\}], \quad G_L(\mathbf{x}, \mathbf{z}) = \lambda_{x,1} G_L^\alpha + \tfrac{1}{2}\lambda_{z,1} G_L^\gamma. \tag{12.40}$$

Cutoff energies are chosen as in Section 12.3.3. The biased energy, $e_b^{(i)}$, for the ith ligand then becomes

$$e_b^{(i)}(\mathbf{x}, \mathbf{z}) = \begin{cases} e^{(i)}(\mathbf{x}, \mathbf{z}) & \text{if } e^{(i)} > e_c \\ e_c(\mathbf{x}, \mathbf{z}) & \text{otherwise.} \end{cases} \tag{12.41}$$

Similarly, the biased energy, $E_b^{(m)}$, for the mth template becomes

$$E_b^{(m)}(\mathbf{x}, \mathbf{z}) = \begin{cases} E^{(m)}(\mathbf{x}, \mathbf{z}) & \text{if } E^{(m)} > E_c \\ E_c(\mathbf{x}, \mathbf{z}) & \text{otherwise.} \end{cases} \tag{12.42}$$

If the ith ligand in the organocation is to be changed at the proposed new zeolite values $\mathbf{x}'$ and $\mathbf{z}'$, the biased probability for selecting ligand l_i' from the library is

$$f[E(p)] = \frac{\exp[-\beta e_b^{(l_i')}(\mathbf{x}', \mathbf{z}')]}{\sum\limits_{j=1}^{N_L} \exp[-\beta e_b^{(j)}(\mathbf{x}', \mathbf{z}')]}. \tag{12.43}$$

To satisfy detailed balance, the bias for the reverse move is needed:

$$f[E(c)] = \frac{\exp[-\beta e_b^{(l_i)}(\mathbf{x}, \mathbf{z})]}{\sum\limits_{j=1}^{N_L} \exp[-\beta e_b^{(j)}(\mathbf{x}, \mathbf{z})]}. \tag{12.44}$$

Similarly, the bias probability for selecting template m' at proposed new zeolite values $\mathbf{x}'$ and $\mathbf{z}'$ is

$$f[E(p)] = \frac{\exp[-\beta E_b^{(m')}(\mathbf{x}', \mathbf{z}')]}{\sum\limits_{j=1}^{N_T} \exp[-\beta E_b^{(j)}(\mathbf{x}', \mathbf{z}')]}. \tag{12.45}$$

The bias for the reverse move is needed to satisfy detailed balance:

$$f[E(c)] = \frac{\exp[-\beta E_b^{(m)}(\mathbf{x}, \mathbf{z})]}{\sum\limits_{j=1}^{N_T} \exp[-\beta E_b^{(j)}(\mathbf{x}, \mathbf{z})]}. \tag{12.46}$$

The acceptance rule in the biased scheme is

$$p_{\text{acc}}(c \to p) = \min\left\{1, \frac{f[E(c)]}{f[E(p)]} \exp(-\beta \, \Delta E)\right\}. \tag{12.47}$$

The parallel tempering move is also considered. The acceptance rule for a parallel tempering exchange move is equation (12.3). As always, this exchange step is experimentally cost-free, and it can be effective at allowing the protocol to escape from local optima.

12.4.3 Effectiveness of the Monte Carlo Strategies

The size of the organocation library is fixed at $N_T = 15$ and $N_L = 1000$. The parameters $\lambda_{x,i}$ and $\lambda_{z,i}$ are adjusted so that the synergistic terms will contribute in the ratio $E_L : E_T : E_{LL} : E_{TL} = 1 : 1 : 2 : 1.2$ in the root-mean-square sense in the composition and noncomposition phases. The relative contributions from the zeolite and molecule are fixed by requiring that they be roughly of the same order for the optimal configurations obtained by the grid or random protocols. So that this occurs, the parameters are found by trial and error, due to the complicated molecular terms. The σ_H values are first adjusted, then the $\lambda_{x,i}$ and $\lambda_{z,i}$ in equations (12.33)–(12.36) are adjusted. As mentioned, the $\lambda_{x,i}$ and $\lambda_{z,i}$ are adjusted so that the total molecular contribution from either type of phase accounts for roughly 15% of the corresponding constant phase term in E_{zeolite} in the root-mean-square sense.

A few short preexperiments are performed to locate optimal parameters for the protocols. The optimal value for the probability of changing a template is $p_{\text{template}} = 0.02$, since the size of the template library is relatively small. The maximum random displacements are $|\Delta \mathbf{x}| = 0.1/\sqrt{d-1}$ and $|\Delta\mathbf{z}| = 0.2$ in the composition space and noncomposition space. The optimal inverse protocol temperature for simple Metropolis Monte Carlo is $\beta = 50$ for $d = 3$ and $d = 4$ and $\beta = 20$ for $d = 5$. The optimal inverse protocol temperature for the biased Monte Carlo schemes is $\beta_b = 500$ for $d = 3$ and $d = 4$ and $\beta_b = 200$ for $d = 5$. It is optimal to have the samples divided into three subsets when biased Monte Carlo is combined with parallel tempering, with 25% of the population at $\beta_1 = \frac{1}{2} \beta_b$, 50% at $\beta_2 = \beta_b$, and 25% at $\beta_3 = 2 \beta_b$. The switching probability, p_{ex}, is 0.1. Determination of these parameters corresponds to gaining familiarity with a new protocol experimentally.

The protocols are tested with a constant organocation library size and increasing numbers of composition and noncomposition variables. Results are shown in Figure 12.9. From this figure, it can be seen that the multiround Monte Carlo protocols are better than single-pass protocols such as grid and random. The simple Metropolis method finds optimum samples that are twice as good as those from the grid or random protocols. The biased Monte Carlo is even more efficient, and the optimum figures of merit from biased Monte Carlo far exceed those from simple Metropolis with even 1000 rounds. The main contribution to the optimal figures of merit is coming from the ability of biased Monte Carlo to find favorable designs for the structure-directing agent. Only for the more complicated $d = 4$ and $d = 5$ systems is parallel tempering noticeably effective. It is more important to keep all 1000 of the samples at the optimum temperature for the relatively simple systems, such as $d = 3$. In either case, since most of the complexity arises from the organocation library variables, the impact of using parallel tempering is small.

Using a model for high throughput zeolite synthesis, multipass Monte Carlo methods are shown to work better than single-pass protocols. Sophisticated biased Monte Carlo schemes are highly efficient and significantly better than simple Metropolis Monte Carlo. For complicated systems with five or more framework chemical compositional variables, parallel tempering is the best method.

Interestingly, the complexity of the chemical space is largely determined by the

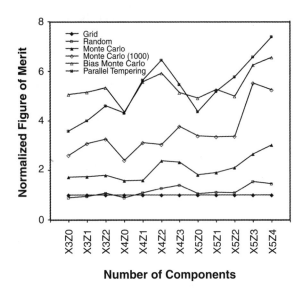

Figure 12.9 The optimal figure of merit found with different protocols with different number of composition (**x**) and noncomposition (**z**) variables. The complexity of the structure-directing agent libraries is identical in all cases. The values are scaled to the optimum found by the grid searching method. Each scaled value is averaged over results from 10 instances. (Reproduced, with permission, from Chen, L. and Deem, M. W., *Molecular Physics,* 2002, 100: 2175–2181, copyright © 2002, Taylor & Francis Ltd., London.)

complexity of the template. The composition and noncomposition variables play a minor role compared to the template. Physically, this means that templates are almost essential to successful zeolite synthesis, which is in accord with experiment. In addition, choice of template is crucial to effective materials performance, because without an appropriate template, no material is made, which is also in accord with experiment. The robotic technology for both template synthesis and zeolite synthesis [21] is available, and it should only be a matter of time before the Monte Carlo protocols are put to use.

SUMMARY

To summarize, there are essentially four main challenges to a successful HTE program. The first challenge is the design of sufficiently robust and transferable chemistries so that parallel synthesis is possible. The second challenge is the design of rapid and accurate robotic screens. The third challenge is the design and maintenance of the databases necessary to track the experimental data produced. The fourth and final challenge is deciding what materials to make, given the ability to synthesize and screen in high throughput. This last challenge of library design and redesign has been addressed by this chapter, and by this book as a whole. It is this last topic to which theory seems most likely to make a contribution.

This chapter has described the Monte Carlo approach to library design and redesign. An alternative approach, with which some comparison was made, is use of GAs. The difference between Monte Carlo and GAs is not in the proposal of new samples, often called the mutation event in GAs. Any mutation event possible in GAs is also a possible move in a Monte Carlo strategy. The difference between the two approaches lies in the selection step. Both GAs and Monte Carlo can perform well. Only Monte Carlo, however, is guaranteed to sample from the figure of merit, the guarantee being provided by Markov chain theory. The benefit of sampling becomes apparent when secondary and tertiary performance screens are performed on hits identified by the primary screen.

Despite the many successes of HTE, the method has been criticized as a simple machinery, lacking incorporation of a priori knowledge when compared with the traditional synthetic approach. A priori knowledge, such as chemical intuition, previous database or experimental information, well-known theory, patentability, or other specific constraints, is indispensable to an efficient library design and is the traditional province of the synthetic chemist. It is fascinating that the Monte Carlo approach to HTE can naturally incorporate such knowledge in the experimental design through the technique of biased Monte Carlo.

The Monte Carlo strategies discussed in this chapter are complementary to other approaches to library design and redesign. For example, while the random construction is superior to the grid approach, even more sophisticated methods that take into account concerns such as diversity or a priori chemical knowledge can be used for the initial library design. Similarly, if a purely spatially addressable synthesis is too expensive in an application, the Monte Carlo protocol can be used to design a few freely addressable samples, and a less expensive combinatorial method can be used to produce a cloud of samples about these few points. If the figure of merit is locally smooth, but globally rough, interpolation procedures can be used to provide a good experimental estimate for the bias that will assist a Monte Carlo protocol to sample the chemical and experimental space. Finally, when promising hits are identified by a Monte Carlo experiment, gradient-type optimization methods can be used to refine the final process and material variables.

ACKNOWLEDGMENT

It is a pleasure to acknowledge the contributions of my collaborators Marco Falcioni and Ligang Chen. This work was supported by the National Science Foundation.

APPENDIX 12.A: THE SIMPLEX OF ALLOWED COMPOSITIONS

In any high throughput experiment involving elemental compositions as variables, some means of searching the composition space must be developed. Typically, the composition space is defined by the mole fractions of each element. If there are d mole fractions, there are only $d - 1$ variables to be searched, as the mole fractions sum to unity. Since there are only $d - 1$ variables among the d mole fractions, x_i, the changes to the mole fractions that occur during the library redesign must be made in a correlated fashion. In particular, the changes to the mole fractions must be made in the $d - 1$ dimensional hyperplane that is consistent with the constraint $\Sigma_{i=1}^{d} x_i = 1$. Mathematically, this constraint can be written as as $\mathbf{x} \cdot \mathbf{u}_d = 1/\sqrt{d}$, where the vector $\mathbf{u}_d = (1/\sqrt{d}, 1/\sqrt{d}, \ldots, 1/\sqrt{d})$.

It is convenient to perform a change of variables from the mole fractions to variables that lie in the allowed hyperplane. This change of variables then allows each of the new variables to be changed independently. What this amounts to mathematically is a rotation of the coordinate system describing the mole fractions to a coordinate system in which the first $d - 1$ components are orthogonal to $\mathbf{u}_d$. Defining the original coordinate system by the unit vectors

$$\mathbf{e}_1 = (1, 0, \ldots, 0, 0)$$

$$\mathbf{e}_2 = (0, 1, \ldots, 0, 0)$$

$$\cdot$$
$$\cdot$$
$$\cdot$$

$$\mathbf{e}_{d-1} = (0, 0, \ldots, 1, 0), \tag{12.A.1}$$

the new coordinate system is identified by the Gram–Schmidt procedure:

$$\mathbf{u}_1 = \frac{\mathbf{e}_1 - (\mathbf{e}_1 \cdot \mathbf{u}_d)\, \mathbf{u}_d}{|\mathbf{e}_1 - (\mathbf{e}_1 \cdot \mathbf{u}_d)\, \mathbf{u}_d|}$$

$$\mathbf{u}_2 = \frac{\mathbf{e}_2 - (\mathbf{e}_2 \cdot \mathbf{u}_d)\mathbf{u}_d - (\mathbf{e}_2 \cdot \mathbf{u}_1)\mathbf{u}_1}{|\mathbf{e}_2 - (\mathbf{e}_2 \cdot \mathbf{u}_d)\mathbf{u}_d - (\mathbf{e}_2 \cdot \mathbf{u}_1)\mathbf{u}_1|}$$

$$\cdot$$
$$\cdot$$

$$\mathbf{u}_i = \frac{\mathbf{e}_i - (\mathbf{e}_i \cdot \mathbf{u}_d)\mathbf{u}_d - \sum_{j=1}^{i-1} (\mathbf{e}_i \cdot \mathbf{u}_j)\, \mathbf{u}_j}{|\mathbf{e}_i - (\mathbf{e}_i \cdot \mathbf{u}_d)\mathbf{u}_d - \sum_{j=1}^{i-1} (\mathbf{e}_i \cdot \mathbf{u}_j)\, \mathbf{u}_j|}, \qquad i < d. \tag{12.A.2}$$

The relationship between the original mole fractions and the new variables, w_i, is given by a simple rotation:

$$\mathbf{x} = R\mathbf{w}, \tag{12.A.2}$$

where R is the rotation matrix, and the value of R_{ij} is given by the ith component of $\mathbf{u}_j$. How this coordinate system rotation works for a three-component system is shown in Figure 12.10. To ensure that the mole fractions sum to unity, $w_d = 1/\sqrt{d}$. These new variables, w_i, $i < d$, can be changed independently. Any changes made to

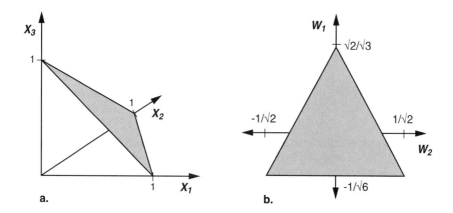

Figure 12.10 The allowed composition range of a three-component system is shown in the (a) original composition variables, x_i, and (b) Gram-Schmidt variables, w_i. (Reproduced, with permission, from Deem, M. W., *Advances in Chemical Engineering*, Vol. 28, A. Chakraborty, Ed., 81–121; copyright © 2001. Academic Press, San Diego.)

w_i, $i < d$, when the mole fractions are calculated from equation (12.A.3), will satisfy the constraint that the mole fractions sum to one. Some changes, however, may violate the constraint that the mole fractions must all be positive. In such cases, the procedure described in Appendix 12.B is used to map the invalid mole fractions to valid mole fractions.

APPENDIX 12.B: REFLECTING BOUNDARY CONDITIONS

As shown in Figure 12.10, the valid mole fractions form a simplex. Not only do the mole fractions sum to one, but they also are all nonnegative. If a particular sample has a composition that is near the edge of the allowed simplex, very often the random change to the composition that occurs in the Monte Carlo protocol will lead to a new composition with invalid values. To remedy this situation, it is best to have a mapping of such invalid values back to the allowed composition simplex. With such a mapping, all of the attempted Monte Carlo moves will at least be valid, and experimental effort will not be wasted.

There is a procedure that ensures all Monte Carlo moves lead to valid new composition points. The procedure involves reflecting an invalid composition point about the $d - 2$-dimensional hyperplanes that define the edges of the allowed simplex. Figure 12.11 shows the procedure for a move in a three-component system that requires only a single reflection. The procedure requires the unit normals, $\mathbf{n}_i$, to each of the faces of the simplex as well as the constants, c_i, that define the faces by the equation $\mathbf{x} \cdot \mathbf{n}_i = c_i$. As shown in Figure 12.11, in the case of a three-component system there are three such unit normals and three such constants. We initially define $\mathbf{y} = \mathbf{x}(c)$. The procedure executes the following steps:

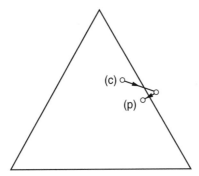

Figure 12.11 Schematic of the reflecting boundary conditions used to obtain valid composition variables. (Reproduced, with permission, from Chen, L. and Deem M. W., *Molecular Physics,* 2002, 100: 2175–2181, copyright © 2002, Taylor & Francis Ltd., London.)

- Determine whether the proposed new composition, $\mathbf{x}$, is in the allowed region. If so, define the new composition $\mathbf{x}(p)$ to be $\mathbf{x}$ and stop.
- If not, find the face i for which the quantity $t_i = (c_i - \mathbf{y}\cdot\mathbf{n}_i)/[(\mathbf{x}-\mathbf{y})\cdot\mathbf{n}_i]$ is minimal, taking into account only those faces for which both t_i and $(\mathbf{x}-\mathbf{y})\cdot\mathbf{n}_i$ are positive.
- Define $\mathbf{y} \leftarrow \mathbf{y} + t_i(\mathbf{x}-\mathbf{y})$.
- Reflect the composition point through face i by the equation $\mathbf{x} \leftarrow \mathbf{x} - 2(\mathbf{x}\cdot\mathbf{n}_i - c_i)\mathbf{n}_i$.
- Go back to the first step.

At each step of this algorithm, the magnitude of the composition vector decreases by a finite amount, and so this procedure converges. The procedure is also reversible. In other words, for each forward move $\mathbf{x}(c) \to \mathbf{x}$ that leads to the new composition $\mathbf{x}(p)$, there is always a backward move $\mathbf{x}(p) \to \mathbf{x}'$ that leads to the original composition $\mathbf{x}(c)$. Furthermore, when the original Monte Carlo move is in the allowed move sphere $|\mathbf{x}(c) - \mathbf{x}| < \Delta x$, the reverse Monte Carlo move is as well: $|\mathbf{x}(p) - \mathbf{x}'| < \Delta x$. This reversibility is important, because it allows the procedure to satisfy detailed balance. These reflecting boundary conditions provide a simple modification of the spherical move so that new compositions are always within the allowed simplex. These boundary conditions are analogous to the periodic boundary conditions that are more typically used in atomistic simulations.

These reflecting boundary conditions can be given a geometrical interpretation. Essentially, they are a form of billiards in the $d-1$-dimensional simplex. In each move, we chose a displacement $\mathbf{x} - \mathbf{x}(c)$. We can imagine the composition variable as a small ball that moves at a constant speed along this trajectory. The motion is continued until either a boundary is reached or the entire length of the move has been traveled. If a boundary of the simplex is reached, the ball reflects off the hyperplane by Newtonian mechanics, and the trajectory is continued in the new direction. The reflections continue until the ball has traveled a distance equal to the chosen length of the displacement, $|\mathbf{x} - \mathbf{x}(c)|$. The new point $\mathbf{x}(p)$ is then given by the location of the ball at the end of the trajectory.

REFERENCES

1. Manousiouthakis, V.; Deem, M. W. *J. Chem. Phys.,* 1999, 110: 2735–2756.

2. Kauffman, S.; Levin, S. *J. Theor. Biol.,* 1987, 128: 11–45.

3. Perelson, A. S.; Macken, A. *Proc. Natl. Acad. Sci. USA,* 1995, 92: 9657–9661.

4. Bogarad, L. D.; Deem, M. W. *Proc. Natl. Acad. Sci. USA,* 1999, 96: 2591–2595.

5. Falcioni, M.; Deem, M. W. *Phys. Rev. E,* 2000, 61: 5948–5952.

6. Chen, L.; Deem, M. W. *J. Chem. Inf. Comput. Sci.,* 2001, 41: 950–957.

7. Deem, M. W. In A. Chakraborty, Ed., *Advances in Chemical Engineering,* Vol. 28, Academic Press, San Diego, 2001, 81–121.

8. Pirrung, M. C. *Chem. Rev.,* 1997, 97: 473–488.

9. Weinberg, W. H.; Jandeleit, B.; Self, K.; Turner, H. *Curr. Opin. Chem. Bio.,* 1998, 3: 104–110.

10. McFarland, E. W., Weinberg, W. H. *TIBTECH,* 1999, 17: 107–115.

11. Xiang, X.-D.; Sun, X.; Briceño, G.; Lou, Y.; Wang, K. A.; Chang, H.; Wallace-Freedman, G.; Chang, S.-W.; Schultz, P. G. *Science,* 1995, 268: 1738–1740.

12. Briceño, W. G.; Chang, H.; Sun, X.; Schultz, P. G.; and Xiang, X.-D. *Science,* 1995, 270: 273–275.

13. Danielson, E.; Golden, J. H.; McFarland, E. W.; Reaves, C. M.; Weinberg, W. H.; Wu, X. D. *Nature,* 1997, 389: 944–948.

14. Danielson, E.; Devenney, M.; Giaquinta, D. M.; Golden, J. H.; Haushalter, R. C.; McFarland, E. W.; Poojary, D. M.; Reaves, C. M.; Weinberg, W. H.; Wu, X. D. *Science,* 1998, 279: 837–839.

15. Wang, J.; Yoo, Y.; Gao, C.; Takeuchi, I.; Sun, X.; Chang, H.; Xiang, X.-D.; Schultz, P. G. *Science,* 1998, 279: 1712–1714.

16. Francis, M. B.; Jamison, T. F.; Jacobsen, E. N. *Curr. Opin. Chem. Biol.,* 1998, 2: 422–428.

17. Dickinson, T. A.; Walt, D. R.; White, J.; and Kauer, J. S. *Anal. Chem.,* 1997, 69: 3413–3418.

18. Menger, F. M.; Eliseev, A. V.; Migulin, V. A. *J. Organic Chem.,* 1995, 60: 6666–6667.

19. Burgess, K,; Lim, H.-J.; Porte, A. M.; and Sulikowski, G. A. *Angew. Chem. Int. Ed.,* 1996, 35: 220–222.

20. Cole, B. M.; Shimizu, K. D.; Krueger, C. A.; Harrity, J. P. A.; Snapper, M. L.; Hoveyda, A. H. *Angew. Chem. Int. Ed.,* 1996, 35: 1668–1671.

21. Akporiaye, D. E.; Dahl, I. M.; Karlsson, A.; Wendelbo, R. *Angew. Chem. Int. Ed.,* 1998, 37: 609–611.

22. Reddington, E.; Sapienza, A.; Gurau, B.; Viswanathan, R.; Sarangapani, S.; Smotkin, E. S.; Mallouk, T. E. *Science,* 1998, 280: 1735–1737.

23. Cong, P.; Doolen, R. D.; Fan, Q.; Giaquinta, D. M.; Guan, S.; McFarland, E. W.; Poo-

jary, D. M.; Self, K.; Turner, H.; Weinberg, W. H. *Angew. Chem. Int. Ed.*, 1999, 38: 483–488.

24. Van Dover, R. B.; Schneemeyer, L. F.; Fleming, R. M. *Nature*, 1998, 392: 162–164.

25. Novet, T.; Johnson, D. C.; Fister, L. *Adv. Chem. Ser.*, 1995, 245: 425–469.

26. Zones, S. I.; Nakagawa, Y.; Lee, G. S.; Chen, C. Y.; Yuen, L. T. *Microporous Mesoporous Mat.*, 1998, 21: 199–211.

27. Niederreiter, H. *Random Number Generation and Quasi-Monte Carlo Methods,* Society for Industrial and Applied Mathematics, Philadelphia, 1992.

28. Bratley, P.; Fox, B. L.; Niederreiter, H. *ACM Trans. Math. Softw.*, 1994, 20: 494–495.

29. Geyer, C. J. In *Computing Science and Statistics: Proceedings of the 23rd Symposium on the Interface*, American Statistical Association, New York, 1991, 156–163.

30. Swendsen, R. H.; Wang, J. S. *Phys. Rev. Lett.*, 1986, 57: 2607–2609.

31. Marinari, E.; Parisi, G.; Ruiz-Lorenzo, J. In *Spin Glasses and Random Fields*, Vol. 12, *Directions in Condensed Matter Physics*, A. P. Young, Ed., World Scientific, Singapore, 1998, 59–98.

32. Falcioni, M.; Deem, M. W. *J. Chem. Phys.*, 1999, 110: 1754–1766.

33. Sedgewick, R. *Algorithms,* 2nd ed., Addison-Wesley, New York, 1988.

34. Balkenhohl, F.; von dem Bussche-Hünnefeld, C.; Lansky, A.; Zechel, C. *Angew. Chem. Int. Ed. Engl.*, 1996, 35: 2288–2377.

35. Jandeleit, B.; Schaefer, D. J.; Powers, T. S.; Turner, H. W.; Weinberg, W. H. *Angew. Chem. Int. Ed.*, 1999, 38: 2494–2532.

36. McFarland, E. W.; Weinberg, W. H. *Trends Biotechnol.*, 1999, 17: 107–115.

37. Furka, A.; Sebestyen, F.; Asgedom, M.; Dibo, G. *Int. J. Peptide Protein Res.*, 1991, 37: 487–493.

38. Ohlmeyer, M. H. J.; Swanson, R. N.; Dillard, L. W.; Reader, J. C.; Asouline, G.; Kobayashi, G. A. R.; Wigler, M.; Still, W. C. *Proc. Natl. Acad. Sci. USA*, 1993, 90: 10922–10926.

39. Francis, M. B.; Jamison, T. F.; Jacobsen, E. N. *Curr. Opin. Chem. Biol.*, 1998, 2: 422–428.

40. Tan, D. S.; Foley, M. A.; Stockwell, B. R.; Shair, M. D.; Schreiber, S. L. *J. Am. Chem. Soc.*, 1999, 121: 9073–9087.

41. Weinberg, W. H.; Jandeleit, B.; Self, K.; Turner, H. *Curr. Opin. Chem. Bio.*, 1998, 3: 104–110.

42. Schreiber, S. L. *Science*, 2000, 287: 1964–1969.

43. Nazarpack-Kandlousy, N.; Zweigenbaum, J.; Henion, J.; Eliseev, A. V. *J. Comb. Chem.*, 1999, 1: 199–206.

44. Desai, M. C.; Zuckermann, R. N.; Moos, W. H. *Drug Dev. Res.*, 1994, 33: 174–188.

45. Ajay, W.; Walters, P.; Murcko, M. A. *J. Med. Chem.*, 1998, 41: 3314–3324.

46. Bures, M. G.; Martin, Y. C. *Curr. Opin. Chem. Biol.*, 1998, 2: 376–380.

47. Drewry, D. H.; Young, S. S. *Chemometrics Intell. Lab.*, 1999, 48: 1–20.

48. Hassan, M.; Bielawski, J. P.; Hempel, J. C.; Waldman, M. *Mol. Diversity*, 1996, 2: 64–74.

49. Zheng, W.; Cho, S. J.; Tropsha, A. *J. Chem. Inf. Comput. Sci.*, 1998, 38: 251–258.

50. Zheng, W.; Cho, S. J.; Tropsha, A. *J. Chem. Inf. Comput. Sci.*, 1998, 38: 259–268.

51. Zheng, W.; Cho, S. J.; Waller, C. L.; Tropsha, A. *J. Chem. Inf. Comput. Sci.*, 1999, 39: 738–746.

52. Reynolds, C. H. *J. Comb. Chem.*, 1999, 1: 297–306.

53. Linusson, A.; Wold, S.; Nordén, B. *Mol. Diversity*, 1999, 4: 103–114.

54. Gobbi, A.; Poppinger, D. *Biotechnol. Bioeng.*, 1998, 61: 47–54.

55. Weber, L.; Wallbaum, S.; Broger, C.; Gubernator, K. *Angew. Chem. Int. Ed. Engl.*, 1995, 34: 2280–2282.

56. Sheridan, R. P.; Kearsley, S. K. *J. Chem. Inf. Comput. Sci.*, 1995, 35: 310–320.

57. Singh, J.; Ator, M. A.; Jaeger, E. P.; Allen, M. P.; Whipple, D. A.; Soloweij, J. E.; Chowdhary, S.; and Treasurywala, A. M. *J. Am. Chem. Soc.*, 1996, 118: 1669–1676.

58. Brennan, M. B. *Chemical Engineering News*, June 5, 2000, 78(23): 63–73.

59. Brown, R. D.; Clark, D. E. *Expert Opin. Therap. Pat.*, 1997, 8: 1447–1459.

60. Brown, R. D.; Martin, Y. C. *J. Med. Chem.*, 1998, 40: 2304–2313.

61. Katritzky, A. R.; Kiely, J. S.; Hébert, N.; Chassaing, C. *J. Comb. Chem.*, 2000, 2: 2–5.

62. Frenkel, D.; Smit, B. *Understanding Molecular Simulation: From Algorithms to Applications,* 2nd edition, Academic Press, San Diego, 2002.

63. Shuker, S. B.; Hajduk, P. J.; Meadows, R. P.; Fesik, S. W. *Science*, 1996, 274: 1531–1534.

64. Fejzo, J.; Lepre, C. A.; Peng, J. W.; Bemis, G. W.; Ajay; Murcko, M. A.; Moore, J. M. *Chem. Biol.*, 1999, 6: 755–769.

65. Maly, D. J.; Choong, I. C.; Ellman, J. A. *Proc. Natl. Acad. Sci. USA*, 2000, 97: 2419–2424.

66. Griffey, R. H.; Hofstadler, S. A.; Sannes-Lowery, K. A.; Ecker, D. J.; Crooke, S. T. *Proc. Natl. Acad. Sci. USA*, 1999, 96: 10129–10133.

67. Davis, M. E.; Zones, S. I. In *Synthesis of Porous Materials: Zeolites, Clays, and Nanostructures*, Occelli, M. L.; Kessler, H., Eds., Marcel Dekker, New York, 1997, 1–34.

68. Burkett, S. L.; Davis, M. E. *Chem. Mater.*, 1995, 7: 920–928.

69. Burkett, S. L.; Davis, M. E. *Chem. Mater.*, 1995, 7: 1453–1463.

70. Chen, L.; Deem, M. W. *Mol. Phys.*, 2002, 100: 2175–2181.

CHAPTER 13

EXPLORING A SPACE OF MATERIALS: SPATIAL SAMPLING DESIGN AND SUBSET SELECTION

FRED A. HAMPRECHT
University of Heidelberg

ERIK AGRELL
Chalmers University of Technology

13.1 INTRODUCTION

We assume you are facing the following situation: you wish to find a material that features the best possible value for some property that you can measure quantitatively. This is your *response*. You can control the composition as well as the formulation and synthesis conditions. These variables form the basis of your *experimental space*. We define the *experimental region* as that part of experimental space that you decide to explore in your quest for a better material.

Previous chapters have dealt with optimization of the response. This works fine once you have found an area in your experimental region in which the response is greater than zero. What, however, if you have not? Also, you may already have explored an area of your experimental region until you have found the local response maximum, but you may be dissatisfied with what you have found: you now wish to screen the remainder of your experimental region for other areas that show a nonzero response.

In both cases, you are obliged to sample your experimental region systematically, at least until you have found a new active area. A similar approach may be useful even to locate the response maximum within a known active area, in cases when the experiments are too time-consuming to admit a sequential design, e.g., if long annealing is required. This entire chapter is about how to perform this sampling most effectively in the sense of gaining the clearest possible picture of your response surface.

13.2 MODELING THE RESPONSE SURFACE

To prevent wasting of time and money, we wish to sample space systematically in an optimal fashion. Fine, but what are "systematic" and "optimal"? Intuitively, we understand that, in the absence of detailed prior knowledge about the response surface, systematic means the samples should be distributed uniformly throughout the experimental region. However, different systematic sampling schemes are conceivable, and a large variety of these have been proposed in the literature (see references in [1]). Many have been brought forward with eloquent arguments, but as long as these are purely verbal, it is difficult to weigh one against the other. We advocate the use of methods that are derived from clearly (and mathematically) stated assumptions concerning the *response surface*. The name of the latter derives from the case of a two-dimensional experimental region. If the response is plotted in the third dimension for each point in the experimental region, the impression of a surface results. In this analogy, a steepest-ascent algorithm can be compared to a hiker who chooses the steepest path to the nearest peak, which corresponds to a local response maximum. We will always talk of "surfaces," even in higher dimensions.

13.2.1 Polynomial Approximation

The *design* is the sampling scheme, or the set of all points at which experiments are to be performed. The theory of optimal experiments [2] assumes that the response can be approximated with an analytical model. Once a model is fixed, those points that have the greatest statistical leverage are selected for the design, that is, those points that minimize the uncertainty in the estimates of the model parameters. Such optimality criteria go under the name of alphabetic criteria; namely, A-optimality (minimizing the trace of the inverse of the information matrix), D-optimality (maximizing the determinant of the information matrix), and G-optimality (minimizing the maximum prediction variance) [3]. By far the most popular models are polynomials, that is, the sum of a constant plus a plane plus a paraboloid and so on. The difficulty with most implementations is that they consider only low-order polynomials as models. This is problematic if the response surface can be expected to be irregular, for instance, featuring multiple peaks (statisticians refer to these as modes) and valleys. A second-order polynomial can provide a good model for a single peak, but not for a chain of mountains. One consequence is that, especially in higher dimensions, most of the design points placed optimally by the preceding criteria lie on the hull of the experimental region, while few points are selected within. This lack of representation of the inner region has started a quest for more space-filling algorithms.

In the following sections, we describe a different model of the response surface that provides a lot of flexibility. Optimal sampling strategies will then be derived under that model.

13.2.2 Stochastic Processes

Not knowing what the actual response surface looks like, we can model it using a stochastic approach. For starters, we could draw numbers from a random-number generator, one for each of the points in the experimental region. The resultant surface would be pure noise. If you have reason to believe that this is a good model for the particular response surface you are studying, you may skip this chapter: the best you can then do is to avoid sampling the same point in experimental space twice, but other than that, all conceivable designs will be equally good, on average.

In reality, however, many response surfaces exhibit spatial correlation. That is, one can expect two proximate points in the experimental region to feature similar responses. Exceptions do certainly occur, for instance, in the case of sharp phase transitions, but even then the remainder of the experimental region will usually show some spatial correlation. This "similar property principle" applies to a wide range of systems and we assume it to hold in the following.

What we want, then, is a mathematical tool that can exhibit a range of spatial correlations while being random in character, because the true response surface is not known. In other words, in our model we wish to represent the expected *smoothness* of the unknown response surface (which may be guessed or inferred from previous experiments on similar systems), but not the exact location of the response minima and maxima. Once such a model is found, we can derive designs that are optimal for it.

A suitable technique that satisfies our requirements is a *stochastic process*, also known as random field or random process. Such processes arise in the description of systems that evolve in space subject to probabilistic laws. We do not claim that a response surface is probabilistic in nature, but we do assume that a response surface looks similar to a suitably chosen stochastic process.

Before we illustrate these concepts graphically, we need a little more terminology: consider a random-number generator from which random numbers can be drawn, one at a time. These individual numbers are called *realizations*. In the case of a particular stochastic process, an entire random surface (and it can generate infinitely many of these) is just *one* realization. These realizations will have their peaks and valleys at different positions, but they will also have one thing in common: the smoothness.

Figure 13.1 shows two realizations of each of four different stochastic processes. The surfaces are obviously all different, but pairs of them appear similar: they share the same smoothness. The smoothness of the surfaces is governed by the range[1] of the covariance functions as well as by the behavior of the covariance functions near the origin. A cusp, as in the exponential covariance function, leads to surfaces that are much rougher on a short scale.

---

[1]The range indicates the interval over which the covariance function differs "substantially" from zero. This definition is admittedly vague and conventions in the literature differ. In the calculations shown here, the exponential and Gaussian covariance functions with range ρ are given by
$$\text{Cov}(x_1, x_2) \sim \exp(-\|x_1 - x_2\|/\rho) \text{ and } \text{Cov}(x_1, x_2) \sim \exp(-\|x_1 - x_2\|^2/\rho^2),$$
respectively.

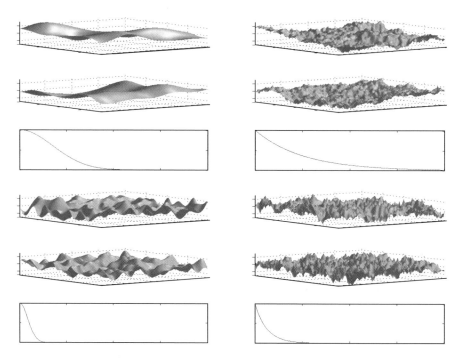

Figure 13.1 Stochastic models for response surfaces. Shown are two realizations each of four stochastic processes with their corresponding isotropic (circularly symmetric) covariance functions. Left: processes with a Gaussian covariance function; right: processes with an exponential covariance function. Top: long range; bottom: short range. The smoothness is governed by the covariance function: its degree of differentiability around the origin determines the smoothness at a microscopic scale, its decay with distance governs the appearance of mountains and valleys at a large scale. All simulations performed with gstat [4]. (Reprinted, with permission, from Hamprecht, F. A., *J. Chemical Information and Computer Science,* 2002, 42: 414–428; copyright © 2001. American Chemical Society, Washington, DC.)

It is now time to quantify this smoothness and the class of stochastic processes we use. First, consider the mean of the stochastic process $Z(x)$ at position x,

$$\bar{Z}(x) = E[Z(x)],$$

where E denotes expectation. One obtains the mean as one averages the response at a specific position, x, over all possible realizations of a stochastic process. In the following development, we will assume this mean to be a constant, $\bar{Z}$, over the experimental region. Next, consider the *covariance* between two points x_1, x_2, given by

$$\mathrm{Cov}(x_1, x_2) = E[(Z(x_1) - \overline{Z})(Z(x_2) - \overline{Z})].$$

The covariance function describes how similar the response is at positions x_1, x_2, averaged over all possible realizations. To make things easier, we assume that the covariance function depends only on the length and orientation of the vector connecting x_1 and x_2. This assumption and the one for a constant mean basically express the belief that the response surface does not change in character throughout the experimental region. It can still exhibit wild cliffs, deep canyons, and vertiginous peaks, but the character of the response surface landscape should be homogeneous throughout. Mathematically, these assumptions are denoted *second-order* (or weak or wide-sense) *stationarity*.[2] As a final simplification, we can assume that the covariance function depends only on the length, but not on the orientation, of the vector connecting x_1 and x_2. Such covariances are called *isotropic* and the phenomena they describe admit, on average, no discernible orientation.

Summarizing this section, we recommend following only those arguments for the selection of experimental designs that are directly derived from clearly stated models for the response surface. We have quoted two such models: the polynomial approximation, leading to the alphabetic criteria; and stochastic processes, for which we will derive optimality criteria in the following sections.

13.3 KRIGING

The central assumption now is that the response surface can be modeled as the realization of a second-order stationary stochastic process. Bear with us while we introduce an interpolation method that will, in turn, lead to criteria for optimal designs.

Actually, an interpolator has some interest in its own right: given a number of measured responses at some points in the experimental region, it will give an estimate of the response in between these points. Reflecting the importance of this task, a large number of methods have been proposed: polynomials, splines, B-splines, thin-plate splines, etc. The *kriging*[3] or *best linear unbiased estimator* stands out because it provides not only an estimate of the response surface throughout the experimental region, but also the uncertainty of that estimate, as a function of space: the uncertainty is low around the available data and grows with distance from these. Just how quickly the uncertainty grows with distance depends on the smoothness of the surface (see Figure 13.2).

---

[2]Fixing the first two moments, the mean and the covariance, is not very restrictive in the sense that there is an infinite number of different stochastic processes with the same first two moments. Generally, all finite-dimensional moments [5] are required to characterize a stochastic process exhaustively. Gaussian processes (which were used in Figure 13.1) are an exception, because they are completely determined by their first two moments.

[3]This interpolator has been developed independently in different disciplines, but has probably gained the greatest popularity in the mining community that is compelled to extract a maximum of information from each of their expensive samples. The name honors the mining engineer D. G. Krige [6].

The simple kriging estimator $\hat{Z}(x)$ of the true response surface $Z(x)$ is given by

$$\hat{Z}(x) = \sum_{i=1}^{n} \lambda_i(x) Z(y_i),$$

where y_i are those points at which experiments have been performed, that is, the design, and $\lambda_i(x)$ are optimal weights, one for each of the n measurements $Z(y_i)$.[4] Note that these weights vary over space [otherwise, the estimate $\hat{Z}(x)$ would be a constant]. They are found as the solution of a set of n linear equations

$$\sum_{i=1}^{n} \lambda_i(x) \, \text{Cov}(y_i, y_j) = \text{Cov}(x, y_j) \qquad j = 1, \ldots, n \tag{13.1}$$

depending on the covariance function of the stochastic process, and are optimal in the sense of minimizing the mean-square error (MSE) [1,5] between the true and the estimated response surfaces,

$$MSE(x) = E[(\hat{Z}(x) - Z(x))^2].$$

Again, the MSE of an interpolator should be minimal on *average*, that is, considering all possible realizations of the stochastic process.

We have stated previously that the growth of the uncertainty of the interpolation with the distance from design points depends on the smoothness of the surface. This smoothness enters through the covariance function in equation (13.1) for the optimal weights $\lambda_i(x)$.

The MSE of the kriging estimator, or the uncertainty of the interpolation as a function of space, is given [1,5] by

$$MSE(x) = \text{Cov}(x, x) - \sum_{i=1}^{n} \lambda_i(x) \, \text{Cov}(y_i, x). \tag{13.2}$$

What does this equation depend on? On the covariance function, the design y_i, and the optimal weights which, in turn, depend only on the covariance function and y_i. If you ponder this, the all-important conclusion is that the uncertainty in the interpolation depends on the measurement sites (the design) and the covariance function, but not on the measured values $Z(y_i)$.

As a consequence, the design can now be optimized so as to minimize the MSE, or the uncertainty in the interpolation, *before* performing any experiments. The catch is in the assumptions: the estimated uncertainty is only correct when the "true" covariance function is used.

---

[4]In the following we will assume that $\bar{Z} = 0$. To estimate a process $Y(x)$ with a nonzero, but constant and known, mean, simply subtract the mean from every measurement, apply the regular kriging estimator to the obtained zero-mean process $Z(x) = Y(x) - \bar{Y}$, and add $\bar{Y}$ again to $\hat{Z}(x)$ to obtain an unbiased estimate $\hat{Y}(x)$.

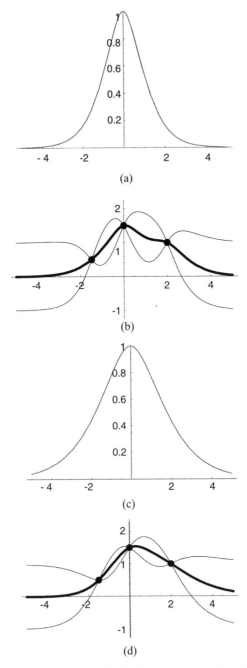

Figure 13.2 Simple kriging example. (a), (c) Two covariance functions that differ in their range; (b), (d) an interpolation (bold line) of the three bold points for the corresponding functions. Far from the data, the interpolation goes to the assumed mean, in this case, zero. The thin lines indicate the uncertainty of the interpolation in terms of $\pm\sqrt{MSE(x)}$, given by equation (13.2). The estimates lie within the marked intervals with a likelihood of 68%. The uncertainty grows faster for the covariance function with the shorter range.

13.4   OPTIMALITY CRITERIA

Equation (13.2) gives the expected error for location x. The design can then be modified to control the uncertainty at that specific location. We wish, however, to find a design that is optimal for the entire experimental region $\mathcal{R}$. We will introduce two natural measures of global performance that have been proposed [7], then illustrate in what sense they differ fundamentally (Sections 13.5.2, 13.5.3), and finally argue which of the two should be used (Section 13.5.4).

13.4.1   Integrated Mean-Square Error

The integrated MSE aims to minimize the total MSE for the entire experimental region:

$$\min \int_{\mathcal{R}} MSE(x)\, dx. \tag{13.3}$$

Given a particular experimental region (see Section 13.1) and covariance function, this criterion or objective function can be optimized with any deterministic (e.g., steepest descent) or stochastic (e.g., simulated annealing) optimization method.[5] Algorithmic details and several examples of such an optimization are given in [1].

Figure 13.3 shows a square experimental region and the design of 121 points that has been optimized under equation (13.3) using an exponential covariance function with a range of one-fifth of the square's side length. The design obviously reflects the particular shape of the experimental region under study, but its inner points seem to have a will of their own: they are arranged in a hexagonal pattern that does not match the shape of the boundary! This might lead us to hypothesize that in two dimensions, the intrinsic structure of a design that minimizes the integrated MSE may be hexagonal.

To find out whether this hypothesis is of any value, we should make the experimental region larger and larger and add more and more points to it to keep the density of the design constant. In this way, the points at the interior can be expected to be influenced less and less by the particular shape of the experimental region.

Ultimately, and to answer the question in full generality, the experimental region and the number of points should become infinitely large. This assumption greatly facilitates a mathematical analysis. We state, without proof, the following results:

- If the response surface becomes infinitely smooth or infinitely rough, it does not matter which sampling design is used;
- If the response surface is very rough, the best design is given [8] by the densest sphere packing (see Section 13.5.1);

---

[5]If an opinion exists as to which parts of the experimental region should be sampled most intensely, this prior belief can be expressed by a spatial weighting of the mean square error [1].

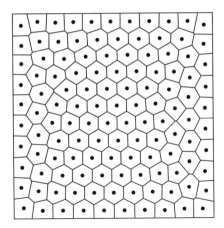

Figure 13.3 Design of 121 points for a square region, optimized under the integrated MSE.

- If the response surface is very smooth, the best design is given by the Fourier transform of the densest sphere packing [9].

These notions are introduced in Section 13.5.1, but as a preview, we have disclosed the result that in two dimensions, both the best sphere packing and its Fourier transform are indeed given by a hexagonal arrangement and our inspection of Figure 13.3 thus has allowed us to anticipate some deep results about the inherent characteristics of optimal designs in general.

13.4.2 Maximum Mean Square Error

Whereas the integrated MSE criterion discussed earlier strives to minimize the *average* ignorance, the maximum MSE criterion aims to minimize the *maximum* interpolation error in the experimental region. A brute-force recipe to find a design that minimizes the maximum MSE consists of identifying that location in space that has the greatest error, then perturbing the design, finding the location that now has the greatest error, etc.

If the stochastic process is assumed Gaussian and the response surface is very rough, this criterion becomes almost equivalent [10] to the problem of thinnest sphere covering, which is introduced in the next section.

13.5 INFINITE EXPERIMENTAL REGIONS

In an earlier thought experiment, we have considered an infinitely large experimental region, to be explored with an infinite number of points. This is a trick to make boundary effects negligible and learn something about the inherent character of the different design criteria. We use this trick in the present section to accentuate the differences between using integrated and maximum MSE.

In Section 13.6, we will turn our attention back to the practically more relevant case of a finite experimental region.

13.5.1 Uniformly Spaced Points: Lattices

Intuitively, we can expect an optimal design on a perfectly homogeneous infinite space to be very homogeneous itself. The most regular arrangement of points conceivable is a *lattice*. Mathematically, a lattice consists of all linear combinations of some basis vectors, with all coefficients assuming integer values only (see Figure 13.4). This yields a highly regular structure with the property that, with an often cited observation, "if you sit on one lattice point and view the surrounding set of lattice points, you will see the identical environment regardless of which point you are sitting on" [11]. In this sense, all lattices are perfectly uniform arrangements: there is not one part of space where the points lie closer together than in any other part.

The first dimension does not offer much variety when it comes to lattices: there is only one way of arranging points at equal intervals, and the resultant lattice is called Z_1. (We follow approximately the terminology of [12], where the first letter indicates the family and the number gives the dimension.) In contrast, an infinite number of lattices can be conceived for any dimension higher than one. In two dimensions, two fundamental lattices have emerged as particularly interesting, the *cubic* lattice, Z_2, and the *hexagonal* lattice, A_2 (see Figure 13.4).

Figure 13.4 also displays a set of basis vectors for each lattice. A lattice is fully specified through its basis vectors, but the converse is not true: the same lattice can be generated from different sets of basis vectors. If the basis vectors are collected

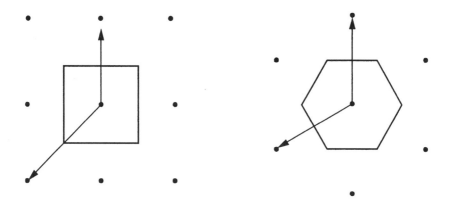

Figure 13.4 The cubic lattice, Z_2, and the hexagonal lattice, A_2, along with their Voronoi cells and one possible choice of basis vectors.

into a matrix, we obtain a *generator matrix* for the lattice. Examples of generator matrices for Z_2 and A_2 are

$$\begin{bmatrix} 0 & 2 \\ -2 & -2 \end{bmatrix} \quad \text{and} \quad \begin{bmatrix} 0 & 2 \\ -\sqrt{3} & -1 \end{bmatrix},$$

where the rows represent the same basis vectors as those shown in Figure 13.4.

Also shown in the figure are the so-called *Voronoi cells* of the lattices [13,14]. The Voronoi cell of a point in a point set makes up the part of space that is closer to this point than to any other. Contrary to the case of general point sets, there is only one type of Voronoi cells in a lattice, i.e., each Voronoi cell can be superimposed onto any other by translation only, without rotation or reflection.

When we move on to three-dimensional lattices, three basic lattices deserve particular attention, all of them well known from crystallography. They are illustrated in Figure 13.5. The cubic lattice, Z_3, simply consists of all three-dimensional points with integer coordinates. If we remove every second point of the cubic lattice as in Figure 13.5b, the *face-centered cubic* (fcc) lattice is obtained. (Alternatively, it can be constructed by centering a point on each of the six faces of a cube and stacking such cubes, hence the name.) And if instead we remove three fourths of the points in a cubic lattice, leaving only those with either all even or all odd coordinates, we obtain the *body-centered cubic* (bcc) lattice depicted in Figure 13.5c. Even though the point sets in themselves may not look very different, their geometrical properties differ substantially, as illustrated by the shapes of their Voronoi cells. In standard lattice notation, the fcc lattice is denoted A_3 and the bcc lattice, A_3^*. Their generator matrices, as well as their generalizations and other lattices in higher dimensions, are deferred to the Appendix.

13.5.2 Packing, Quantizing, and Covering Problems

Packing   In Section 13.4.1 we claimed that, in the case of an infinite design region and a rough response surface, the integrated MSE is minimized by a design that corresponds to a best sphere packing. The *sphere-packing* problem asks for the densest possible way of arranging nonoverlapping spheres of equal size. The density becomes maximal when the volume of the interstitial regions is minimized. In lattices, the multiple translations of a single Voronoi cell tile all of space. As a consequence, it is sufficient to study a single Voronoi cell to understand the properties of the entire lattice. To reduce interstitial space, it is advantageous to have a Voronoi cell that is as similar as possible to a sphere. Since spheres do not tile space in dimensions greater than one, the best approximation is sought. If we scale all lattices such that their Voronoi cells have unit volume, the best packing lattice will be the one with the largest inscribed sphere[6]. Its radius, the *packing radius, r,* becomes

---

[6]Lattices are not the only possible method of stacking spheres densely, but the best lattices are generally either the best packing method known or not much worse than the best known.

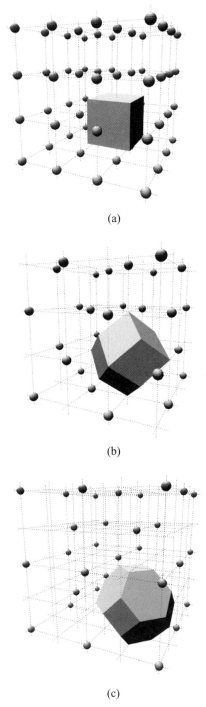

Figure 13.5 The three basic three-dimensional lattices and their Voronoi cells. (a) The cubic lattice, whose Voronoi cell is simply a cube. (b) The fcc lattice. Its Voronoi cell is a rhombic dodecahedron and has 12 faces. (c) The bcc lattice. Its Voronoi cell is a truncated octahedron and has 14 faces.

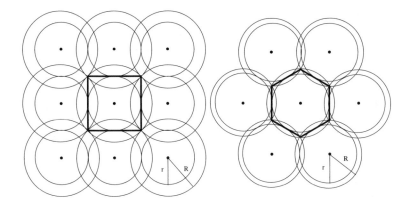

Figure 13.6   The cubic and the hexagonal lattice with the largest inscribed and smallest circumscribed spheres of their Voronoi cells. If both lattices are scaled such that their Voronoi cells have unit volume, the hexagonal lattice has the greater packing radius r and smaller covering radius R, cf. Section 13.5.2 and Figure 13.7.

the figure of merit for the packing problem: the greater r, the denser a lattice can pack spheres (see Figure 13.6).

As we go to higher dimensions, we encounter an ever increasing number of potentially useful lattices, while at the same time the evaluation of their characteristics becomes more complex. Fortunately, a large amount of research has been devoted to the search for good lattices, and the results are tabulated for several quality measures and for all dimensions that are relevant for our purposes [12]. The best packing, quantizing, and covering lattices (see the following sections) known for dimensions up to eight are summarized in Table 13.1. Definitions of all lattices mentioned in the table, in terms of their generator matrices, are found in the Appendix.

Quantizing   Work on the optimal sampling problem in signal processing [9] has led to the conclusion that for very smooth response surfaces in an infinite experimental region, the integrated MSE is minimized by a design that is the dual[7] of the best sphere packer.

A glance at Table 13.1 shows that the duals of the best packing lattices can be found[8] in the column entitled Quantizing. The *quantizing problem* (or vector quantization), which owes its name to an application in digital communications, seeks an arrangement of points such that their Voronoi cells become as compact as possible, where compactness is measured by their average second moment. Indeed, the intriguing fact that the best known lattices for packing and quantizing are duals of each other has been observed also outside the range shown in the table, and it was

---

[7]The dual of a lattice is essentially its Fourier transform, if the point set is regarded as a multidimensional sum of Dirac impulse functions.

[8]The lattices Z_1, A_2, D_4, and E_8 are self- or isodual.

Table 13.1 The Best Known Lattices According to Three Criteria

Dimension	Packing	Quantizing	Covering
1	Z_1	Z_1	Z_1
2	A_2	A_2	A_2
3	A_3 (fcc)	A_3^* (bcc)	A_3^*
4	D_4	D_4	A_4^*
5	D_5	D_5^*	A_5^*
6	E_6	E_6^*	A_6^*
7	E_7	E_7^*	A_7^*
8	E_8	E_8	A_8^*

Note: A lattice L^* is the dual, or Fourier transform, of the lattice L. Z_d is the d-dimensional cubic lattice, and A_d is the generalized hexagonal lattice. These and the other lattice families are defined in the Appendix.

earlier conjectured to be a generally valid relation between optimal lattices in any dimension [12]. This conjecture does not appear to be true for arbitrary dimensions [15], but the general tendency is still valid: the dual of a good packing lattice is good for quantizing and vice versa [16]. We will have reason to recall this property in Section 13.6.

As before, if we restrict ourselves to lattices, it is sufficient to study a single Voronoi cell; and also as before, the most compact body of a given volume according to this measure is the sphere. Yet again, tiling space with spherical cells is not possible, and thus one seeks a lattice whose Voronoi cells are as nearly spherical as possible, by the previously used measure, while still tiling space. The figure of merit now is $\bar{r}$, which gives the root of the mean-squared Euclidean distance of all points in a Voronoi cell to its center.

Covering Finally, we noted (Section 13.4.2) that, at least for rough Gaussian[9] stochastic processes, for an infinite experimental region, the maximum MSE is minimized by a design that offers a *thinnest covering*. In this problem, the aim is to seek an arrangement of spheres of equal size such that each point in space is covered by at least one sphere, while the average number of spheres covering a point in space should be as low as possible. Arguing in terms of lattices, the radius of the smallest circumscribed sphere around the Voronoi cell, the *covering radius, R*, should become minimal (see Figure 13.6).

We have seen that all criteria, packing, quantizing, and covering, seek to find lattices with as spherical a Voronoi cell as possible; however, different definitions of sphericity are used and the following section shows that these lead to wildly differing lattices.

---

[9]Gaussian here refers to the finite-dimensional distribution, not to the covariance function.

13.5.3 The Geometry of Voronoi Cells

By far the most popular multidimensional lattice in experimental design, as well as in a wide range of other applications, is the cubic lattice, Z_d. It is what one obtains by simply allowing the same discrete set of equidistant values for all variables, independently of each other. While this property makes it extremely easy to generate and utilize this lattice, the lattice is by no means a good lattice in the sense presented in the previous section. Note that it does not appear in Table 13.1 for dimensions higher than one. Figures 13.4–13.6 suggest why: its Voronoi cell is a square, in general a hypercube, which is not a very good approximation to a sphere.

The most important characteristic of a point in a Voronoi cell is its "radius," i.e., its distance to the center of the cell—the direction is irrelevant for the performance, as long as a rotation-invariant figure of merit is used. Hence, the performance of a lattice is fully determined by the distribution of the radii in the Voronoi cell. We define the *radial distribution* of a body as the percentage of the volume that lies outside a circle of a certain radius. For lattice Voronoi cells, this is equivalent to the likelihood that the distance between a random point in multidimensional space and its closest lattice point is greater than a certain value. A uniform distribution is assumed over a large enough region to make boundary effects negligible.

In preparation for considering higher dimensions, we begin with the radial distributions of the Voronoi cells of the two-dimensional lattices of Figure 13.6. The corresponding functions are shown in Figure 13.7. The density of both lattices have been normalized to one lattice point per unit volume. The three quality measures r, $\bar{r}$, and R (see Section 13.5.2) are marked in the diagrams for both lattices.

If we look for a d-dimensional body of unit volume with greatest inscribed sphere (largest r), smallest moment of inertia (smallest $\bar{r}$), or smallest circumscribed sphere (smallest R) without requiring that it allow a tiling of space, we find that the sphere is optimal by all criteria. Hence we include in the diagrams the corresponding curve for the distance between points in a sphere and its center. This curve serves as a lower bound in the diagrams. A good lattice, in the sense that it has as much of the Voronoi cell as possible located close to its center, would in these diagrams be identified by its proximity to the spherical lower bound. As we can see in Figure 13.7, the curves for the two lattices differ only in their tails. They follow the same shape down to the packing radius r of Z_2. From there on, A_2 has a steeper slope, reflecting the rounder shape of its Voronoi cell.

Figures 13.8–13.10 show the same type of curves for higher dimensions. For dimensions 4, 8, and 16, the best lattices for packing, quantizing, and covering, respectively, are illustrated, along with the cubic lattice. Again, the best lattice would have a curve that in some sense lies as close as possible to that of the sphere. The curves coincide with those of the sphere for distances below their packing radii r and above their covering radii R, respectively. They deviate between these points, where the cubic lattice always shows the most prominent tail. The ratio R/r is equal to $\sqrt{d}$ for Z_d and thus tends to infinity as the dimension increases. "Good" lattices,

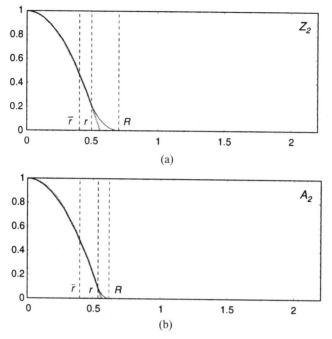

Figure 13.7 The solid curves show the radial distribution, i.e., the percentage of the Voronoi cells of the two-dimensional lattices (a) Z_2 and (b) A_2 (see Figure 13.4) that lies outside a given radius. Dashed lines indicate the figures of merit for packing (packing radius r), quantizing (root-mean-square distance $\bar{r}$), and covering (covering radius R) problems. The gray curve illustrates the radial distribution of a circle with the same area as the Voronoi cells considered.

on the other hand, typically have a ratio less than two for any dimension. In particular, this is always true for the optimal packing lattice.[10]

In all cases, and in particular for the cubic lattice, the curves approach zero rapidly long before the theoretical zero, which occurs at the covering radius R. This may be somewhat surprising in view of the large number of vertices that a typical Voronoi cell possesses.[11] Apparently, the volume close to such a vertex is so small that even the sum of them accounts for only a negligible volume.

One might think of a high-dimensional Voronoi cell as being similar to a sea urchin: a spherical shape with a large number of thin needles on its surface. Such a shape is illustrated in Figure 13.11, which is a nonlinear but radius-preserving mapping from an eight-dimensional hypercube to three-dimensional space. We observe

---

[10]Because for any lattice with $R \geq 2r$, a new lattice with the same r, but higher point density, can be created by inserting points in the void between points in the original lattice.

[11]The Voronoi cell of any d-dimensional lattice has between 2^d and $(d + 1)!$ vertices, inclusively, where the lowest value is attained by Z_d and the highest by A_d^* [17].

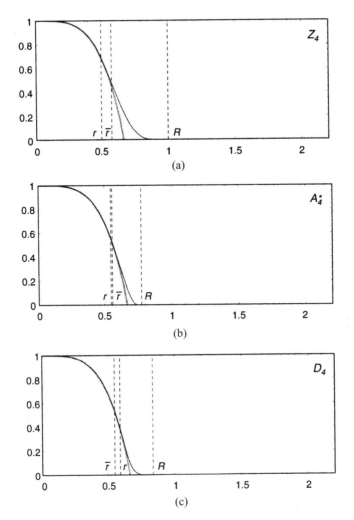

Figure 13.8 The radial distribution of (a) Z_4, (b) A^*_4 (the best known lattice for covering), and (c) D_4 (the best known lattice for packing and quantizing). The gray curve represents a four-dimensional sphere.

that the spines are extremely thin, indicating that the space close to a vertex accounts for a negligible fraction of the volume of the hypercube. On the other hand, the shape is also almost empty in the center. This behavior is typical for high-dimensional polytopes, not just hypercubes. Most of the mass is concentrated on a shell with relatively small radial variation.[12]

---

[12]For d-dimensional hypercubes, the radius of the shell approaches $R/\sqrt{3}$ as d increases.

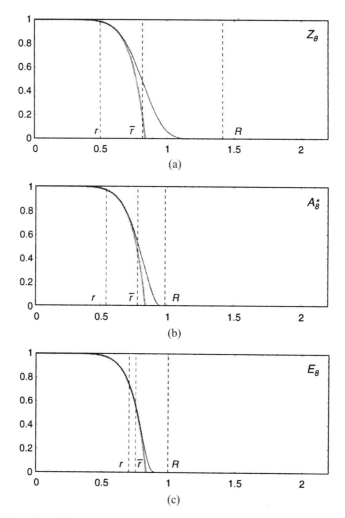

Figure 13.9 The radial distribution of (a) Z_8, (b) A_8^* (the best known lattice for covering), and (c) E_8 (the best known lattice for packing and quantizing). The gray curve represents an eight-dimensional sphere.

13.5.4 Discussion of Optimality Criteria

We have characterized the integrated and maximum MSE criteria by means of the lattices that they lead to when the experimental region becomes infinitely large and the response surface is very smooth or very rough. Figures 13.8–13.10 show that almost all of the volume of a Voronoi cell is confined to a radius that is much smaller than the distance of the vertices from the center. Repeating our metaphor (illustrated in Figure 13.11), the corners of a Voronoi cell in a higher dimension take the shape of needles and their volume in experimental space is small. If the probability

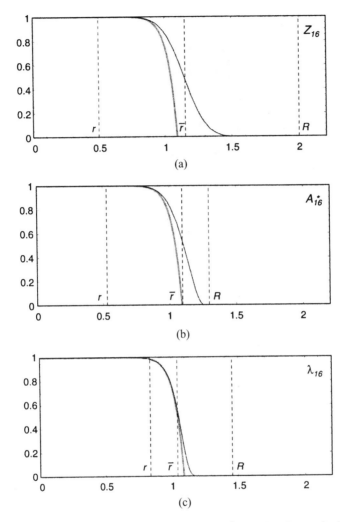

Figure 13.10   The radial distribution of (a) Z_{16}, (b) A_{16}^* (the best known lattice for covering), and (c) Λ_{16} (the best known lattice for packing and quantizing). The gray curve represents a 16-dimensional sphere.

of finding an interesting area of high response is uniform throughout the experimental region, it is unlikely that it will come to lie in one of the needles. For covering (and thus the maximum MSE criterion), however, all that matters is the distance of the tips of these needles from the center, irrespective of what little volume they occupy. The integrated MSE criterion leads to Voronoi cells with longer needles. This criterion pays no heed to the large interpolation uncertainty near the tip of these needles, simply because their statistical weight is so low.

Also, in the sense of the integrated MSE criterion, *any* additional design point

Figure 13.11 A three-dimensional representation of an eight-dimensional hypercube. The body has the same radial distribution and the same number of vertices as the hypercube. A very small fraction of the mass lies near the vertex. Also, most of the interior is void.

will reduce the uncertainty of the experimenter (as long as it does not coincide with a previous point). By the maximum MSE criterion, however, only *optimally* placed points reduce the uncertainty if the range of the covariance function is short.[13]

For these reasons, and assuming that the top priority in most applications is the best overall performance throughout the experimental region, we encourage the use of the integrated MSE criterion. High-dimensional lattices are, however, rarely useful in experimental design unless an extremely large number of experiments can be performed, as explained in the next section.

13.6 FINITE EXPERIMENTAL REGIONS

Having learned more about the quintessential differences between the integrated and maximum MSE criteria in the admittedly somewhat academic setting of infinite experimental regions, we are ready to study the practically relevant case of finite experimental regions. Owing to the "curse of dimensionality," these regions have their peculiarities in higher dimensions.

If you can modify all of your experimental variables, such as composition, formulation, and synthesis conditions independently, and if you set upper and lower limits for each of these variables individually, your experimental region is a rectan-

---

[13]The reason being that for short ranges, the maximum MSE criterion becomes similar to the sphere covering problem. The maximum MSE then depends on the point (or points) that is farthest from its closest design point. Only additional design points that reduce this greatest minimal distance lead to a reduction in maximum MSE.

gular body. For instance, if you can choose between 0 and 1 unit (milligram, mol, etc.) of any of the three components c_1, c_2, and c_3, the associated experimental region is a cube in the first octant of experimental space, see Figure 13.12. If dependencies between the variables exist, the shape of the experimental space changes. The most common case is the so-called *mixture problem* in which all the components must add up to one unit. Geometrically, the condition $c_1 + c_2 + c_3 = 1$ is a plane that passes through the points $(0, 0, 1)$, $(0, 1, 0)$, and $(1, 0, 0)$. If one intersects this plane with the cube $0 \leq c_1, c_2, c_3 \leq 1$, the resultant experimental region is an equilateral triangle (see Figure 13.12), often used for summarizing data on ternary mixtures. If one has a mixture problem with four components, the experimental region becomes a regular tetrahedron, and so on. The geometrical entities line segment (in dimension $d = 1$), equilateral triangle ($d = 2$), regular tetrahedron ($d = 3$), etc., are summarily denoted as d-dimensional *simplices*.

How do these polyhedra generalize to higher dimensions? We will look at cubic and simplicial experimental regions in higher dimensions.

For starters, imagine that you can set d different reaction parameters (temperature, pressure, etc.) to k different values each. You can then choose from k^d distinct sets of reaction conditions (see Figure 13.13). For example, $d = 8$ parameters that are set to only $k = 3$ values each result in 6561 distinct combinations. How many of these lie on the surface of your experimental region? The answer is 6560, because only one point, namely, the one where all eight parameters assume their central value, lies in the interior.

Similar properties hold for the simplices: if d components can be varied under the condition that they must sum to some constant, all possible combinations lie in a $(d-1)$-dimensional simplex. Assuming that the individual components can be varied in k steps, the total number of discrete combinations is $(k + d - 1)!/((k-1)!d!)$. Again, most points will lie on the surface: in an 8-dimensional simplex, the resolu-

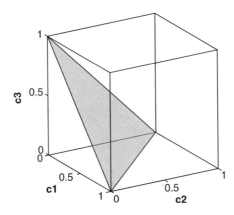

Figure 13.12 A cubic experimental region $0 \leq c_1, c_2, c_3 \leq 1$ and the regular triangle containing all ternary mixtures, with $c_1 + c_2 + c_3 = 1$.

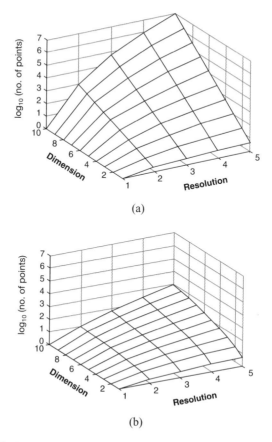

(a)

(b)

Figure 13.13   The logarithm of the number of points in (a) cubes and (b) simplices of various dimensions with the number of points on one edge given by "Resolution." Points are arranged on the cubic lattice, Z_d, in the cubes and on the generalized hexagonal lattice, A_d, in the simplices.

tion k must be 10 (or higher) to ensure that at least one point lies in the interior of the simplex. At this stage (8-dimensional simplex with 10 points along each edge) the simplex already comprises 24,310 points. In this example, the points have been packed in the generalized hexagonal lattice, A_d, which has the right symmetry for a simplex [12].

Two conclusions that can be drawn from this section are that the volume of high-dimensional spaces is vast, and most of the volume is concentrated on the surface. Unless special measures are taken, most of the design points will lie equally on (or near) the hull of the experimental region, (almost) equaling one or more of the constraints that define the experimental region.

Another implication is that if one can perform only a few experiments compared to the volume of the experimental region, the shape of the experimental region will have a heavy impact on the optimal design. To explain this in terms of Figure 13.3, if you put just as many points in a higher-dimensional experimental region, most points will make contact with one of the surfaces, and the structure of the optimal design will have to deviate strongly from the nice lattice that it would really "like" to be.

As a consequence, if you have a large number of points (say, more than 10^d in dimension d), an extract from a lattice will be a good approximation to an optimal design. If you have fewer points, you may want to think about optimizing your design numerically.

If you can make a reasonable guess at the covariance structure of your response surface, use equation (13.3) as a criterion in conjunction with your favorite optimizer. If not, you should make a guess as to whether your response surface is more likely very rough or very smooth on the scale of your sampling density. Judging by the toy examples from this section, in case of doubt, you are probably in the rough regime and should then use some sphere-packing algorithm, e.g., [18]. If you have reason to believe that your response surface is very smooth, you will want to use the dual of a sphere packing. Since that is not defined for finite experimental regions, nor for nonlattice point sets, you might want to put some trust in the conjecture mentioned in Section 13.5.2 and decide to design a point set that is good for vector quantization. This problem has been studied extensively in connection with the encoding of analog signals for transmission over a digital network. (For an excellent overview, see [19].) Design algorithms for this purpose also are readily applicable to experimental design, for sufficiently smooth surfaces, with or without a weight function. Available algorithms fall into one of two main classes: block-iterative [20] or sample-iterative [21].

13.7 EXTENSIONS

The following subsections cover further applications based on the kriging formalism (Sections 13.7.1 and 13.7.2), compare the kriging approach to popular algorithms for the generation of space-filling designs (Section 13.7.3), and discuss the repercussions of some of the assumptions made and how they can be dealt with (Section 13.7.4).

13.7.1 Application to Subset Selection

In the foregoing, we have assumed that each point in the experimental region is eligible as a design point. This need not be true in all applications.

For instance, the finest resolution in the amount of deposited component that a pipetting robot can achieve may be an entire droplet. This leads to a discretization of the experimental region.

In other applications, for instance, in pharmacological screening, a small number of compounds should be selected from a larger library. As in the continuous case, this selection should be optimal in the sense of allowing us to learn a maximum about the response in the experimental region. This is the *subset selection* problem. We assume that the compounds have been embedded in a property space (see Chapter 1, Table 1) such that the response varies as smoothly as possible from one compound to the next. Each compound is represented by one point in property space, and as in the continuous case, points that are close in the experimental space should evoke a similar response.

In both these examples, the experimental region now consists of a cloud of points in space. In cases such as the screening example, the density of the cloud will not be uniform, there may be a clutter of points in some areas of the experimental region, while others may be nearly empty.

Now, since the continuous experimental region has been replaced by a cloud of points, and assuming the integrated MSE criterion has been chosen, the integral in equation (13.3) can be replaced by a summation of the MSE over all points. However, the nonuniformity of the point density needs to be taken into account, otherwise the resultant design will have most of its points in areas that are highly populated by the library [1]. Such a design is called a *representative* subset, and is not compatible with the assumption of uniformly distributed probability of high response, which in turn asks for *diverse* subsets.

If, instead, the maximum MSE has been chosen as the design criterion, the optimization process is simplified in a similar way. It then suffices to calculate, for a given trial design, the MSE at each point in the cloud, and retain the largest value found. The trial design can then be perturbed and the new maximum MSE found, etc.

In summary, modeling the response surface as a stochastic process and design optimization based on the best linear unbiased estimator can be applied to the subset selection problem, with the computational simplification consisting of having to consider only the given library members as candidate points for the design rather than the entire experimental region.

13.7.2 Adaptive Sampling

All previous arguments were based on the assumption that a design is sought first, and then all experiments determining the responses are performed. Such experiments are called *simultaneous*. What if you have been lucky enough to locate an area of promising response, and if you now wish to switch from exploration to optimization? You could define a narrower experimental region and proceed as before, first optimizing a design and then performing measurements. If you do not wish to waste experiments, you can construct a design around the sampling points that you already have in that narrower experimental region (for details, see [1]). If you do have a fair idea of the smoothness of the response surface, however, you can do even better than that. You can construct a design that directly uses the information from the previous measurements. The basic idea in such an *adaptive* (also called sequential) experiment is to put your next design point where the response (as predicted by the kriging interpolator) plus the estimated uncertainty of your prediction are maximal. For a highly readable account of this approach, see [22].

13.7.3 Relation to Other Algorithms

The algorithm you employ should mirror your assumptions concerning

1. The response surface;
2. Your experimental strategy.

Concerning point 1, modeling the response surface as the realization of a stationary stochastic process allows you to lay open the implicit assumptions that many published algorithms rely on.

For instance, the maximin algorithm (which seeks to maximize the minimum distance between any two design points) implicitly tries to find a good sphere packing, and as such relies on the assumption of a rough response surface in combination with the aim of minimizing the average uncertainty. Be aware that the popular maximin algorithm pushes design points into the hull of the experimental region. This can be avoided by using a definition of sphere packing that relies on the sphere packing density in the experimental region rather than on the interpoint distance [1].

If, instead, an algorithm tries to minimize the maximum distance between any point in the experimental region and its closest design point ("minimax"), it seeks to solve the covering problem, which in the current framework can be interpreted as trying to minimize the maximum uncertainty on a rough response surface, as in Section 13.4.2. Some additional correspondences are discussed in [1].

Concerning point 2, an algorithm that proposes design points in a sequential fashion is not compatible with a simultaneous experimental setup. If all experiments are performed without taking previous measurements into account, then an optimal design should also be found by varying all design points simultaneously, rather than by augmenting an initial small design, point by point.

13.7.4 Technical Details

The expected uncertainty in a kriging prediction depends on knowledge of the true covariance function. If a guessed or fitted covariance function is used instead, the uncertainty is underestimated. An uncorrelated random error component can be included to account for random measurement error. The scaling of the coordinate axes is vital; an isotropic covariance function assumes that all spatial directions are of equal importance. If one dimension of the experimental region represents the quantity of a particular additive and its units are changed from, say, millimol to mol while keeping all other things equal, the experimental region is compressed a thousandfold in that direction and many fewer points will be available to illuminate the effect of the additive on the response. In other words, prior knowledge must be used to scale the axes of the experimental region or, equivalently, parameterize a non-isotropic covariance function.[14] Also, less stringent assumptions concerning the na-

---

[14]Rescaling of the experimental space means replacing a covariance matrix that is constant along its diagonal and zero elsewhere with a diagonal covariance matrix with differing diagonal elements. This is equivalent to making a shift from Euclidean to weighted Euclidean distance.

ture of the response surface can be made [1,5], in particular the response can be modeled as the sum of a trend (given, e.g., by a polynomial) and a stochastic process. Depending on the relative importance of the two, the resultant design will vary between one that is optimal according to the "alphabetic" criteria (Section 13.2.1) and one that is optimal for the pure stochastic model.

13.8 CONCLUSIONS AND PRACTICAL RECOMMENDATIONS

Before anything else, you are required to define your experimental region (defined in Section 13.1). If your experimental space has no reasonable natural limits, you need to fix constraints arbitrarily. Upper and lower bounds on individual dimensions are the most convenient constraints to formulate and implement, but you should generally try to define your experimental region as tightly as possible. When working in multiple dimensions you are almost always data starved, and if you can afford only so many design points, better not waste them in an overly generous margin.

In exploring an unknown space of materials, we think that modeling the response surface as a realization of a stochastic process offers more flexibility in the incorporation of prior knowledge through weight and covariance functions. If nothing is known about the covariance structure of the response surface, at the very least a guess is required as to whether it is smooth or rough on the scale of the interpoint distance. In light of Section 13.5.4, we will discuss only the integrated MSE criterion in this section.

Remembering that the volume of a multidimensional space is vast, the few thousand design points you can afford may easily look forlorn. Their isolation will usually lead to a situation in which your response surface can vary significantly between design points. In this case, if you have relatively few[15] design points, you should, for the reasons given in Sections 13.4.1 and 13.6, consider numerical optimization of your design with a sphere-packing algorithm, as in [18]. If you have many design points available relative to the dimensionality of your problem, you may ignore boundary effects and excise part of the packing lattice for your design.

If you can assume that your response surface is smooth, you are in a fortunate position: you can then hope to gain a fairly accurate picture of the true response surface from your experiments. Also, if you have too few design points to simply use the dual of a packing lattice (see Sections 13.4.1 and 13.5.2), you can rely on efficient and robust algorithms for vector quantization (Section 13.6).

If you are in a position to make an educated guess at the covariance structure of the response surface, you can optimize your design under equation (13.3) using any local or global optimizer. Also, if for some reason you want more design

---

[15]It is difficult to give numbers that are useful for a wide range of situations, but if hard pressed for one, we might say that "few" means $\ll 10^d$ in dimension d.

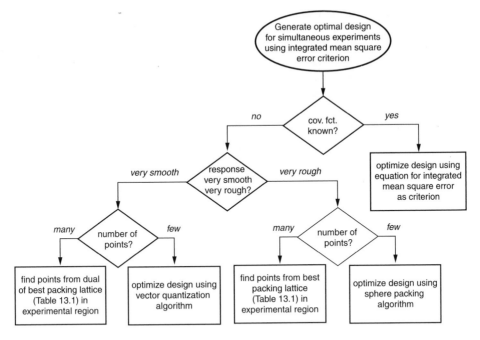

Figure 13.14   Flowchart for determining a suitable design method.

points in a particular area, you can introduce this bias through a weight function [1].

In essence, you can get away with very little knowledge about your response surface, but if you do know more, you have ways of putting this knowledge to work in a consistent framework. The preceding remarks are condensed in the flowchart in Figure 13.14.

In summary, if you can afford many experiments and wish to invest little into optimization of the experimental design, you can sample your experimental region on a lattice—though preferably not on the cubic. Switching to a more suitable lattice will improve the performance significantly for the same number of experiments, still assuming that the number of experiments is large. If you need to make the most out of a limited number of experiments, you will be forced to invest time and expertise in numerical optimization of the design.

APPENDIX 13.A   GENERATOR MATRICES FOR LATTICES

13.A.1   Generating Lattice Points

As discussed in Section 13.5.1, each point in an infinite lattice can be specified by a linear combination of basis vectors, where all coefficients take integer values only.

The basis vectors can be collected, row by row, into a generator matrix. In this Appendix, we give one generator matrix for each lattice mentioned in the chapter (the Z, A, D, and E families, and Λ_{16}).

To span a d-dimensional lattice, d basis vectors are required. However, each of these vectors can have more than d components. The two-dimensional hexagonal lattice A_2, for instance, can be defined as the intersection of the three-dimensional cubic lattice (x, y, z) with x, y, z integer, and of the plane $x + y + z = 0$. The resultant generator matrix can be written in the form

$$\begin{bmatrix} -1 & 1 & 0 \\ -1 & 0 & 1 \end{bmatrix},$$

which, unlike the matrices given in Sections 13.5.1 and 13.A.2, contains no square root. In general, generator matrices are often easier to write down using basis vectors with more than d components, and textbooks usually give these matrices. However, these lattices first need to be rotated, thus eliminating the redundant dimensions, before they can be used in an experimental design. All generator matrices tabulated in the following have exactly d columns for d-dimensional lattices. Most of them are taken from the appendix in [15], where further details can be found.

For every lattice, there is a dual (denoted by a star). One way to define it is through Fourier analysis. Consider the sum of Dirac impulses, one at each point of a particular lattice. The Fourier transform of the sum is another sum of Dirac impulses. The locations of these impulses are the dual of the original lattice. If B is a square generator matrix for a lattice, then $(B^{-1})^T$ is a generator matrix for its dual.

For any particular lattice, basis vectors can be chosen in a multitude of ways. Furthermore, lattices that can be obtained from each other just by rotation, scaling, and possibly reflection are defined to be *equivalent* (denoted by $\cong$), because, even if such lattices do not contain the same points, they have analogous properties as far as Euclidean distance is concerned. Useful equivalence relations include $Z_1 \cong Z_1^* \cong A_1 \cong A_1^*$, $A_2 \cong A_2^*$ (hexagonal), $A_3 \cong D_3$ (fcc), $A_3^* \cong D_3^*$ (bcc), $D_4 \cong D_4^*$, $E_8 \cong E_8^*$, and $\Lambda_{16} \cong \Lambda_{16}^*$.

Once an appropriate lattice has been chosen, we must also be able to determine the points of the lattice that fall within a given experimental region $\mathcal{R}$. The general procedure can be outlined as follows:

1. Project the region $\mathcal{R}$ onto the first basis vector. This defines an interval in the direction of the basis vector; find the endpoints of this interval. Repeat for all d basis vectors and call the intervals $J_1, \ldots, J_d$.

2. For $i = 1, \ldots, d$, generate all integer multiples of basis vector i that lie within J_i and call this set of vectors S_i.

3. Compute the set of all points that can be obtained as the sum of i vectors, one from each S_i. These points belong to a parallelepiped with sides parallel to the basis vectors. The parallelepiped encloses $\mathcal{R}$.

4. Discard all points in the set that lie outside $\mathcal{R}$.

13.A.2 Generator Matrices

$$Z_d, \quad d \ge 1: \begin{bmatrix} 1 & 0 & 0 & \cdots & 0 \\ 0 & 1 & 0 & \cdots & 0 \\ 0 & 0 & 1 & \cdots & 0 \\ \vdots & \vdots & \vdots & \ddots & \vdots \\ 0 & 0 & 0 & \cdots & 1 \end{bmatrix}$$

$$A_d, \quad d \ge 1: \begin{bmatrix} \alpha & 1 & \cdots & 1 \\ 1 & \alpha & \cdots & 1 \\ \vdots & \vdots & \ddots & \vdots \\ 1 & 1 & \cdots & \alpha \end{bmatrix} \quad \text{with } \alpha = \sqrt{d+1} + 2$$

$A_d^*, \quad d \ge 1$: as above   with $\alpha = \sqrt{d+1} - d$

$$D_d, \quad d \ge 3: \begin{bmatrix} 2 & 0 & 0 & \cdots & 0 \\ 1 & 1 & 0 & \cdots & 0 \\ 1 & 0 & 1 & \cdots & 0 \\ \vdots & \vdots & \vdots & \ddots & \vdots \\ 1 & 0 & 0 & \cdots & 1 \end{bmatrix}$$

$$D_d^*, \quad d \ge 3: \begin{bmatrix} 1 & 0 & \cdots & 0 & 0 \\ 0 & 1 & \cdots & 0 & 0 \\ \vdots & \vdots & \ddots & \vdots & \vdots \\ 0 & 0 & \cdots & 1 & 0 \\ \frac{1}{2} & \frac{1}{2} & \cdots & \frac{1}{2} & \frac{1}{2} \end{bmatrix}$$

$$E_6: \begin{bmatrix} 1 & 0 & 0 & 0 & 0 & \alpha \\ 0 & 1 & 0 & 0 & 0 & \alpha \\ 0 & 0 & 1 & 0 & 0 & \alpha \\ 0 & 0 & 0 & 1 & 0 & \alpha \\ 0 & 0 & 0 & 0 & 1 & \alpha \\ \frac{1}{2} & \frac{1}{2} & \frac{1}{2} & \frac{1}{2} & \frac{1}{2} & \frac{3\alpha}{2} \end{bmatrix} \quad \text{with } \alpha = \sqrt{3}$$

E_6^*: as above   with $\alpha = 1/\sqrt{3}$

$$E_7: \begin{bmatrix} 2 & 0 & 0 & 0 & 0 & 0 & 0 \\ 0 & 2 & 0 & 0 & 0 & 0 & 0 \\ 0 & 0 & 2 & 0 & 0 & 0 & 0 \\ 0 & 0 & 0 & 2 & 0 & 0 & 0 \\ 1 & 1 & 1 & 0 & 1 & 0 & 0 \\ 0 & 1 & 1 & 1 & 0 & 1 & 0 \\ 0 & 0 & 1 & 1 & 1 & 0 & 1 \end{bmatrix}$$

$$E_7^*: \begin{bmatrix} 2 & 0 & 0 & 0 & 0 & 0 & 0 \\ 0 & 2 & 0 & 0 & 0 & 0 & 0 \\ 0 & 0 & 2 & 0 & 0 & 0 & 0 \\ 1 & 1 & 0 & 1 & 0 & 0 & 0 \\ 0 & 1 & 1 & 0 & 1 & 0 & 0 \\ 0 & 0 & 1 & 1 & 0 & 1 & 0 \\ 0 & 0 & 0 & 1 & 1 & 0 & 1 \end{bmatrix}$$

$$E_8: \begin{bmatrix} 2 & 0 & 0 & 0 & 0 & 0 & 0 & 0 \\ 1 & 1 & 0 & 0 & 0 & 0 & 0 & 0 \\ 1 & 0 & 1 & 0 & 0 & 0 & 0 & 0 \\ 1 & 0 & 0 & 1 & 0 & 0 & 0 & 0 \\ 1 & 0 & 0 & 0 & 1 & 0 & 0 & 0 \\ 1 & 0 & 0 & 0 & 0 & 1 & 0 & 0 \\ 1 & 0 & 0 & 0 & 0 & 0 & 1 & 0 \\ \frac{1}{2} & \frac{1}{2} & \frac{1}{2} & \frac{1}{2} & \frac{1}{2} & \frac{1}{2} & \frac{1}{2} & \frac{1}{2} \end{bmatrix}.$$

$$\Lambda_{16}: \begin{bmatrix} 4 & 0 & 0 & 0 & 0 & 0 & 0 & 0 & 0 & 0 & 0 & 0 & 0 & 0 & 0 & 0 \\ 2 & 2 & 0 & 0 & 0 & 0 & 0 & 0 & 0 & 0 & 0 & 0 & 0 & 0 & 0 & 0 \\ 2 & 0 & 2 & 0 & 0 & 0 & 0 & 0 & 0 & 0 & 0 & 0 & 0 & 0 & 0 & 0 \\ 2 & 0 & 0 & 2 & 0 & 0 & 0 & 0 & 0 & 0 & 0 & 0 & 0 & 0 & 0 & 0 \\ 2 & 0 & 0 & 0 & 2 & 0 & 0 & 0 & 0 & 0 & 0 & 0 & 0 & 0 & 0 & 0 \\ 2 & 0 & 0 & 0 & 0 & 2 & 0 & 0 & 0 & 0 & 0 & 0 & 0 & 0 & 0 & 0 \\ 2 & 0 & 0 & 0 & 0 & 0 & 2 & 0 & 0 & 0 & 0 & 0 & 0 & 0 & 0 & 0 \\ 1 & 1 & 1 & 1 & 1 & 1 & 1 & 1 & 0 & 0 & 0 & 0 & 0 & 0 & 0 & 0 \\ 2 & 0 & 0 & 0 & 0 & 0 & 0 & 0 & 2 & 0 & 0 & 0 & 0 & 0 & 0 & 0 \\ 2 & 0 & 0 & 0 & 0 & 0 & 0 & 0 & 0 & 2 & 0 & 0 & 0 & 0 & 0 & 0 \\ 2 & 0 & 0 & 0 & 0 & 0 & 0 & 0 & 0 & 0 & 2 & 0 & 0 & 0 & 0 & 0 \\ 1 & 1 & 1 & 1 & 0 & 0 & 0 & 0 & 1 & 1 & 1 & 1 & 0 & 0 & 0 & 0 \\ 2 & 0 & 0 & 0 & 0 & 0 & 0 & 0 & 0 & 0 & 0 & 0 & 2 & 0 & 0 & 0 \\ 1 & 1 & 0 & 0 & 1 & 1 & 0 & 0 & 1 & 1 & 0 & 0 & 1 & 1 & 0 & 0 \\ 1 & 0 & 1 & 0 & 1 & 0 & 1 & 0 & 1 & 0 & 1 & 0 & 1 & 0 & 1 & 0 \\ 1 & 0 & 0 & 1 & 1 & 0 & 0 & 1 & 1 & 0 & 0 & 1 & 1 & 0 & 0 & 1 \end{bmatrix}$$

REFERENCES

1. Hamprecht, F. A.; Thiel, W.; van Gunsteren, W. F. *J. Chem. Inf. Comput. Sci.,* 200, 42: 414–428.

2. Fedorov, V. V. *Theory of Optimal Experiments,* Academic Press, London, 1972.

3. Atkinson, A. C.; Donev, A. N. *Optimum Experimental Designs,* Oxford University Press, Oxford, 1992.

4. Pebesma, E. J.; Heuveling, G. B. M. *Technometrics,* 1999. 41: 303–312. http://www. gstat.org.

5. Chilès, J.-P.; Delfiner, P. *Geostatistics: Modeling Spatial Uncertainty.* Wiley Series in Probability and Statistics, Wiley, New York, 1999.

6. Cressie, N. A. C. *Math. Geol.,* 1990, 22: 239–252.

7. Sacks, J.; Welch, W. J.; Mitchell, T. J.; Wynn, H. P. *Stat. Sci., 1989,* 4: 409–435.

8. Agrell, E.; Hamprecht, F. A.; Künsch, H. in preparation.

9. Petersen, D. P.; Middleton, D. *Inf. Control,* 1962, 5: 279–323.

10. Johnson, M. E.; Moore, L. M.; Ylvisaker, D. *J. Stat. Plann. Inf.,* 1990, 26: 131–148.

11. Gersho, A. *IEEE Trans. Inform. Theory,* 1982, IT-28: 157–166.

12. Conway, J. H.; Sloane, N. J. A. *Sphere Packings, Lattices and Groups,* 2nd ed., Vol. 290, *Grundlehren der Mathematischen Wissenschaften.* Springer-Verlag, New York, 1993.

13. Okabe, A.; Boots, B.; Sugihara, K. *Spatial Tessellations. Concepts and Applications of Voronoi Diagrams.* Wiley, Chichester, UK, 1992.

14. Agrell, E. *Voronoi-Based Coding.* Ph.D. Thesis, Chalmers University of Technology, Göteborg, Sweden, 1997.

15. Agrell, E.; Eriksson, T. *IEEE Trans. Inform. Theory,* 1998, 44(5): 1814–1828.

16. Forney, G. D., Jr. In *Coding and Quantization,* Vol. 14, *DIMACS Series in Discrete Mathematics and Theoretical Computer Science,* Calderbank, R.; Forney, G. D., Jr.; Moayeri, N., eds., American Mathematical Society, Providence, RI, 1993.

17. Agrell, E.; Eriksson, T.; Vardy, A.; Zeger, K. *IEEE Trans. Inform. Theory,* 2000, 48(8): 2201–2214.

18. Hardin, R. H.; Sloane, N. J. A. *Operating Manual for Gosset: A General-Purpose Program for Constructing Experimental Designs,* 2nd ed., AT&T Bell Labs, Murray Hill, NJ, 1994. http://www.research.att.com/~njas/gosset/.

19. Gray, R. M.; Neuhoff, D. L. *IEEE Trans. Inform. Theory,* 1998, IT-44(6): 2325–2383.

20. Linde, Y.; Buzo, A.; Gray, R. M. *IEEE Trans. Commun.,* 1980, COM-28: 84–95.

21. Yair, E.; Zeger, K.; Gersho, A. *IEEE Trans. Inform. Theory,* 1992, 40(2): 294–309.

22. Jones, D. R.; Schonlau, M.; Welch, W. J. *J. Global Optim.,* 1998, 13: 455–492.

INDEX